D0455374

COAL EXPLORATION

Proceedings of the first
International Coal Exploration Symposium
London, England, May 18-21, 1976

Edited by William L.G. Muir

Other books, directories, and marketing aids
for the mining and minerals processing industries:

Waste Production and Disposal in Mining, Milling, and Metallurgical Industries
by Roy E. Williams

Copper Porphyries by Alexander Sutulov

Minerals Transportation Volume 1: Proceedings of the First International Symposium on Transport and Handling of Minerals, Vancouver, British Columbia, Canada, October 1971

Minerals Transportation Volume 2: Proceedings of the Second International Symposium on Transport and Handling of Minerals, Rotterdam, Netherlands, October 1973

Tailing Disposal Today: Proceedings of the First International Tailing Symposium, Tucson, Arizona, USA, October-November 1972

Review of the World Coal Industry to 1990 by W.L.G. Muir

World Mining Glossary of Mining, Processing, and Geological Terms
Revised edition of over 10,000 terms and phrases in five languages—English, Swedish, German, French, and Spanish—with alphabetized, cross-referenced indexes.
Edited by R.J.M. Wyllie and George O. Argall, Jr.

World Mines Register: International directory of active mining companies, operations, and key personnel

World Mining Copper Map: Shows every known important porphyry copper and molybdenum mining operation in the world

Map of World Coal Resources and Major Trade Routes

Map of the Coal Fields of the World
Published by the USSR Ministries of the Coal Industry, Geology, and Higher and Specialized Education.
Edited by Alexander K. Matveev

TABLE OF CONTENTS

Section One

Section Two

Section Three

Section Four

Section Five

Section Six

Section Seven

Section Eight

LIST OF ILLUSTRATIONS

---

* unnumbered photograph

LIST OF TABLES

FOREWORD BY THE EDITOR

The first International Coal Exploration Symposium was a response to the new importance that coal has assumed in the international energy field. The movement was triggered and given a strong impetus by the fourfold increase in the price of oil in late 1973, but it would have taken place without that, although more slowly and less dramatically. The basic fact is that the world's resources of coal far exceed those of any other energy mineral. Coal has, therefore, a long-term future both in steel-making and direct thermal power generation and as a source of liquid and gaseous hydrocarbons when the reserves of the natural products begin to wane.

The new importance of coal has brought many relative newcomers to the search for, and development of, new coalfields. Many major oil companies and many international mining houses, which have until recently been engaged mainly in metal mining, are now involved in the coal industry. The new price levels of coal make remote deposits in the developing world of possible economic importance and open another new field of activity.

In these circumstances the sponsors of the symposium believed that a conference that brought together these diverse interests to study and discuss the present state of the art of coal exploration would serve a useful purpose. If some of the newcomers lacked experience in coal, others brought their own expertise from other fields and adapted it to the special characteristics of coal exploration. In so doing, important new techniques were, and are still being, developed. Some of the most interesting papers presented dealt with the contribution of oil technology in improving the technology of coal exploration.

The delegates attending the conference represented 34 countries and many different interests and technologies. To provide something for everyone was not within the scope of a 3½ day meeting and was not attempted. Nor could all the papers be of equal interest to those who had long experience and expertise in the field of coal exploration. The Programme Committee believe that the mix of papers presented served the main purpose of the meeting, and that this volume of proceedings brings together a body of knowledge which defines the state of present technology and indicates the direction of new developments.

The Committee take this opportunity to thank the authors of the papers, all busy men occupying senior positions each in his own branch of the industry. Particular thanks are extended to the National Coal Board for their participation: to Sir Derek Ezra for his keynote address, and to Mr. A.M. Clarke and Mr. Keith Shaw, two very senior members of the staff, who presented papers. The Committee was helped during the preparation of the programme by discussions with the National Coal Board staff. The cooperation of the United States Geological Survey in granting permission for the paper given by Dr. Wilfred P. Hasbrouck of their staff and Professor Frank A. Hadsell of the Colorado School of Mines is also gratefully acknowledged. It is unnecessary to list all the authors by name, but all are equally thanked.

Thanks are also given to the San Francisco staff of Miller Freeman Publications, particularly Mr. George O. Argall, Jr., and Mr. George H. Roman, for their advice and help, especially in arranging the North American contribution to the Symposium.

W.L.G. Muir
Program Chairman
August 1976

New Zealand drillers examine coal cuttings from a test hole in the Huntly area, which has revealed a much larger coal base than suspected. Photo courtesy World Coal

AUTHORS & ABSTRACTS

A.F. LESSARD

SYMPOSIUM CHAIRMAN

Arnold F. Lessard, Resources Engineering & Management International, Inc., Denver, Colorado, USA

Arnold Lessard has a diploma in accounting and B.S. and M.A. degrees in social sciences from Boston and Columbia Universities. In addition, he has done advanced studies in economics and Middle East area studies.

Following U.S. Government service as Head of the Personnel Development Division of an independent agency, he joined Booz Allen & Hamilton in 1956. He is presently with Resources Engineering and Management International Ltd.

A. Lessard's primary fields of interest are the associated problems of planning and executing industrial development projects, and related recruitment, training and development of personnel. He is a counselor to top managements of private and public organizations in Europe, Africa and the Middle East. In addition, he is the author of several publications in the field of supervisory and executive development.

W.L.G. MUIR

PROGRAM CHAIRMAN

William L.G. Muir, Mining Consultant, High Wycombe, Buckinghamshire, UK

Mr. W.L.G. Muir graduated in mining from Glasgow University in 1931 and took his Master's degree at McGill University in 1934. He obtained the British First Class Colliery Manager's Certificate in 1936. He had experience in mine operations, with periods of prospecting, in Canada, the British Coalfield and Zambia.

Much of his management experience was gained in West Africa where he was General Manager and later Mines Director of the Sierra Leone Development Company Ltd. which operated an open-cast iron mine with its own railway and seaport.

He is now an independent mining consultant, and has carried out assignments in Algeria, Austria, Greece, Iran, Liberia, and the United Kingdom. These assignments included designing and installing management control systems, and corporate organizations for state-owned and privately owned mining companies, as well as technical and economic evaluations.

He is the author of "Review of the World Coal Industry to 1990" published by Miller Freeman Publications in 1975, and the 1976 Supplement to it.

SYMPOSIUM COORDINATOR

George O. Argall, Miller Freeman Publications, San Francisco, California, USA

George O. Argall, internationally known mining engineer, has been Editor of *World Mining* for 25 years. In 1974 he also became Publisher of *World Coal*.

He is a graduate of the Colorado School of Mines and worked in gold, silver, tungsten, molybdenum, and uranium mines before joining the *World Mining* staff.

G.O. ARGALL
He has inspected mines in more than 50 countries and written several hundreds of articles about these mines and mineral processing plants. He won the Jesse H. Neal award for editorial excellence for his special issue "Japan and the World of Mining" in 1969.

NORTH AMERICAN CHAIRMAN

George H. Roman, Miller Freeman Publications, San Francisco, California, USA

George H. Roman, Editor of *World Coal*, is a graduate of the University of Illinois with a B.S. degree in Mining Engineering.

After graduation in 1961, he worked as an underground miner with Freeman Coal Company, a Mine Engineer with Consolidation Coal Company and an Associate Engineer with the firm of mining consultants,

G.H. ROMAN
Paul Weir Company. Prior to joining Miller Freeman Publications he worked for several years as an Assistant Editor of McGraw Hill's *Coal Age* magazine. In addition to his editorial duties, Mr. Roman is also General Manager of *World Coal* and *World Mining*.

He is a registered Professional Engineer and a member of the American Institute of Mining Engineers (AIME) and the Society of Mining Engineers (SME).

K. WHITWORTH

ASSISTANT SYMPOSIUM COORDINATOR

Keith Whitworth, Miller Freeman Publications, London, UK

Keith Whitworth, Managing Editor of *World Coal*, began his mining career with the Blackwell Colliery Company during World War II. In 1954, he joined the Tothill Press Limited as an Editorial Assistant on *Colliery Engineering* until he was appointed Editor of that publication in 1957. He held this position until 1965 when he was named Editor of *Colliery Guardian*, a post he held for nine years until he joined *World Coal*.

Whitworth is an associate member of the Institute of Mining Engineers (UK) and a T. Eng. (CEI). He has lectured at Nottingham University and has served as Honorary President of the University of Nottingham Mining Society and as Honorary Secretary of the former National Association of Colliery Managers.

SIR DEREK EZRA

SPECIAL KEYNOTE SPEAKER

Sir Derek Ezra M.B.E., Chairman, National Coal Board, London, UK

Sir Derek (created a Knight Bachelor in Birthday Honours, June 1974) was born on 23rd February, 1919, and educated at Monmouth School and Magdalene College, Cambridge.

He joined the Army as a gunner in 1939. He was demobilized in 1947 with the rank of Lt. Colonel and joined the Marketing Department of the National Coal Board.

From 1952 to 1956 he represented the Board on the United Kingdom Delegation to the High Authority of the European Coal and Steel Community in Luxembourg. In 1956 he was appointed Deputy Manager of the Inland Branch of Marketing Department at national headquarters and later transferred to the equivalent post at the London and Southern Regional Sales Office.

He became Manager of the Board's London and Southern Regional Sales Office on 1st July, 1958 and was appointed Director-General of Marketing in August, 1960 in succession to Mr. F. Wilkinson.

Sir Derek was appointed to be a member of the National Coal Board in July, 1965 and was appointed Deputy Chairman in May, 1967; his appointment as Chairman Designate of the National Coal Board was announced on 8th April, 1971, and he became Chairman on 3rd July, 1971.

Sir Derek is also:
—Chairman of Associated Heat Services and a Di-

rector of the British Fuel Company and of J.H. Sankey and Son Ltd.—all National Coal Board Associated companies

—President of CEPCEO—the organization of the European common market coal producers

—Chairman-Designate of the British Institute of Management, succeeding Sir Frederick Catherwood from 6th October, 1976 (Sir Derek is currently Chairman of the BIM's Board of Fellows)

—Chairman of the Confederation of British Industry's Europe Committee

—A member of the British Overseas Trade Board and Chairman of the BOTB European Trade Committee

—A member of the Government's Advisory Council for Energy Conservation

—A Governor of the London Business School.

1 WORLD COAL SUPPLY AND DEMAND

By Henry E. Collins, Consulting Mining Engineer, London, UK

H.E. COLLINS

Henry Edward Collins is a chartered engineer with a Master of Engineering degree from the University of Sheffield, where he was a Senior Lecturer of Mining Engineering from 1935-1939.

From 1945 Mr. Collins was British Chairman of the UK/US Coal Control Group (later Combined Coal Control Group) in West Germany. In 1948 he was awarded the C.B.E. On his return to the British coal industry he occupied senior positions in the National Coal Board and became Board Member for Production. He was then consultant to the NCB until 1967. He is now a private consultant. He is vice-president of the World Mining Congress and past-president of the Institution of Mining Engineers.

He is the author of "The Revitalised Coal Industries of the World" and of many papers.

ABSTRACT: The paper presents the current world energy balance and indicates its likely pattern by the turn of this century. It is suggested that the present exponential growth rate in energy usage cannot be maintained into the 21st century. There will have to be a deceleration in the energy growth rate so that the curve of annual energy consumption becomes asymptotic to a line representing the maximum maintainable annual world energy demand. The important role in the future of fossil fuels and particularly of coal, is indicated. The consequential importance of exploration of coal deposits throughout the world is stressed.

A.K. MATVEEV

2 DISTRIBUTION AND RESOURCES OF WORLD COAL

By Prof. Alexander K. Matveev,
Professor, Moscow State University, USSR

Prof. Alexander K. Matveev, mining engineer and geologist, is Professor at Moscow State University where he presently heads the laboratory of geology and coal petrography.

Prof. Matveev has authored many papers on theoretical work in coal geology, and has written several monographies: Geology of Coal Fields of the USSR; Coal Fields of Foreign Countries, in four volumes. He is also the editor of the "Map of World Coal Fields."

ABSTRACT: The report is illustrated by the Map of Coal Fields of the World. It presents the principles of compilation of the map, regularities of the distribution of coal fields of the world and approximate total geological reserves. The necessity for coordinated international calculation of the reserves is also suggested.

D.H. ADAMS

3 OBJECTIVES AND ORGANISATION OF COAL EXPLORATION PROJECTS

By Dell H. Adams, Vice-President—Exploration,
Consolidation Coal Co.,
McMurray, Pennsylvania, USA

Dell H. Adams graduated from the University of Kentucky in 1964 with a B.S. degree in civil engineering. Following graduation he was employed for a year by Island Creek Coal Company as Project Engineer. In 1965 he joined Peabody Coal Company as Project Engineer, later serving as Assistant Chief Engineer and then as Chief Engineer of Peabody's Alabama and Kentucky divisions. Before joining Consolidation Coal Company, Adams was a Consulting Engineer in his own engineering firm of Adams and Fuller Engineering Service in Madisonville, Kentucky. He joined Consolidation Coal in December 1968 as an Engineer in the Western Division. Since that time Adams has served as Chief Engineer of the Mid-Western Division, Director of Engineering and Land, Vice President of Operations in the Western Division, and Director of Exploration and Development activities. In his current position Adams is responsible for Consolidation's overall coal exploration activities. No abstract of Mr. Adams' paper was provided.

4 APPLICATION OF WIRELINE LOGGING TECHNIQUES TO COAL EXPLORATION

By Donald R. Reeves, Managing Director,
BPB Industries (Instruments) Ltd., East Leake,
Leicestershire, UK

D.R. Reeves graduated from London University in Geology in 1956 and, after carrying out a year's geophysical research at Cambridge University, worked for Shell Oil and the National Coal Board before joining BPB Industries Ltd. in 1960. He was responsible for initiating developments in borehole logging in 1965 and became Director and General Manager of BPB Industries (Instruments) Ltd. when this company was formed in 1970, and Managing Director in 1974.

ABSTRACT: Wireline Logging techniques have been used in coal exploration for over twenty years but until recently only with limited success. Early logs were run using oilfield equipment or simply resistivity measurements, but with the introduction of specially designed density tools, more accurate and important information could be obtained, stimulating the importance of the technique.

Current applications utilise wireline techniques in two fields: the original application, related to acquiring more geological information in both cored and non-cored holes for correlation, lithology and coal thickness.

With the improvements in tool design and calibration techniques, a second field of petrophysical analysis of the coal itself is being developed, and although still at an early stage, regular analysis of ash content and, in some favourable cases, rank, can already be made.

In the future, Field Digitisation and Computer Analysis will become standard and will enhance all present applications. Slimline formation resistivity measurement devices are just being introduced and will give an important additional parameter for petrophysical analysis, whilst the introduction in the near future of special sonic logs will open a third field of application: rock strength.

B.A. LAVERS

5 RECENT DEVELOPMENTS IN COAL PETROPHYSICS

By Brian A. Lavers and Lambert J.M. Smits,
Shell International Petroleum Co., The Hague,
The Netherlands

B.A. Lavers graduated from the Royal School of Mines, Imperial College. London, in 1956 and served with the Royal Engineers in Malaysia prior to joining Shell in 1959.

After assignments in the Netherlands and Libya, he went to Venezuela in 1965 where he became successively Senior Petrophysical Engineer, Senior Lake Reservoir Engineer, Lake Development Superintendent.

L.J.M. SMITS

In 1973 he became Head of Petrophysical Engineering for Shell International Petroleum Co. in The Hague. In January 1976, he took his present position as Production Manager of Production Development, Oman.

L.J.M. Smits joined Shell in 1953 after receiving a degree in Physical Engineering at the College of Advanced Technology in Dordrecht. He then joined the Koninklijke Shell Laboratory, Amsterdam and was transferred in 1961 to the Koninklijke Shell Exploration and Production Laboratory, Rijswijk, as a Research Scientist. In 1970, he was transferred to the Nederlandse Aardolie Co., Assen, as a Senior Petrophysical Engineer, head of the Petrophysics Section.

In 1972, he became Senior Petrophysical Engineer, second to the Head of the Department, at the Shell International Petroleum Co. in The Hague.

ABSTRACT: In 1973 a major programme of coal exploration was initiated by Shell Companies operating in various parts of the world. While it was recognized that Coal Petrophysics was still a fairly new and inexact science, a policy decision was made at the outset to log every borehole.

Routine logging for coal requires a suitable package of wireline instrumentation, purpose designed to do the necessary job at low cost. The tools must be slim hole (2 OD), have good vertical resolution and provide well-calibrated and repeatable measurements of formation density, hydrogen index (neutron), natural gamma ray, resistivity and caliper; the pressure and temperature ratings of the tools can be very low.

Operations have become quite extensive and there are now 12 logging units working for Shell in South Africa, Botswana, Swaziland, Indonesia, Sarawak, Australia and Canada. Several of the units have been built to fit into two 700 pound packages for transport by light helicopter. This level of activity is supported by production department petrophysical engineers.

The logs are used as qualitative geological tools for coal seam identification, correlation and characterization; and quantitatively to estimate the commercial properties of the coals.

The paper also discusses an integrated computer system developed to handle coal core analyses and petrophysical evaluation.

A.M. CLARKE

6 SEISMIC SURVEYING AND MINE PLANNING: THEIR RELATIONSHIPS AND APPLICATION

By Anthony M. Clarke, Head of Mining Geotechnology and Chief Geologist, National Coal Board, UK

After serving from 1953 to 1955 with the Seismograph Service Corp. in the Middle East, India and Pakistan as Seismologist, Geologist and Permitman, A.M. Clarke worked for two years as Research Officer, Mapping, Fuel Survey.

He has been with the National Coal Board since

1957, starting as Assistant Divisional Geologist, Coal and Metaliferous Mining, in the Durham Division. In 1963 he moved to National Coal Board headquarters as Assistant Chief Geologist, then to Deputy Chief Geologist, Research and Development in 1967. In 1974 he became Head of Mining Geotechnology and Chief Geologist, the position he presently holds.

Mr. Clarke was graduated from King's College, London in 1953 with a B.Sc. in geology.

ABSTRACT: Reasons are given for the extended history (until the last three years) of misunderstanding of needs and capabilities on the part of a technically sophisticated international coal mining industry, long in need of results which have been technically feasible from seismic surveying, and a similarly large and sophisticated international seismic surveying industry, often in need of contract work. (Illustrations are taken from the situation in the UK.) As part of the main theme of the paper, and illustration of when, where and to what extent seismic surveying is called for in coal mining, a description is given of the place of seismic surveying in the three-phase logical sequence of exploration in the search for new mine sites.

It is then shown that, in existing mines, consideration of the best balance of mining effort applied on a moving annual basis between (a) exploration for disturbances to the continuity of production, (b) insurance against them and (c) the loss of revenue and the effort likely to be expended in recovering from unexplored uninsured disturbances—provides the best forum for integration of seismic exploration into normal practice at existing mines. The long established Geosimplan approach to test these balances within a mine is briefly outlined.

Finally, it is pointed out that after three years' continuous use on a multicrew basis in the UK, coal mining seismic surveying is still in its infancy. The paper concludes with a brief review of the current development objectives in coal mining seismic surveying in the UK.

T.E. DALY

7 SEISMIC METHODS FOR THE DELINEATION OF COAL DEPOSITS

By Thomas E. Daly, Director of Exploration Services, and Richard F. Hagemann, Senior Geophysicist, Resources Engineering & Management International, Denver, Colorado, USA

Tom Daly is a geologist-geophysicist specializing in petroleum exploration. He holds a B.S. degree in geology from Missouri School of Mines.

Upon graduation in 1949, Mr. Daly spent five years with Phillips Petroleum Co. and four years with Continental Geophysical Co. in seismic operations. He then was with Sinclair Oil & Gas Co. from 1957 to 1966 and was Chief Geophysicist and Exploration Manager of an international subsidiary of that company.

Mr. Daly joined SONATRACH in 1966 as a member of a four man team retained to assist in the organization and start-up of that company. While employed by SONATRACH he was the Chief Geophysical Adviser and Assistant Manager of Exploration. He left SONATRACH in 1970 to become a partner in Resources Engineering & Management International Inc. and director of their exploration operations. In his present position he has been directly associated with exploration activities in Scandinavia, United Kingdom, Ireland, Germany, Nigeria, Algeria, Libya, and Egypt. He has served as international exploration adviser to governments, national companies and numerous private companies.

R.F. HAGEMANN

Richard F. Hagemann holds a degree in geological engineering (geophysics) from the Colorado School of Mines and has followed special studies at Washington University, Harvard University and M.I.T. in meteorology and radar electronics

He was a research Geophysicist for Sun Oil Co., conducting research in data processing and interpretation in seismology, and a Technical Adviser and Geophysical Consultant in instrumentation, field techniques and interpretation for the Compagnie Générale de Géophysique for twelve years. Following this, he established his own company for the specialized purchase of geological, oil field, and electronic equipment and components for export throughout the world.

Mr. Hagemann, until recently, was assigned by REMI as Senior Supervisory Geophysicist on loan to SONATRACH (the national oil co.) in Algeria. His responsibilities included the preparation of geophysical programs designed to evaluate specific areas, supervision of data acquisition and processing, interpretation of the data with recommendation for further exploratory actions, training of Algerian personnel, and acting as adviser to the Chief Geophysicist.

Mr. Hagemann has been supervising, as a REMI Senior Geophysicist, seismic data acquisition and interpretation for the UK National Coal Board for the past year.

ABSTRACT: The paper deals with the present use of seismic surveys in the coal industry and explains how data acquisition techniques are adapted.

The processing and the problems of interpreting the data are then covered. The paper concludes with a look at areas needing further research in order to make the seismograph a more useful tool to the coal industry.

T.C. KREY

8 IN-SEAM SEISMIC EXPLORATION TECHNIQUES

By Prof. Theodor C. Krey, Retired Chief Supervisor, Prakla-Seismos GmbH, Hannover, Fed. Rep. of Germany, Professor at University of Hamburg

Prof. Theodor Krey received his Doctorate in Natural Sciences at the University of Munich in 1965, and has been since 1968 a professor at the University of Hamburg. Prof. Krey, now retired, was Chief Supervisor with Prakla-Seismos GmbH, and Technical Manager of Prakla-Seismos Geomechanik GmbH. His professional activities center around applied geophysics, mainly seismic, including refraction and mining seismics.

ABSTRACT: Surface seismic methods do normally not succeed in detecting faults with a throw of a few meters only on account of the rapid attenuation of high frequencies. Therefore the seam wave seismic method was developed, which uses waves guided by the seam. After a short presentation of the theoretical background of seam waves the practical execution of surveys using these waves underground will be reported.

Sources and receivers may not only be planted in the galleries and faces but also in drilled holes. Both transmitted and reflected seam waves are observed and evaluated. With transmitted seam waves, useful information could be obtained up to a distance of more than one kilometer, whereas the recognition of reflected waves seems to be limited to a distance of up to about 100 times the seam thickness. Within these ranges the seam wave seismic method has very often successfully assisted mine planning in western and central Europe during the last decade. Various results are discussed.

There are still some technical and physical impediments which impair the new method to a certain degree, and which call for improvements as in the case more or less with all methods of exploration geophysics. With seam waves the main problems arise from the following reasons: equipment must be firedamp-proof; the desired type of seam waves should be produced exclusively or at least to a high percentage in order to avoid undesired noise. These and other problems concerning the possible progress in the promising seam wave seismic method are briefly discussed.

W.P. HASBROUCK

9 GEOPHYSICAL EXPLORATION TECHNIQUES APPLIED TO WESTERN UNITED STATES COAL DEPOSITS

By Wilfred P. Hasbrouck, Coordinator of Coal Geophysics Research Project, Denver Federal Center, Denver, Colorado and Frank A. Hadsell, Professor Geophysics, Colorado School of Mines, Golden, Colorado, USA

Wilfred P. Hasbrouck received his Geophysical Engineer degree from the Colorado School of Mines in

1950. After working for Mountain Geophysical Co. and Stanolind Oil and Gas Co., he returned to the Colorado School of Mines in 1953 as an instructor and graduate student, receiving his D.Sc. in 1964. During this time he also consulted for several oil and mining firms, and served as Secretary-Treasurer, Vice President, and President of the Denver Geophysical Society. In 1962 he joined the U.S. Geological Survey where he worked in magnetotellurics and seismomagnetics. From 1966 to 1973, as a member of the Earth Sciences Laboratories of the National Oceanic and Atmospheric Administration, he conducted research in geopiezomagnetics and worked toward developing magnetic exploration field techniques for use in lava-covered areas. He returned to the U.S. Geological Survey in 1973, and is presently coordinator of their Coal Geophysics Research project.

F.A. HADSELL

F.A. Hadsell earned a B.S. in Petroleum Engineering and a M.S. in Physics from the University of Wyoming. In 1961 he was awarded a D.Sc. degree in Geophysical Engineering by the Colorado School of Mines. No sooner had he stepped off the platform than he joined the faculty, and has been at "Mines" ever since, except for a recent academic leave at Mobil's Field Research Laboratory, summers with the U.S. Geological Survey, a little consulting for Sun Oil Co. and a Fullbright Lecture Tour of Australia. Currently, in addition to teaching, he is the Principal Investigator on a grant: "Studies of seismic waves within coal seams."

ABSTRACT: Coals at strippable depths in the Western United States generally are thick, low grade, often severely parted and split, and variable in thickness within short distances. Also, outcrops over vast stretches of the Western Plains are either concealed or burned. For deposits such as these, drilling and stratigraphic correlation have been the mainstays of most pre-mining programs. Now, however, geophysical techniques present an economically attractive means of augmenting geologic and borehole data: by reducing the number of drill holes required to delineate and evaluate a deposit, geophysical techniques can decrease overall costs of coal exploration. Results from the joint coal geophysics research program of the U.S. Geological Survey and the Colorado School of Mines indicate that for strippable coals of the Western United States: (1) high-precision gravity surveys can be used to locate cutouts of thick coal seams; (2) burn facies can be mapped effectively and quickly with magnetic methods; (3) seismic seam waves can be observed when seam boundaries are well defined; (4) combination of borehole logging, seismic seam-wave certification, and shallow seismic reflection techniques is the preferred geophysical exploration method when precise mapping is required.

R. VOIGT

10 EVALUATION OF GEOLOGIC, HYDROLOGIC, AND GEOMECHANIC PROPERTIES CONTROLLING FUTURE LIGNITE OPEN PIT MINING

By Dr. Rudolf Voigt, Hydrogeologist, Rheinbraun Consulting, Cologne, Fed. Rep. of Germany

Dr. Rudolf Voigt obtained a diploma in geology from the Jlena University in the German Democratic Republic and a Doctorate in Natural Sciences from the Technical University of Aachen, Fed. Rep. Germany.

From 1961 to 1969 he worked on the hydrogeological mapping of the Lower Rhine Basin Area with a research department of the Aachen University.

Dr. Voigt worked on a post doctorate with the Hydrology and Water Resources Office of the University of Arizona, specializing in ground water hydrology, in 1969 and 1970. He then applied his expertise in hydrology and geology to dewatering of open pit mines in Germany and Turkey.

Since 1974, Dr. Voigt is part-time lecturer with the Heidelberg European Division, University of Maryland.

ABSTRACT: Geological properties, such as stratigraphy and tectonics, determine the basic design of a lignite open pit mine. Dewatering measures, either by means of tube wells or by means of drainage ditches and galleries, require the knowledge of the spatial distribution of the hydrologic parameters of the water bearing strata top and bottom of the seam(s). Geomechanical data of the overburden must be known in assessing the stability of slopes at the working face and at the dumping side of a mine.

The paper describes methods to obtain pertinent data during the exploration stage and how they are processed and used during the planning phase.

R.D. ELLISON

11 GEOTECHNOLOGY: AN INTEGRAL PART OF MINE PLANNING

By Richard D. Ellison, Vice-President, and Allen G. Thurman, Manager Rocky Mountain Operations, D'Appolonia Consulting Engineers, Inc., Pittsburgh, Pennsylvania, USA

Dr. Ellison is a civil engineering graduate of Pittsburgh's Carnegie-Mellon University. He is a vice president of D'Appolonia Consulting Engineers and has directed this firm's mining and environmental activities for the past decade. His interdisciplinary group has been involved in over 200 projects throughout the United States and abroad, including work for many of the major coal mining operators.

Dr. Thurman is the manager of D'Appolonia's Rocky Mountain operations. His civil engineering education and experience have been obtained in various western states and institutions, while his Ph.D. was obtained from Carnegie-Mellon University in Pittsburgh. He has served as Chairman of the University of Denver's Department of Civil Engineering, Acting Dean of this institution's College of Engineering and was also an Assistant Director of the Denver Research Institute. He is professionally registered in Colorado, Oregon, Wyoming, Montana and Utah.

A.G. THURMAN

Dr. Thurman joined D'Appolonia in 1974 to direct the firm's consulting services to the rapidly expanding western coal mining industry.

ABSTRACT: Worldwide energy demands are causing rapid expansions of coal mining activity in all of the major United States coal areas. New, high-production operations, often in previously unmined and difficult seams, entail large capital expenditures for mine equipment and development. Cost considerations and a trend towards increased mechanization are gradually causing industry to place more emphasis on premining investigation and analysis.

Currently, the collection of most premining data is accomplished through expensive boring and laboratory testing programs, aimed primarily towards evaluating only the coal quality and quantity. There often is too little expert geotechnical involvement in the planning or interpretation of these core boring or laboratory testing programs. Also, appropriate geological interpretation and geophysical exploration (surface and borehole) techniques are often not used to efficiently maximize knowledge from costly boring programs.

This paper discusses:

- The relative importance and justification for expert geotechnical involvement in premining investigations and design, and the relationship of this expert involvement with traditional exploration and mine planning;

- Relationship between geotechnical factors, geological conditions and mining considerations;

- Efficient means of implementing geotechnical premining programs for a variety of conditions and mining methods;

- Procedures for getting maximum information from each boring, without increasing costs;

- Data evaluation techniques that vary from simple experience judgements to complex computer analyses; and

- Examples of applications in major mining projects.

An end product of each premining investigation program is the prediction of mine behavior in order to select the best mining methods, minimize operating difficulties and eliminate costly surprises.

R.G. WILSON

12 ESTIMATING THE POTENTIAL OF A COAL BASIN

By Rodney G. Wilson, Consultant Geologist, CRA Exploration Pty. Ltd., Melbourne, Australia

Rod Wilson was born in the Yorkshire Coalfield and educated at Manchester University where he gained the degree of BSc (Hons) Geology in 1959. Subsequently he worked for the National Coal Board Coal Survey and Opencast Executive until 1971. During this period Mr. Wilson was involved in coal exploration activities in the majority of British Coalfields. In 1971 he joined the Rio Tinto Group first as Supervising Geologist for the coal exploration activities in Australia, then as Chief Geologist (Coal and Uranium) and more recently Consultant Geologist of CRA Exploration in Melbourne, Australia. During this period Mr. Wilson has visited coal, uranium and base metal mines and exploration camps in some 20 different countries.

ABSTRACT: The task of estimating the potential of a coal basin includes four major elements viz. Regional geological appraisal; Engineering and reverse economic studies; Detailed evaluation of coal deposits; and Feasibility studies. This paper restricts itself to the Regional geological aspects of the problem. It proposes the formulation of a predictive model of a coal depository based upon an understanding of coal as a sedimentary rock, a knowledge of the geological factors influencing coal properties and an understanding of the depositional environment of coal bearing strata. The principles on which the model is based i.e. the genesis of coal type and coal rank and the effects of the tectonic setting of a coal basin are discussed, and the paper concludes with a brief methodology.

P.G. STRAUSS

13 COAL PETROGRAPHY AS AN EXPLORATION AID IN THE WEST CIRCUM-PACIFIC

By Peter G. Strauss, Chief Coal Geologist, Robertson Research (Australia) Pty. Limited, Sydney, Nigel J. Russell, Coal Petrologist, Robertson Research (Singapore) Private Limited, Allan J.R. Bennett, Coal Petrologist, Commonwealth Scientific and Industrial Research Organisation, Division of Mineralogy, Sydney, and C. Michael Atkinson, Senior Coal Geologist, Robertson Research (Australia) Pty. Limited, Sydney

Mr. P.G. Strauss obtained a degree in Geology from the University of London in 1954. He worked for fifteen years with the National Coal Board in the East Midlands, UK, which included service with the Opencast Executive, the Coal Survey, and with the Regional Geological Services as Area Geologist.

In 1970 he joined the Robertson Research Group in Sydney, Australia and, as Chief Coal Geologist, has been technically responsible for the group's consultancy interests world wide, the most widely ranging involvement being in the west Circum-Pacific region.

Mr. Nigel J. Russell graduated in Geology with first class honours at St. Andrews University, Fife, in 1966. He was stationed in Australia from 1968 to 1975; the first year as teaching fellow at the University of New England, followed by two years as a petrologist/mineralogist with Planet Management & Research Pty. Limited.

He joined Robertson Research (Australia) Pty. Ltd. in 1972 where he has built up a coal microscopy section, undertaking petrographics analyses and supporting interpretation studies of coals especially from the west Circum-Pacific region. Early in 1976 he was transferred to Singapore to continue his work and to set up a geochemistry section.

A.J.R. BENNETT

Mr. A.J.R. Bennett studied mining at the Camborne School of Mines, mining geology at the Royal School of Mines in London, and geology at Chelsea College, London. Mr. Bennett spent ten years on research of coal preparation and coal petrology at the National Coal Board, Coal Research Establishment, Stoke Orchard, England.

In 1963 he joined CSIRO in the Division of Mineralogy at North Ryde, N.S.W. and is currently studying the origin and diagenesis of coal and carbonaceous material in Australian sedimentary basins.

Mr. C.M. Atkinson graduated in geology at the University of Durham in 1966 and has since been based in Australia. From 1966-1970 he was engaged with John Taylor & Sons in regional geological compilation and appraisal work in a variety of minerals, including sedimentary uranium deposits.

Since 1970 he has been working with the Robertson Research Group, and over one year as a petroleum geologist in Indonesia. From 1973, as Senior Coal Geologist, he has been involved in numerous sedimentary basin studies, coal property assessments and exploration projects principally in Australia and the Indonesian Archipelago.

C.M. ATKINSON

ABSTRACT: The advantages of utilizing microscopy techniques on coal sampled in the field are often not sufficiently appreciated. Normally the results are valid for weathered samples in contrast to more conventional chemical analyses.

The techniques involved in field sampling and in the determination of the rank and type of coal are briefly reviewed. A range of examples are presented, taken from initial exploration surveys in the west Circum-Pacific region. The interpretation of the petrographic results assists in a variety of ways in providing a preliminary geological and economic evaluation of a coal deposit.

An attempt is made to summarize the coal-type provinces throughout the region.

W.H. SMITH

14 COMPUTER EVALUATION AND CLASSIFICATION OF COAL RESERVES

By William H. Smith, Consulting Geologist, Champaign, Illinois, USA

William H. Smith received his B.S. and M.S. degrees in Geology from Ohio State University with a minor in Cartography. He served as an officer in the U.S. Navy during World War II, following which he joined the Ohio Geological Survey where he was in charge of the Coal Geology Division. In 1955 he joined the Illinois Geological Survey where his principal research interests have been in mapping coal resources and in stratigraphy of coal-bearing rocks. He is a registered Professional Geologist, a Fellow of the Geological Society of America, and a member of the American Institute of Mining and Metallurgical Engineers, Society of Economic Paleontologists and Mineralogists, Society of Sigma Xi, and the American Congress on Surveying and Mapping. He is the author of numerous publications on coal geology and related fields.

ABSTRACT: It is now possible to store in a data bank and later retrieve in map format all of the data concerning a mining operation that is required for land management, geological evaluations and mining engineering.

This mapping requires a coal data base containing all the specific information obtained during the exploration and evaluation program. The build-up of this base is described with its storage and retrieval system.

The calculation of coal reserves by computer is described and the method of excluding areas in which the seam is too thin or absent. Reserves can be classified, for example, by ash or sulphur content.

Conversion of existing records to establish a fully integrated coal data base capable of producing structure contour maps, coal isopach maps, and complete reserves estimates, will take some time but can be built up a module at a time.

W.W. SVENDSEN

15 COAL EXPLORATION TECHNIQUES AND TOOLS TO MEET THE DEMAND OF THE COAL INDUSTRY

By Walter W. Svendsen, Technical Director, Longyear Co., Minneapolis, Minnesota, USA

Walter W. Svendsen joined the Longyear Co. in 1946 as a Design Engineer and was instrumental in the development of a number of new products.

He was the co-inventor of Longyear's Wireline Core Barrel and holds patents on other items such as a shut off system for core barrels, a semi-automatic chucking device, a long feed drill with articulating mast as well as patents on several other devices used in core drilling.

In 1958 he became part owner of a contracting and manufacturing concern and in 1967 President of Drill Tech. Inc., a research and development group.

In 1970 Mr. Svendsen returned to Longyear as Manager of International Engineering and in 1971 was appointed Manager of Corporate Engineering. In 1975 he was named Technical Director.

Mr. Svendsen has served as President of the Diamond Core Drill Manufacturers Assoc., Chairman of the Technical Committee for Core Drilling Equipment, U.S.A. representative to the I.S.O., and at present is Chairman of the International Standards Committee of D.C.D.M.A.

ABSTRACT: Strip mining and auger mining in the United States have expanded while underground mining has declined. This has changed exploration to increase the mobility, speed and instrumentation of the drilling rig and to produce larger core samples. Types of data needed to decide what mining method to use are discussed. Degassing techniques and methane gas recovery and techniques are reviewed. Wireline equipment, operating techniques, applications and advantages are discussed. Improved equipment for deep, difficult explorations to reduce costs and time are described. Global field and offshore drilling experience and techniques are compared. Instrumentation needs and advantages in reducing costs and improving information are presented.

K. SHAW

16 DEVELOPMENT AND ADAPTATION OF DRILLING EQUIPMENT TO COAL EXPLORATION

By Keith Shaw, Chief Exploration Engineer, National Coal Board, Huthwaite, Sutton-in-Ashfield, Nottinghamshire, UK

Keith Shaw, C. Eng., F.I. Min. E., apprenticed as a mining surveyor with Manchester Collieries Ltd. in 1939 and was educated at Wigan Mining & Technology College to qualify in 1945 as a mine surveyor.

On nationalisation of the British Coal Mining Industry in 1947, he joined the Planning Department of the newly constituted North Western Divisional Headquarters, subsequently transferring as Planning Engineer to the St. Helens Area of that Division.

In 1951, Mr. Shaw was appointed Assistant Divisional Planning Engineer at the East Midlands Divisional Headquarters and in 1963 was promoted to Divisional Planning Department.

On the dissolution of Divisions in 1967, he was appointed Deputy Chief Mining Engineer of the North Nottinghamshire Area, a post which he retained until his recent appointment in September 1975 as the Board's Chief Exploration Engineer.

ABSTRACT: The paper discusses the modifications made to exploration drills and equipment to fit them to the special conditions to be found in the exploration for coal. It considers the subject under the following headings:

Underground drilling in which the machines and equipment have to be adapted to confined spaces and stringent safety regulations. Wire line equipment has been adapted to drill upwards.

In-seam drilling in which the drilling is done for methane drainage, and to prove the continuity of the seam or locate discontinuities.

Surface drilling which is classified as shallow/medium and deep drilling. A recent departure in using oil well equipment has proved successful and given much improved drill rates.

Off-shore drilling which describes equipment used in the Wimpey Sealab.

17 MAIN PRINCIPLES OF EXPLORATION OF COAL DEPOSITS IN THE USSR AND NEW PROBLEMS OF EXPLORATION METHODS AT THE PRESENT STAGE

By E.V. Terentyev, Candidate of Geological and Mineralogical Sciences, Ministry for the Coal Industry of the USSR

E.V. TERENTYEV

E.V. Terentyev is Candidate of Geological and Mineralogical Sciences, Ministry for the Coal Industry of the USSR, and Chief Engineer of the ministry's Geological Department. He has extensive field experience, particularly in the Pechora Basin, which is within the Arctic Circle, and Donets Basin, which is the largest coal-producing basin in the Soviet Union and is in the Ukraine, where he was born and raised. In his present position Terentyev travels to all Soviet coal basins, including those in Siberia, to provide geological investigative services on behalf of the Coal Ministry. He has published a number of articles and was a member of the editorial board for the "Map of the Coal Fields of the World." No abstract of his paper was furnished.

18 EXPLORING COAL DEPOSITS FOR SURFACE MINING

C.E. WIER

By Charles E. Wier, Exploration Manager,
International Coal, AMAX Coal Co.,
Indianapolis, Indiana, USA

Charles E. Wier was born in Jasonville, Indiana, a coal mining town, on May 15, 1921. He received A.B., A.M. and Ph. D. degrees in economic geology at Indiana University—the latter in 1955. Concurrently, he taught geology at Indiana University and was head of coal research at the Indiana Geological Survey until 1974. Since then, he has worked on coal exploration projects for AMAX International Group in Botswana, Southwest Africa, Mozambique, Madagascar, Sarawak, and Australia.

He has published about 50 articles on coal geology and remote sensing. Memberships in professional organizations include: The Geological Society of America, Society of Economic Geologists, American Institute of Mining Engineers, and American Association of Petroleum Geologists.

ABSTRACT: Exploration for coal consists of four stages: 1) Preliminary investigation, 2) Reconnaissance of favorable area, 3) Detailed exploration of project area, and 4) Evaluation. There are many procedures and decisions in each stage and each stage, after the first, depends on information obtained in previous stages.

The first task in preliminary investigation is to do thorough library research. Many bibliographies on coal are available to aid in this search for information. After available information is digested, the geologist visits the area and does a regional reconnaissance. This should include seeking information from members of local government agencies.

After a favorable area is chosen, depending on availability of geologic information, a reconnaissance geologic mapping program may be desirable. A few drill holes may be required to see if coal seams are present or to determine the approximate quality.

When considering a restricted area, further and more detailed geologic mapping may be required. Detailed drilling will be required. A drilling program affords the best opportunity to obtain a large amount of information. The coal seam can be seen in a core, geophysical logs help with correlation and seam evaluations, and samples can be collected for analysis.

Much of the data collected are evaluated by plotting on maps. Iso lines showing thickness of overburden, thickness of coal, overburden to coal ratio, structure, percent ash, percent sulfur, etc., quickly show desirable and undesirable areas. If the report that is written at this stage is favorable, a full scale feasibility study is made. Results of this study determine if the project is carried forward into development.

19 EXPLORATION AND GEOLOGICAL STRUCTURE OF COAL MEASURES IN WESTERN CANADA

C.W. BALL

By Clive W. Ball, Chief Geologist, Canex Placer Ltd., Vancouver B.C., Canada

Clive Ball, born in Australia, was educated at the Brisbane Grammar School and University of Queensland receiving a Master of Science degree in Geology and Mineralogy.

As a field geologist his work has taken him through the greater part of Australia on assignments for the Broken Hill Proprietary Company, The Bureau of Mineral Resources, Canberra, and the Geological Survey of Queensland.

For the past twenty-eight years Clive Ball has worked for Placer Development Limited with head office in Vancouver, British Columbia and has been responsible for field examinations in Canada, the U.S.A. and Columbia. During his career he has completed field examinations for the evaluation of gold, silver, lead, zinc, copper, molybdenum, tin, tantalite and tungsten ore deposits, and a variety of non-metallic mineral occurrences including coal, potash, magnesite, fluorite, and bauxite.

ABSTRACT: Exploration for coal in Western Canada is concentrated in Jurassic, Cretaceous and Tertiary sub-basins each of which tends to be characterized by coal of a specific rank.

With the exception of the Plains Region of Alberta and Saskatchewan, the main coal measures have suffered intense folding and thrust faulting with consequent distortion and disruption of the individual seams. Further complications are caused by normal faulting.

The seams often show steep pitch and correlation is difficult on account of facies changes and lensing out of the coal seams. Exploration involves diamond drilling, rotary test wells and the use of electro-logs. Adit entries are driven in order to obtain samples for extensive laboratory testing. Areas of coking coal in the Foothills and Rocky Mountain regions require close grid testing for correlation and gauging the coal quality.

On account of recent demands for thermal coal, the lower rank lignitic and sub-bituminous coals in the Tertiary basins are being brought closer to the production stage.

The potential reserves of coal for coking, thermal and chemical uses are considered to be quite adequate for future requirements. However, the structural complexity of the geology will continue to pose a challenge for the mining engineers in continuing to meet rising costs.

J.K. HAMMES

20 FINANCIAL CONSIDERATIONS IN EVALUATING NEWLY DISCOVERED COAL DEPOSITS AND VENTURES

By John K. Hammes, Vice-President, Metals and Mining Department, Citibank, N.A., New York City, New York, USA

John K. Hammes, Vice-President of the Mining Division of the First National City Bank since 1972 holds a B.S. degree in Mathematics and Geology from the University of Missouri. He also graduated with an M.S. in Mining Engineering and a Ph. D. in Mineral Engineering and Microeconomics (B.A.) from the University of Minnesota, School of Mines.

In his present capacity, John Hammes provides professional mining engineering advice to the Bank. His previous experience has been in the field of economic and technical evaluation of mining ventures with Kennecott Copper Corp., and with the U.S. Bureau of Mines.

Mr. Hammes has published for AIME's "Copper Ore Mining," a chapter of "Surface Mining Volume" and various technical articles on mine finance.

ABSTRACT: The paper follows a discussion of financial considerations in the exploration, and evaluation phase is presented. Emphasis is on international aspects such as problems introduced by exchange rate fluctuations, and questions of financing exploration. Increased complexity of evaluation process for international projects is reviewed. Financial considerations which should be taken into account in the provisions of coal sales agreements and work contracts negotiated during this phase are covered.

R.A. SCHMIDT

21 CONSUMER COAL CRITERIA AS A GUIDE TO EXPLORATION

By Richard A. Schmidt, Technical Manager, Fossil Fuel Resources, Fossil Fuel Department, Electric Power Research Institute, Palo Alto, California, USA

Dr. Schmidt is Manager of the Resource Extraction and Preparation Program at the Electric Power Research Institute.

Prior to May 1, 1974, Dr. Schmidt was Senior Geologist and Program Manager at Stanford Research Institute, where he participated in research in economic geology, mineral economics, energy R&D, environmental assessments, and support of legislative and executive decision-making on behalf of Federal and State governments, as well as private industry.

Dr. Schmidt carried out numerous studies on the technology and costs of coal production for combustion and conversion. He led SRI's Study of Surface Coal Mining in West Virginia, performed for the legislature of that State, and was a member of the National Academy of Sciences' Committee on the Rehabilitation Potential of Western Coal Lands. He also served as a consultant to the Department of Defense in connection with several studies and assessments dealing with energy-related matters.

Dr. Schmidt received his formal geological education at Franklin and Marshall College, and his advanced degrees were earned at the University of Wisconsin (Madison). He is a member of the Society of Mining Engineers.

ABSTRACT: Exploration for coal resources and reserves will fall short of meeting consumer requirements unless coal quality factors are considered together with those pertaining to quantity. Standard coal analyses are useful and convenient, but are not universally available; even when available, such data do not necessarily represent the behavior of coals under conditions of use. As a result, deposits of coal-bearing rocks that otherwise may be targets for intensive exploration could prove to be of limited potential. An assessment of resource/reserve data from the standpoint of criteria for optimum use in electric utility combustion and conversion is presented as a guide to future coal exploration.

●

The presentations by A. Papadopoulos, E. Doganis, J. Nunes Neto, and P.K. Ghosh that constitute chapters 22, 23, and 24 were not part of the original Symposium schedule. These chapters record these unscheduled papers as well as major comments offered during the Symposium discussion periods, and are reprinted here because of their substantial nature and relevance to the Symposium theme. The publishers regret that biographies of the authors and abstracts of their papers were unavailable at the time of printing.

SECTION 1

Chairman's Opening Address
by A.F. Lessard

**A Keynote Address on
the Place of Exploration in
the International Coal Mining Industry**
by Sir Derek Ezra

Section Discussion

CHAIRMAN'S OPENING ADDRESS

A. F. LESSARD
Chairman

Resources Engineering and Management International
Denver, Colorado, U.S.A.

I. INTRODUCTION

1. The Purposes and Objectives of the Symposium.

(1) To reassess coal exploration in the light of the future of the industry.

(2) To highlight and summarise state-of-the-art developments in the application of exploration, drilling and mining technology to coal exploration and mine planning.

(3) To review the expanding role of computational techniques and systems in exploration and mine planning.

(4) To assess the relationships of exploration and mine development to marketing, conservation and pollution control considerations.

(5) To provide an international forum for, and an exchange of ideas and experience, among internationa leaders in the field of coal exploration.

2. The Permanent Contribution of the Symposium to the
World Coal Industry.

   (1) The publication of the Proceedings, and their
       distribution on a world wide basis as a means of
       diffusion to practitioners of current developments,
       practices, trends and problems in the area of coal
       exploration and mine planning.

   (2) The use of the Proceedings in university level
       training programmes.

   (3) The publication of follow-up general interest
       articles and press releases to acquaint policy
       makers in government and industry with the reali-
       ties of expanded coal exploration and coal mining
       programmes.

   (4) The establishment of personal contacts among
       leaders in the field in an effort to encourage
       cooperation and cross fertilisation in the use and
       evaluation of new techniques in exploration and
       mine planning.

   (5) The provision of an organisational vehicle for
       future symposiums or special interest seminars in
       the overall field of coal exploration and mine
       planning.

II MAJOR ISSUES AND TRENDS IN INTERNATIONAL ENERGY
ECONOMICS AS A BACKGROUND TO THE SYMPOSIUM'S DISCUSSIONS

1. Introduction

   (1) The world wide energy problem long antedates the
       Arab-Israeli oil embargo crisis.

   (2) Unease and concern over accelerating energy consump-
       tion and growing concentration of supply sources
       was evident much earlier.

   (3) The October 1973 war made the problem less manageable
       in terms of seeking and reaching agreement on
       solutions at national and international levels,
       and helped foster a 'crisis' mentality where all
       too often consuming country cooperation and judi-
       cious reflection flew out of the window.

2.   Some Recent Trends in Energy Consumption

   (1)   The world experienced an ever accelerating rate
         of energy consumption.

         .   Between 1925 and 1950 world energy consumption
             grew at the rate of 2.2%.

         .   Between 1950 and 1960 world energy consumption
             grew at the rate of 5% to $5\frac{1}{2}$%.

         .   By 1970 each 1/10 of a percentage point in
             increased energy consumption was equivalent
             to over one million barrels of oil per day.

   (2)   Major increases in energy consumption are
         concentrated in the industrialised nations.

         .   North America, W. Europe and Japan experienced
             dramatic increases in energy consumption.

         .   North American per capita consumption, as an
             example, is currently 30-40 times levels in
             Africa and developing nations in Asia.

   (3)   Increased total energy consumption was accompanied
         by a dramatic change in energy source materials.

         .   Coal accounted for 55% of energy produced in
             1950, while oil accounted for 29%.

         .   By 1972 coal accounted for 28%, while oil
             accounted for 46%.

         .   Oil supplied 6% of power utility fuel in the
             U.S. in the mid 60's. By 1972, this had
             increased to 16%.

   (4)   Sources of origin of supply of energy became
         highly concentrated.

         .   Oil imports to Western Europe in 1962 provided
             37% of total energy consumption. By 1972,
             oil imports provided 60%.

         .   The Middle East and North Africa provided
             Western Europe with 47% of its energy supply
             by 1972.

. The world's acute dependence on Middle East/
North African sources will remain a fact of
life for many years to come.

(5) The forecasting of energy requirements and trends in
the past proved to be poor guides to policy makers
concerned with the problems of energy supply and
demand.

. Total energy consumption forecasts were
normally overly conservative.

. Gross misjudgements were made in assessing
the role of oil. As late as 1966 the OECD
was forecasting a far greater role for coal
than actually developed.

. The rate of oil exploration - key to continued
reserves discoveries and added production -
particularly in the U.S.A., was highly over-
estimated.

. Comfortable global reserve/production ratios
failed to take into account the practical
problems of converting reserves to production
where and when needed.

3. The Coal Industry World Wide has Undergone Fundamental
Structural Changes.

(1) The industry was generally depressed and had been
for a number of years virtually "written off" in a
number of major coal producing countries as a
matter of long range national policy.

(2) Increasing manpower availability problems, capital
raising difficulties, conservation pressure, and
reducing profit margins seemed to foretell over
the long term the eventual run-down fate of the
industry - particulary in Western Europe.

III THIS INTERNATIONAL COAL EXPLORATION SYMPOSIUM ADDRESSES
ITSELF TO AN EMERGING AND REVITALISED WORLD ROLE FOR
COAL. SEVERAL MAJOR THEMES WILL BE DEVELOPED BY THE
SYMPOSIUM SPEAKERS

1. The Essential Unity of the Total World Energy Scene -
   the Technical, Economic and Political Considerations -
   Determine the Pattern of Energy Production and the
   Eventual Role of Coal.

   (1) Mr. H.E. Collins in his discussion of "World Coal
       Supply and Demand" reviews this world energy
       picture and the probable future role of coal.

   (2) Sir Derek Ezra in his Keynote Speech sets the
       scene by showing the place of coal exploration in
       the international coal mining industry.

2. The Coal Industry Is Seeing the Significant Application
   of New Technology in Every Phase of its Operations. The
   Cross-Fertilisation of the Related Oil and Coal Industry
   Technologies is Becoming Particularly Important in Coal
   Exploration.

   (1) Messrs. Krey, Hasbrouck and Daly bring to the
       Symposium state-of-the-art experience in the
       application of seismic technology to coal
       exploration.

   (2) Messrs. Reeves and Lavers review geophysical
       logging techniques and developments in coal
       petrophysics.

   (3) Mr. K. Shaw discusses the adaptation of drilling
       equipment to coal exploration.

3. Technology must be adapted to Respond Effectively to
   Economic Imperatives. The Highly Mechanised and Capital
   Intensive Nature of the Industry Makes Mandatory Greater
   Precision for Coal Mine Planning, Which in Turn Requires
   Greater Pre-Knowledge of Natural Conditions.

   (1) Dr. R. Voigt emphasises the need to understand
       fully the geological, hydrological and geomechanical
       conditions in mining unconsolidated ground in
       order to minimise waste removal without over-
       stepping and creating relatively dangerous conditions

(2) R.D. Ellison presents his views on the collection,
evaluation and use of geotechnical data in mining
in hard ground - both underground and surface.

(3) Messrs. Krey, Daly, Clarke and Hasbrouck describe
the use of seismic information in determining
location, direction, and displacement of faults in
order to take account of the laying out of working
faces in mine planning.

4. The Computer is Being Increasingly Utilised as a Normal
Tool in Exploration and Mine Planning.

(1) Mr. W.K. Smith summarises programmes designed to
use the computer in evaluating and classifying
coal reserves, and in mine design.

(2) Messrs. Reeves and Lavers mention the use of
computers in processing geophysical data, and Mr.
Ellison in evaluating geotechnical data.

5. Information on the Rank and Composition of Coal Under
Exploration Will be Required as Early as Possible in the
Exploration Process to Determine Marketability of End
Product.

(1) Messrs. Lavers, Wilson, Russell and Schmidt address
various aspects of the impact on the exploration
process of ever more stringent product specifications being imposed by consumers seeking efficient
usage and compliance with anti-pollution laws.

IV SOME THOUGHTS ON THE FUTURE: A LOOK INTO ONE MAN'S
CRYSTAL BALL

1. World Demand for Energy Will be Increasingly Conditioned
by the Ability to Find, Produce, Deliver and Pay For Raw
Materials Required for Energy Generation.

(1) My personal view, however, is that no major supply
problems will develop in the short to medium term.

(2) Prices for oil will establish the pricing standard
for other energy generating materials, and in the
short run oil prices can be expected to increase.

(3)   Exploration activities will continue at a high level of activity, but will be selective and sensitive to longer run political implications in host countries.

(4)   Industrialised nations will continue to consume energy on an ever-increasing rate, and account for the "lion's share" of consumption on both gross and per capita bases.

2.   Middle East and North and West African Supply Sources Will Continue to Dominate as Export Suppliers to the World for a Considerable Period.

(1)   Shut-in capacity in these countries can even now supply world increases in demand which can be forecast over the short to medium term.

(2)   Further exploration in these countries can be expected to increase existing reserves estimates and define major new gas and crude producing structures.

(3)   Improved operating practices will result in higher recovery percentages from reservoirs and almost complete utilisation of associated gas now flared in large quantities.

3.   Coal Will Become in the Long Term Less a Source for Thermal Power Generation and Increasingly a Raw Material Used for Producing Hydrocarbon Substitutes.

(1)   The enormous world reserves of coal when compared to gas and crude oil reserves will dictate such a trend.

(2)   South Africa and Eastern Europe can be expected to lead the way in the practical application of new technology in this field.

4.   Serious and Practical Energy Planning and Coordination will Take Place Increasingly at International, Regional and National Levels.

(1)   Defence needs in a period of heightened international tensions will force the Western nations to coordinate energy policies, ensure stockpiles,

prepare energy contingency plans and seek to
ensure continuity of supply. NATO can be expected
to play a major role in such international planning.

(2) The increasing capabilities of the national oil
     and coal companies and boards in both producing
     and consuming countries - will mean increasing
     nation-to-nation agreements based on long range
     demand and supply planning, which will probably
     be tied to long term technical aid and financial
     assistance agreements on a government-
     to-government basis.

(3) OPEC and OAPEC will become less and less concerned
     with pure income maximisation of their member
     countries, and will increasingly become a serious
     force in stimulating long range planning for
     stabilising demand and supply relationships, main-
     taining order in product pricing, developing
     programmes for recycling oil funds and providing
     grant and loan aid to the non-oil producing develo-
     ping world.

(4) Planning at the national level will increasingly
     stress development of coordinated national energy
     demand and supply policies and programmes. The
     concept of "Energy Ministries" will become accepted
     and the activities of national gas, oil, electricity,
     coal and related government and private organisa-
     tions will increasingly be brought under tighter
     planning, investment and budgetary controls.

The Role and Activities of the International Oil
Companies Will Continue to Change. They Will Increasingly
Become Energy Contractors and Operators on A World Scale
with Interests in all Energy Producing Raw Materials.

(1) The international companies will continue their
     recent trend of becoming involved in the whole
     range of energy raw materials including coal, gas,
     oil, atomic energy, gas, thermal sources, tar
     sands, and oil shales.

(2) The international companies will continue to lose
     their control over crude sources. They will be
     forced to make their profits from crude oil obtained at
     only slightly lower rates than going world market prices
     available to any large consumer.

(3) The international companies will increasingly be subjected to technical and operational regulations and surveillance, and be under frequent investigation on corporate, profit and market practices.

(4) The international oil companies will increasingly become minority joint venture partners with state bodies, and will supply technology, manpower and to a lesser degree, capital.

(5) The international oil companies have the capabilities to become - and may well become - the most important energy contractors to the world. Their services may well range from planning and execution of large scale exploration projects, through design and construction of oil/gas field and mine facilities, to design and construction of process plant, terminals and offshore facilities. They can also be expected to supply contract management services over the full range of energy-related projects. Major oil company - mining company - construction company mergers, acquisitions, and joint ventures, may become a reality in the years ahead.

7. The National Petroleum Company - Now in Evidence in Almost All the Major Producing and Consuming Countries - Will Increasingly Assume the Roles and Responsibilities of the International Oil Companies.

(1) The new national companies now emerging in every oil producing country - each with statutory rights to engage in every phase of international oil, gas and associated activities - are rapidly changing from paper organisations developed in response to a political decision to reasonably effective, business oriented enterprises. They will increasingly attract and develop top men - the "best that money can buy".

(2) The national oil companies in the consuming nations will become their government's vehicle for securing tied sources of energy raw materials anywhere in the world, and for directly participating in internal and foreign ventures.

(3) Direct relations between the producer - consumer national oil companies will increase as a factor of the growing internal capabilities of these companies

(4) The national oil companies, as arms of their central governments, will have access to the enormous amounts of capital required for major energy projects. This capital capability has long been almost the exclusive domain of the international oil companies, whose retained earnings and banking connections have enabled them to commit the massive funds which the energy business requires.

8. "Conservation" will Move Increasingly from the Environmental Enthusiast to the Development of Well Defined and Enforced Conservation Codes and Standards Policed by Qualified Technologists.

A Keynote Address on

THE PLACE OF EXPLORATION IN

THE INTERNATIONAL COALMINING INDUSTRY

Sir Derek Ezra

Introduction

There is little need to emphasise to a gathering such
as this, the renewed interest in coal which has become
evident over the past few years. Indeed the holding of
this Symposium is itself both a product and an indication
of this growth in interest. I in fact was approached about
this event, and asked to give this address, when Mr. George
Roman and I met in Mexico City last October; there we were
attending an International Iron and Steel Institute
Conference - to which I had been invited to speak - where
a whole session was devoted to hearing about and discussing
the international coal situation. Wherever I go, within
Britain, in Europe and beyond, I find enormous interest in
what is happening in the coal industries of the world.

The stimulus for this revival of interest was of
course provided by the abrupt rise in the price of oil in
1973, but it is true to say that even before the events of
that year, the world's coal-producing nations were
reassessing their coal policies. It was already being
realised that approximately exponential growth in world
energy demand, set against the limit in the total reserves
of fossil fuels, particularly of oil and natural gas, could

rise to serious problems before the end of the century, as Mr. Collins has illustrated in his paper.

Exploration is of course an essential first step in the extraction of fuels, and indeed of metals and other minerals. I would like to talk a little today about the role of exploration in coalmining, and to illustrate recent developments in exploration in the United Kingdom.

The role of exploration

It is worth reflecting for a moment on the differences between coal exploration and exploration for other fuels, minerals and metals.

This Symposium began with a paper describing where exploration has already shown most of the world's coal to be. This could not be the case for the world's metal mining and petroleum industries, where the successful discovery of the whereabouts of new reserves by exploration controls their future to a far greater extent. However, the knowledge of the world occurrence of coal is not sufficient to characterise the world's coal reserves in the manner required. The importance of exploration in the coal industry is not so much finding more of it but, knowing where most of the coalfields are, in choosing the place where it can be mined most efficiently.

In fact, because coalfields are rarely completely concealed, unlike ore-bodies or oil-bearing strata, they present a relatively unmissable target to the geologist when the outcrop at the surface, and to the drill hole when they do not outcrop. Finding out in general terms where coal lies is not therefore a very difficult or scientific exercise - and the discovery of the vast majority of British coalfields preceded British geologists (the first to discover themselves as an organised scientific body).

Effect of exploration on the market position

The local results of exploration in coal have not, in recent history, altered the market position of coal vis a vis its substitutes in the market for bulk energy supplies. Unexpected discoveries of very large easily extractable bodies of the more valuable type of metal have altered both the market and the supply position for metals. For instance, the discovery of Broken Hill virtually closed the long

established lead mining industry of Britain. More recently
the exploration leading to the discovery of the super-giant
oilfields of the Middle East such as that at Gawar in Arabia,
led to a complete change in the world's bulk energy markets.
Supplies were well in excess of demand. With this experience,
it is easy to forget that the early discoveries of coal
brought about the industrial revolution, located the industries and created the markets and societies which the oil
and metalliferous mining industries now serve. This is why
most of the coalfields conveniently located for the world's
industrial centres already have a long history.

Why should this be?

(i)   because coalfields commonly occur on a larger scale
        than other mineral occurrences;

(ii)   extraction is necessary serially from the seams
        instead of simultaneously at each vertical point in
        the full thickness of the reserves (because strata
        strengths, pressures and movements would not permit
        the latter in the case of coal).

Coalfields, therefore, once discovered, last longer
than oilfields.

The stimulus for exploration

It follows from my previous remarks that in the
metalliferous mining and oil industries, the possibility of
unexpectedly discovering a really rich 'bonanza' ore body
or a 'super-giant' oilfield in the next 'prospect' (and of
being able to exhaust it relatively quickly) commonly
provides a certain impetus to the maintenance of exploration
at a higher level in these industries than is the case in
coal. A large proportion of you at this symposium are from
these other industries, simply because the demand and supply
position in coal has made it expedient for them to diversify
into coal. In coal, in the absence of the 'bonanza' or
'super-giant' stimulus, exploration considerations are more
influenced by the view taken of future market demand in the
light of currently accessed economically mineable reserves.

Market demand

If market demand is the crucial factor in determining
levels of exploration for coal, let us examine how this has

evidenced itself since the last war, and how current estimates of forward demand are affecting exploration programmes today.

(i) Looking back

Changes in the process of finding oil during the 1950s, particularly changes in exploration technology, brought about a huge increase in supply compared with demand, which lasted through most of the 1960s. Since, once discovered, the process of extracting and utilising the oil is relatively simple and cheap, the relative costs of oil and coal moved very much in favour of oil. However, most of the costs of the coal industry are tied up in the extraction and utilisation processes, and the only way in which coal could respond rapidly to these events was by closure of high-cost mines with access only to the less easily recoverable reserves, and by concentration of output in the most productive portions of the most productive seams of the most productive collieries. This process occurred in particular in Western Europe, where coal production (compared for example with the United States) was relatively high-cost, and where (unlike Eastern Europe) it was not shielded from adverse movements in the supply/demand position.

It is easy to forget however that the world consumption of solid fuels continued to rise throughout this period, but at a slower rate than that of oil and natural gas

(m.t.c.e.)	1950	1960	1970	1974	2000 (NCB est.)
Coal and lignite	1,605	2,191	2,394	2,430	4,600
Oil	700	1,396	3,002	3,740	5,500
Natural gas	261	622	1,436	1,750	3,500
Primary electricity	41	86	157	220	6,400
	2,607	4,295	6,989	8,140	20,000

Round about 1970, the situation began to change: as I said at the beginning, the catalyst was the action of the OPEC nations, but there were other underlying trends which were bringing about this change. Adverse change in the opportunity to find oil, brought about by the early successes and by natural opportunities to exploit the short life of oilfields, and to pay back the now heavier (largely offshore) exploration costs, has made less risk capital available to pursue the world's remaining oilfields.

But more fundamentally, the realisation of the basic imbalance between finite reserves of oil and natural gas on the one hand, and the growth in demand on the other, led to an acceleration of interest in coal exploration. Mr. Collins has reviewed in his paper the growth in world population and industrialisation trends (drawing attention to the uncertainties in coupling this well established pattern with total primary energy demand). He reviewed the position with regard to coal and nuclear energy as alternatives in 'post-oil' period of bulk energy supply and the countries and places where, as a consequence of our knowledge of occurrence, coal exploration is now being concentrated.

(ii) Looking ahead

All experts seem now agreed that coal and nuclear energy are the only two major energy sources which at the end of this century will be able to fill the 'energy gap' created by the rundown in world oil and natural gas resources. Novel forms of energy - solar, wind, tidal, geothermal, etc. power - will be able to make only a marginal contribution during the next thirty or so years, and in any case their application will be limited.

Looking at nuclear power, scarcely a day goes by when we do not read of growing doubts about the feasibility of massive nuclear power programmes; in the United States, public acceptability of nuclear power is lessening rather than growing, as would usually be the case with new technology. Estimates of cost are rising constantly; and there are also doubts about the ability of uranium reserves to sustain the planned growth in non-breeder reactors (the breeder reactor being even now a long way from commercial exploitation).

The challenge facing the coal industry is simply this: its reserves are vast, but it is relatively difficult to extract and difficult to use; this is the opposite of the case of oil, where the reserves are limited, but it is relatively easy (once discovered) to extract and to use.

Overcoming the problems of coal

(i) Difficulty in extraction

We have to overcome the problems of nature (geology) and also to ensure that conditions of work underground constantly improve. In the geologically disturbed coal-fields of the world, the problem is to improve the predictability of the disturbances to the continuity of extraction - one of the main themes of this symposium. Coal industries are now largely mechanised; the next step must be to move from mechanisation to automation - new generation of equipment are required which will reduce the number of men needed underground , and improve performance (e.g., new cutting machines required for hard rock). This is the object of the research programmes of all the mining industries.

(ii) Difficulty in use

In the past coal has been used for bulk steam-raising, for direct heating, and (as coke) for smelting/blast furnaces.

In the future, usage for these purposes will be advanced by new techniques (e.g., fluidised bed combustion) but also coal will be used increasingly as a fuel for conversion into other fuels (oil and SNG) and as a feedstock for chemicals.

Research programmes to these ends have made good progress.

Future demand - conclusion

The best estimates of the likely demand for coal in the year 2000 suggest a near-doubling of present levels, and, provided we can make progress on the problems of extraction and utilisation, I am fairly confident that an expansion of demand of this magnitude will materialise. And it is these estimates which, as I have said before, are giving rise to the growth in exploration activity today.

Coal policy in Europe

EEC policy is to reduce the level of import dependence of the Community from some 62% in 1974 to 50% or if possible 40% by 1985.

As long ago as December, 1974, the Council of Ministers laid down a firm objective of maintaining coal production of 253 million metric tons of coal equivalent in 1985 (270 mmt on a ton for ton basis). This involves a UK figure of some 146 mmt, an increase on present levels, and also an increase in Germany. In these two countries therefore we see a growth in exploration activity.

The situation in the UK

In the United Kingdom, as in the rest of Western Europe, the effects of the changes in the oil/coal price relativity in the 1950s and 1960s were particularly marked. Output from deepmines and opencast sites fell from a postwar peak of over 220 million tons in the mid-1950s to under 150 million tons by 1970. The number of producing collieries fell during the same period from 840 to under 300.

The response of the industry to the changed market situation in terms of efficiency was equally marked, with rapid mechanisation of the cutting and filling operations. Manpower productivity rose from a static 25 cwt. in the mid-1950s to 43 cwt. in 1970, a rise of over 70%.

These developments in oil/coal price relativities had a dramatic effect on potential coal supply, in terms of reserves likely to sustain existing capacity. The economically workable reserves at mines remaining open fell at more than ten times the rate at which reserves were being used up in annual production. Exploration activity during this period was low. But by about 1970 it was clear that planning forecasts showed that, without the creation of new capacity, national production at existing mines would fall markedly after the mid-1980s. So the decision to increase exploration was taken in 1970/71, before the 'oil crisis', and against a background and policy which was still unfavourable to coal. This decision was taken because, looking ahead ten years, it was clear that we were literally running out of access to economically workable reserves. The oil crisis gave an impetus to a programme which was already underway.

Plan for Coal, formulated 1973-74, provided for the investment in 20 million tons a year of capacity at new mines and an increase of 22 million tons in output of existing mines by 1985 - this would (together with 5 million tons of extra opencast output) rather more than offset exhaustions.

To select and prove the coal to support this new
capacity, the National Exploration Programme was born.

The National Exploration Programme

	1972/73	1973/74	1974/75	1975/76	1976/77 (est)
Expenditure £m	0.9	1.9	4.0	11.0	13.2
No. of deep boreholes	45	68	96	178	181
Seismic miles	25	240	381	306	300

(24 m. tons out of the 42 m. tons capacity now approved.)

There are three particularly noteworthy points about this
National Exploration Programme

(i)   It is approximately halfway through proving the
      whereabouts of the reserves needed to support the
      best 20 m. tons per year from new mines and the best
      22 m. tons per year extra from existing mines. Most
      of the reserves proved are in much better coal than
      the average in present mines.

(ii) It is the first time exploration for new sites has
      been planned nationally; it is the first programme to
      investigate totally new prospects in places where,
      because of longstanding coalfield tradition, there was
      thought to be no economically workable coal.

(iii) It benefits from the most up-to-date geological
      reasoning, which analyses trends in seams in ways
      unavailable in the past.

      The results to-date have shown that even in a country
with as mature a coalmining industry as Britain, happy
surprises can still occur in the process of renewed coal
exploration. Not so much in finding new coalfields
(although we hope to) but in finding unexpectedly accessible,
productive and extensive portions of coalfields, as the large
Selby extension of the Yorkshire coalfield. A more recent
example in the discovery of even larger, more accessible and
in general nature potentially even more productive Belvoir-
East Leicestershire extension of the Nottinghamshire
coalfield.

Looking further ahead

The NCB's aim is to get the remaining areas of the existing major prospects fully proved by the end of 1976/77. But that will not be the end of the National Exploration Programme.

The Board are planning now towards the year 2000, and all estimates of demand indicate that more coal will be required than at present. The exploration effort will therefore continue to prove the reserves for further capacity.

Conclusion

The British experience illustrates well, the control the market situation exercises on the levels of exploration activity. It has proved possible to anticipate change in the supply and demand situation, and hence change in the exploration requirement.

I hope these reflections have illuminated some of the characteristics which differentiate coal exploration from exploration in the other bulk energy and mining industries. In the matter of how one explores for coal, I do not presume to address experts in the field who form the body of this first international symposium devoted solely to coal exploration, to which I wish every success.

The future of the international coal industries lie literally in your hands.

Section One Discussion

PROF. PETER O'DELL (Professor of Economic Geography, Netherlands School of Economics): You have given a thought provo-
king survey of the outlook for coal in the United Kingdom.
Obviously the problems of the industry and the challenge and
the opportunities in the industry are far wider than those
which are specifically under study at this conference. This
conference is essentially concerned with exploration. I
think you perhaps pinpointed the economic and political
variables that enter into considerations of exploration when
you pointed out that the explorers of today are really
concerned not with the market of tomorrow, but the markets
of the next decade or the end of the century or even the
following century, and that in order to put into exploration.
the money that is required there has to be some degree of
certainty, some degree of security, about the prospects for
the future. I know one can speculate about how energy
demand and supply will develop after the year 2000. We all
know to our cost that kind of forecasting can be extremely
difficult to make. In the final analysis, in my view, the
equation has to be simplified, constructing the energy
sector of our economy, not as one which has in the past been
open to the variability of market forces and fashion and
habit and pressures of various kinds, but a market which is
regulated within the context of what is reasonable and
realistic for the country for a long time into the future,
not just until the next election or over the next decade,
but over the next half century. In that kind of context of
a regulated economy it seems to me that the exploration
effort can then go ahead with a great deal more certainty
attached to it. It is interesting that in this respect the
problems about which you spoke, seem to have much in common
with the problems that are now arising in connection with
the exploration of expensive oil and gas. The elimination of
uncertainty appears to be as important there as it is in
respect of your industry.

D.C. ION (Consultant on Energy Resources, World Energy
Conference): It is most important that Sir Derek has stressed
that the major task facing the coal industry is to increase
its working inventory or proved reserves from the undoubted
large resources. The point of my first contribution to the
discussion of this Symposium was that high resource figures
encouraged complacency and reduced support for this major
task, and as the expected support to this idea was not then
forthcoming, the appreciation by the speaker of what is the

major task is most encouraging.

SIR DEREK EZRA: We have got to bear in mind that, because of
the increasing contribution that coal will have to make from
now to the end of the century for the reasons developed in
Mr. Collins' paper, the working inventory has got to be
built up over the years. As the demand for coal increases
the knowledge of where you are going to go next has to be
widened and, therefore, I can see an intensification of the
exploration effort. By the end of the century, the coal
demand and hopefully the supply, may be almost double what
it is today. We are currently producing and consuming
something like 2,500 million tons of coal and I would say by
the end of the century it could certainly be of the order of
about 4,500 million tons out of the total requirement then
of between 15/20,000 million tons of coal equivalent, and so
I would see this demand for coal going up for the reasons
Mr. Collins deployed and I would say that with that, unless
we have this inventory of proved workable reserves growing
year by year of course we should not be able to attain that
target.

PROF. P. O'DELL: You spoke in a national context. One of
the principal elements of the organisation of the oil industry
has been its internationalisation. I wonder if you could
tell us something about the way in which the experience of
different national entities is now being developed. I wonder
if the flux now towards an expansion of the exploration
effort and the technology that is required, is becoming
increasingly internationalised as a result of the commonness
of the problems to all the industries concerned.

SIR DEREK EZRA: It is perfectly true that there is a growing
degree of internationalisation about the coal business.
Indeed this conference is an illustration of that, but we
are now sending our mining engineers pretty well all round
the world. We have technical exchange agreements with our
friends in the German, Polish, Russian, American, Canadian
and Australian coal industries, and so on. There is growing
up a greater international awareness and exchange of information including exploration techniques, than has ever existed
before. This is highly desirable and we are exchanging
mining experts with our mining colleagues in different parts
of the world. I also believe we shall see a big increase in
the international trade in coal. It has so far been relati-

vely limited, especially compared with oil. In oil of course, we have had the unique situation for any commodity that its production has been virtually concentrated in one part of the world and its consumption has been virtually concentrated in another part so the vast bulk of the oil produced has had to be moved from one end to another. In the case of coal this hasn't been so because coal resources are more widely spread and coal being a bulkier commodity, the tendency has been to use as much of it near the point of production as possible. With the changing energy situation, I would expect that one of the consequences of that will be to see more and more bulk movement of coal from one part of the world to another. This is something that we are watching with great interest and I hope we will be able to participate in.

Mr. P.K. GHOSH (Ministry of Energy, India): What does a country do which has relatively limited oil resources, but more plentiful coal resources, but the coal is more difficult to exploit?

SIR DEREK EZRA: I think the question can be answered quite simply. The first thing to do is to make absolutely sure that the oil is used in ways which make the best use of it and if it can't be done by pricing, then it has got to be done by some form of physical control. In other words, prevent the oil from being used for example in power station generation. In areas where oil has to be used, for motor transport or specialised industrial processes and so on, that is where it ought to be used. Now as far as the coal is concerned I would say that you ought to have a long term plan for its progressive, expansive exploitation and that means that you have got to start with the exploration that we are talking about. You have got to know where your workable reserves are in relation to your existing pits to start with, and where you could sink new collieries to be getting on with and you will have a progressive plan for expansion, bearing in mind that you can expand your capacity more quickly by expanding existing pits, than by sinking entirely new ones. You want progressively to be doing this using such resources as are available. Now if labour is relatively plentifully available, as I imagine it is in India, then you don't want to spend too much capital to start with on a lot of labour saving devices. What you want to do is get the pits started up and you could follow on later with these more expensive, sophisticated ways of

getting coal out. What you really want to do however, is to
identify the reserves and open them up and get them working,
and then you want of course to have a great care for the
preparation of the product. You can lose a lot of money on
a coal venture if you ignore the preparation aspect and
start shipping coal abroad for example, which is deficient
in quality. You have got to make sure what market it is you
want to go for, whether it is the coking market, the steam
market and so on.   That really would be my answer – you
want a careful constraint on the limited amount of oil and
you want a well defined plan starting with exploration and
then exploitation, and preparation for your coal.

B. ALPERN (Cerchar, France): I would like to raise the
problem of shale. I would like to speak about shales which
are rejected from washeries but which contain some coaly
material, for example from 500 to 3,000 calories per gm. but
which can be used as expanded shales for making aggregates
on buildings and so on. Do you intend to use these shales
which contain a small proportion of organic matter? My
second question is: what do you think about in situ technique
of utilisation of coal because this question is very important
to the estimate of reserves of coal. If we are ever to use
coal by in situ techniques without being obliged to mine the
coal, it will completely change the situation.

SIR DEREK EZRA: On the first question, are we developing
processes here for converting the discarded material into
building materials the answer is yes. We have been progressively developing the use of our shale, lying in the pit
heaps, for road infilling, and other basic building purposes
and we are also developing the use of these shales for
conversion into lightweight building aggregates. We are also
looking at the problem another way, to avoid having these
shales at all. We think there are high hopes of that through
the new Fluidised Bed combustion system that we are developing
over here in conjunction with a number of other countries,
because this has many advantages but the relevant one to
this discussion is that it enables you to use a much more
variable and if necessary, lower grade of coal, than is
presently being used for combustion. If you can burn more
of the product as mined, you avoid the business of having
the shale to discard, you avoid having the costly business
of having the preparation and so on, so we are tackling it
two ways and I think both are promising.

Now as to the in situ exploitation of coal, I presume you
are really referring to underground gasification. We did a
considerable amount of research into this in the early 50's,
indeed we had one of the first pilot projects in the world.
We gave it up for a number of technical reasons but also
because oil then came onto the scene in a very cheap and
plentiful way as you know. We have looked at it all again,
we have had an intensive study made of work done in Russia,
in Belgium, in the United States and elsewhere and we have
still concluded that it is too early to say that this is a
practicable proposition. There are many technical difficulties associated with underground gasification. The main one
to our mind being, that at the end of the day, you produce
from the coal, which could be very valuable if fully utilised,
a fairly low grade of gas which then has to be retreated
and re-formed on the surface, and that therefore one really
isn't making the best use of the coal. At the moment our
strategy for coal is rather different. We visualise a coal
industry in which massive quantities of coal are produced
and conveyed to what can be called major coal refineries,
where it will be converted to different qualities of gas,
oil, chemical feedstock and such solid substances as one
requires. That is how we visualise the future. It will be
an entirely different industry from the present, but we feel
that if one tries to combust the coal in situ, one would be
depriving oneself of all these other qualities that coal
possesses and which will become increasingly valuable. We
would rather bring it to the surface in bulk and then convert
it.

PROF. O'DELL: I think that is probably a somewhat contro-
versial note on which to bring your presentation to an end.
In listening to you and in listening to your answers to the
several questions, I think you have made it quite clear that
the whole question of exploration is basically a research
effort and one wonders if over the past 20/25 years, where
the coal industry might now have been if only a very small
percentage of the research funds that have been devoted to
the oil industry development and to the development of
nuclear power, had in fact been switched to the investigations
necessary to coal resources and to evaluate ways and means
of not only getting it out of the ground, but into the
energy market.

Diamond drill coring provides invaluable information for evaluating coal seams. Photo courtesy Christensen Diamond Products

SECTION 2

1.
World Coal Supply and Demand
by Henry E. Collins

2.
Distribution and Resources of World Coal
by A.K. Matveev

3.
Objectives and Organisation of Coal Exploration Projects
by Dell H. Adams

Section Discussion

1

WORLD COAL SUPPLY AND DEMAND.

HENRY E. COLLINS
Consulting Mining Engineer.
United Kingdom.

In considering the future of world coal supply and demand, it is essential to make some reference to the whole energy balance obtaining in the world and to hazard an opinion on how this is likely to develop in the future. The last quarter of a century has been characterised by a remarkable upsurge in the annual demand for primary energy. In fact, during this period, it is probable that the total consumption of energy was greater than during the whole period of previously recorded history. Industrialised countries had taken steps to intensify their economic development in order to give a consequential improvement in living standards. The so-called developing countries have likewise been aiming at higher standards of living. This world-wide economic and industrial growth, associated with a marked increase in world population, has necessitated a continual rise in the annual consumption of primary energy and, incidentally, of raw materials generally.

As seen in Figure No.1, there has been an exponential growth rate in world population. This has been accompanied by a steady increase

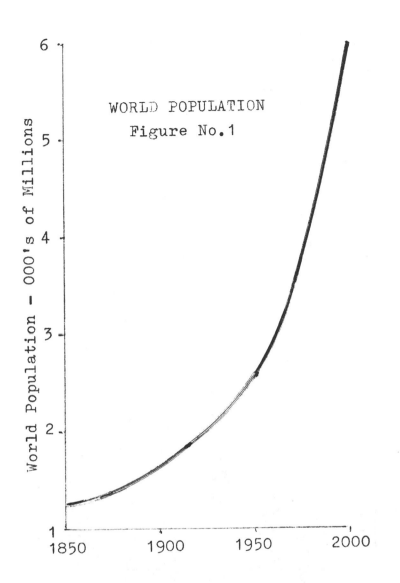

WORLD POPULATION
Figure No.1

World Population - 000's of Millions

in the per capita consumption of energy. Thus,
between 1950 and 1975, the average annual per
capita usage of primary energy has increased
from about 1 tonne to approximately 2 tonnes of
coal equivalent. It is not surprising, therefore,
that the growth rate of primary energy consumption
exhibits an exponential character. In 1950, world
primary energy usage was 2,800 million tonnes of
coal equivalent(m.t.c.e.) rising to approximately
8,000 m.t.c.e. in 1975. Moreover, in this period,

WORLD COAL SUPPLY AND DEMAND 65

the world energy balance changed markedly. World consumption of primary energy by types is shown in Figure No.2.

WORLD CONSUMPTION OF PRIMARY ENERGY

Figure No.2

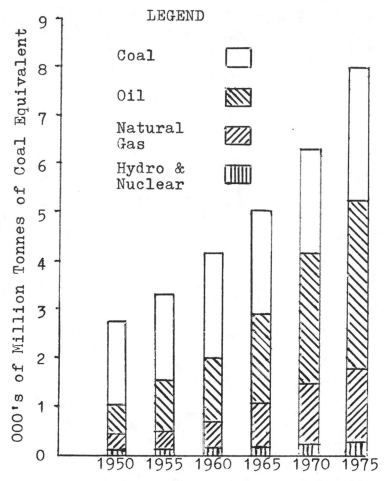

In 1950, coal predominated in the energy balance with a consumption of 1,700 million tonnes, against an oil usage of 600 m.t.c.e. By 1975, these positions had reversed for, although the world consumption of coal increased to around 2,500 million tonnes, that of oil shows almost a six-fold increase to about 3,500 m.t.c.e. Also, in this period, natural gas became a major contributor to the world energy balance with an amount consumed in 1975 estimated at 1,700 m.t.c.e. There is no doubt that, for the greater part of the period under review, the abundance of relatively cheap crude oil had a retarding effect on the growth of coal consumption. In some instances notably in Western Europe, the plentiful supply of cheap crude oil resulted in a serious decline of many coal industries. The energy policies of the Western European countries were, in fact, based on the assumption that cheap crude oil would continue to be available indefinitely. In 1973-74, this naive assumption was proved to be erroneous since in this period there was a four-fold escalation of Middle East crude oil prices. There would appear to be little doubt that the sudden and substantial increase in the price of crude oil has been responsible for a world economic and industrial recession. However, when the world eventually comes to terms with the concept of more expensive energy, with the consequential recovery in industrial production, attention will be more and more directed to the increased production and utilisation of coal.

Primary energy consumed in the world is principally of fossil fuels, coal, oil and natural gas with relatively small contributions from hydro and nuclear power. Other more exotic forms of energy, geothermic, solar, wind and tidal, are at present insignificant. The fossil fuel resources, although considerable and, in the case of coal, very large, are finite in extent and non- renewable in character. Large though these fossil fuel resources might be, they are certainly not sufficient to maintain the current exponential trend in total world energy consumption much beyond the end of this century.

Considering the position of energy usage and demand now and in the future, extrapolation of the curve of actual primary energy consumption gives an estimated annual demand of 10,000 m.t.c.e. by 1980 and approaching double this figure by the year 2000. Even if the growth curve were to be constrained to follow a straight line projection over the next 25 years, the annual energy demand would reach levels of approximately 9,000 m.t.c.e. in 1980 and 15,000 m.t.c.e. at the end of the century. The present world energy position and possible alternative extrapolations of the energy curve are shown in Figure No. 3.

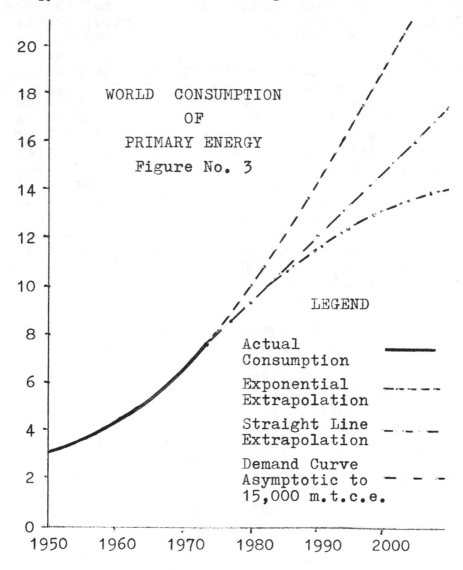

WORLD CONSUMPTION
OF
PRIMARY ENERGY
Figure No. 3

LEGEND

Actual Consumption ⎯⎯⎯

Exponential Extrapolation ⎯ ⎯ ⎯

Straight Line Extrapolation ⎯ ⸱ ⎯

Demand Curve Asymptotic to 15,000 m.t.c.e. ⎯ ⎯ ⎯

The energy sources of the world and the means of exploiting them would be severely strained to meet even the lower levels of demand resulting from the straight line projection - almost twice the present annual consumption - by the year 2000. Moreover, it seems scarcely possible that energy availability could match expanding annual usages far into the next century even if following the straight line projection and still less likely if the current trend were to continue. Therefore, it is very probable that economic growth, which has characterised the past quarter of a century will be slowed down considerably, if not halted altogether, due to lack of energy and the scarcity of essential raw materials.

In recent years, rates of economic growth in the various countries have ranged up to about 10 per cent a year but such rates are unlikely to be sustained in the future. In some countries, not endowed with adequate energy resources, nil growth rates could well obtain and might eventually become negative. In other words, there might have to be a return to a low energy and consequently a much restricted technological mode of existence. A nil growth rate in annual energy consumption would have an adverse effect on the living standards in industrialised countries and would seriously retard improvement in the life styles in the developing countries.

With the advent of high-cost energy, obviously every effort will be made to improve the efficiency in the use of fuels. But increased fuel efficiency would merely have the effect of moving the energy curve forward along the time axis. It would not, of itself, bring about a change in its exponential character. Moreover, greater efficiency in the use of fuels, although very necessary, would not alone be sufficient to meet the future demand consequent upon the continuance of the present trend in consumption.

The continued increased production of primary energy can no longer be relied upon to match excessive increases in demand automatically. Sheer availability of energy will eventually determine world economic growth. Thus, the projected graph of future energy consumption, reflecting the restraint which would be imposed by an anticipated limited availability, will have to follow a constrained curve. It is suggested that this would be asymtotic to a line denoting the maximum sustainable annual energy demand. In Figure No.3 this level is shown as 15,000 m.t.c.e. Although this is an arbitary selection, it is unlikely that the level could be much higher - it might well be considerably lower.

Already, there are signs of a slackening in
the rate of increase in annual energy consumption.
However, this is due more to the enhanced price of
crude oil and the consequential world economic
recession than to any shortfall in supplies.
Indeed, at present, energy is available in excess
of demand. In 1974, Western Europe used some 7 per
cent less energy than had been estimated and in
1975, the differential between expected and actual
consumption is likely to be double that of the
previous year. At present, world figures for these
periods are not available but a useful "barometer"
of world economy and of particular significance
in the level of energy consumption is the annual
production of steel. Except for the Soviet Union,
the Comecon countries and the People's Republic of
China where steel production has continued to rise,
all major steel-producing countries showed very
substantial reductions in steel outputs in 1975
compared with 1974. Total world output of steel
in 1975 at 652 million tonnes was considerably
below the previous year when the figure was 710
million tonnes. This shortfall means a substantial
reduction in the consumption of coking coal and
of other forms of energy used in steel making.

Now, what are the possibilities of being able
to meet the postulated future energy demand even
assuming a ceiling of 15,000 m.t.c.e. per annum?
Reliance will have to be placed principally on
the continued and increased availability of the
fossil fuels. Nuclear energy will eventually no
doubt make a substantial contribution. Indeed,
much is expected of nuclear power in respect of
future primary energy supplies but at present it
meets only a little over 1 per cent of total world
energy consumption. It is possible that the nuclear
contribution could reach around 5 per cent by 1980
and probably 15 per cent by the turn of the century.
But existing commercial nuclear electricity generating stations incorporate reactors which require
natural or enriched uranium. The economically
mineable resources of uranium could prove a limiting factor in the expansion of nuclear generation
of electricity in plants similar to existing types

Much will therefore depend upon the success attending the commercial development of the Fast Breeder Reactor. This type uses plutonium and its utilisation of the energy potential of natural uranium is some sixty times that of present commercial non-breeder reactors. Already, prototypes of the Fast Breeder Reactor are on-stream in the U.K., France and the U.S.S.R. However, it is not until the late 1980's or 1990's that plants incorporating this type of reactor are expected to produce electricity on a commercial scale.

Hydro power will continue to supply around 1 per cent of the total demand for energy but, as for its increased availability, most of the untapped hydro potential is in areas around the world remote from industrial centres.

Of the various exotic sources of primary energy perhaps geothermic is the most promising. Up to the present time, the exploitation of this form of energy has been virtually confined to sites situated in the earth's seismic belt. Prospecting for further useful sources of geothermal energy within this belt is proceeding in a number of countries, including, New Zealand, Greece, Italy and Turkey. However, there are hot rock formations outside the seismic belt but at much greater depths, which could yield geothermal energy. This would require very deep drilling and the creation of permeable conditions within the hot strata to allow of the flow of a working fluid as a means of extracting the heat energy. There is the possibility of using controlled nuclear explosions at depth to induce permeability within the hot strata. At present, geothermal energy contributes but a small part of 1 per cent of world energy usage and it is not likely to become significant in the foreseeable future.

With regard to fossil fuels, the resources of the various forms vary considerably and they are very unevenly distributed throughout the world. The recent publication of the World Energy Conference, "Survey of Energy Resources, 1974" gives the total estimated resources of solid fuels, including, hard coal, brown coal, lignite, and peat, as twelve million megatonnes. Of this vast total, only 1.1 million megatonnes are regarded as proved and recoverable reserves. With improved production techniques and the dire necessity of meeting the energy demand throughout the world, it is almost certain that a much greater proportion of the total resources of solid fuels will be exploited. Recoverable reserves of natural crude oil are recorded as 91,600 megatonnes and of natural gas, 37,400 megatonnes. Oil shales and bituminous sands are considered to have recoverable reserves of oil and kerogen of 230,000 megatonnes but total resources will be very much in excess of this figure. This Survey confirms that the resources of solid fuels far exceed those of any other fossil fuels. It follows that coal will have a continued importance in the world energy balance well into the next century.

The question must be posed: to what extent can solid fuels contribute to the world energy supplies in the foreseeable future? In 1975, out of an estimated total world consumption of primary energy of 8,000 million tonnes of coal equivalent, solid fuels accounted for some 2,500 million tonne and all other forms, 5,500 m.t.c.e. It is possible that annual world coal production could gradually increase to reach 4,300 million tonnes by the year 2000. The straight line projection of the curve of annual world energy consumption gives a total primary energy demand of 15,000 m.t.c.e. at the turn of the century. Thus, in the year 2000, energy sources other than those of solid fuels would have to provide 10,700 m.t.c.e., nearly double the present figure. Assuming a 15 per cent contribution from nuclear energy and say 2 per cent from renewable resources, then oil and natural gas would have to provide over 8,000 m.t.c.e. This figure represents almost double the 1975 consumption and the production of this vast quantity, although not impossible, would present great difficulties. In order to approach such a level, it would be necessary to have large-scale exploitation of the oil shales and bituminou sands to produce substitute crude oil. Similarly, the production of synthetic crude oil and substitute natural gas, presumably by the processing of solid fuels, would have to be accelerated. Even if it proves possible to meet the above-postulated annual demand for energy in the year 2000, it seems extremely unlikely that such a level could be sustained very far into the 21st. century. It is more improbable that energy availability beyond the turn of the present century would be able to support a rising total demand. Thus, as already suggested, the curve of annual energy demand will have to bend over.

In order to achieve the large increase in annual coal production suggested above, extensive prospecting and exploration in various parts of the world would be essential. Potential for the increased coal production and development of new coalfields exists in North America, the U.S.S.R., and the People's Republic of China with possible contributions from Poland, other Eastern European countries, South America, Africa, India and Australia.

Already, forecasts of coal production in the U.S.A. within the next 25 years suggest ultimate increases in annual output of 500 to 600 million tonnes or almost double the present level by the year 2000. This will require considerable development of the coalfields in the Western States where there are large resources of low-sulphur bituminous coal. This might need a change of policy in accordance with which vast areas of Federal lands have been and are being withdrawn from mineral exploration and mining. So far as further development of coal mining in the Eastern States is concerned, de-sulphurisation of the coals will be necessary to meet environmental standards. To this end, there is much to commend fluidised bed combustion for boiler plants in the public utilities. In this system, sulphur can be absorbed by controlled slagging; thus eliminating atmospheric pollution.

In Western Canada, there are extensive bituminous coal resources in the Provinces of Alberta and British Columbia remaining to be more fully exploited. Although these Rocky Mountain coal-bearing areas are some 700 miles from the Pacific coal-loading ports, nevertheless, because of the prime coking quality of much of the coal deposits, more extensive working of the resources is commercially feasible. Also, there are large deposits of lignite, at strip mining depths, in the Province of Saskatchewan. Exploitation of the lignite is already being expanded to provide fuel for thermal electricity generating stations.

Annual output of coal in the U.S.S.R., at around 700 million tonnes, is comparable with that of the U.S.A. and a similar range of additional production may be expected. This would result in almost double the present level of output by the year 2000.

There is no doubt that considerable potential exists for increased coal production in the People' Republic of China. Already, development of new mines is taking place in the coalfields to the south of the Yangtse River where hitherto the deposits have been virtually unworked. No official information is available regarding current annual output of coal and no forecasts of future production have been published. However, it may be assumed that output of coal is running between 400 and 450 million tonnes a year and that a doubling of this level by the year 2000 is not impossible.

Coal production in the Eastern European countries has steadily increased during the last 25 years and this trend could continue. However, much of the additional output is likely to be absorbed within the countries concerned,

In South America, there are large, untapped resources of bituminous and sub-bituminous coal, particularly in Venezuela, Colombia and Peru. In the last two countries, the coalfields are situated in very difficult, mountainous terrain and the development of much infrastructure would be required to allow of coal exports. However, not all the additional world coal production in the foreseeable future will be needed for consumption in its raw state. It is very likely that considerable quantities of coal will be processed into synthetic crude oil and substitute natural gas.

If such processing could be done virtually on the coalfields, the transportation of the resultant liquid and gaseous fuels would present fewer problems than the movement of the coal itself to the ports.

Again, in the African continent, in addition to the present substantial producers of coal, many countries, including Botswana, Mozambique, Swaziland, Tanzania and Zaire, have relatively large coal resources awaiting exploration and subsequent exploitation. Here also, the processing of the coal within easy reach of the mines would solve the transportation problem. Such coal processing would not be cheap but, if the oil requirement of the forward energy demand is to be met, it might be essential. Currently, a project is well advanced for the building of a further large oil-from-coal plant in the Eastern Transvaal. This new venture will have ten times the capacity of the existing SASOL plant in South Africa.

In India, annual coal production is a little over 80 million tonnes and this could be doubled by the turn of the century. Certainly the resources of coal are sufficient to support this enhanced production.

Australia is another country where development of coal production has been continuous over many years. The annual output of bituminous and sub-bituminous coal has reached a little over 60 million tonnes, principally from New South Wales and Queensland. It is in the latter State where the potential for increased production is the greater. The necessary infrastructure has been constructed in Queensland to meet the needs of coal exports. It would seem possible to step up Australian production of hard coal to around 100 million tonnes a year by the end of this century. Similarly, there will be an expansion of lignite production in the State of Victoria in which the present annual output of 24 million tonnes could be almost doubled. This lignite is consumed locally in electricity generating stations.

This brief review of world potential for a

substantial increase in coal production supports
the view that, by the end of this century, the
solid fuel contribution to the likely world energy
demand could exceed 4,000 million tonnes a year.
It is very appropriate that this International
Symposium should be devoted to Coal Exploration.
Drilling and all the usual forms of prospecting
for coal deposits will be required on a most
extensive scale throughout the world if coal
production is to match the likely future demand.

In conclusion, even if there is the considerable increase in world coal production which has
been postulated, the continued availability on a
reasonably large scale of crude oil and natural
gas, augmented by the processing of oil shales
and bituminous sands and possibly by the liquefac-
tion and complete gasification of coal, as well as
a much inhanced contribution from nuclear energy,
the portents are that the rate of industrial and
economic growth will have to decelerate between
now and the turn of the century and continue
thereafter to keep below the maximum maintainable
level of energy demand. Although this might well
seem a pessimistic conclusion, it is really quite
realistic in the light of present scientific and
technical knowledge. However, within the period
reviewed above, it is not impossible that other
forms of energy, such as nuclear fusion, might
be developed.

2

Distribution and Resources of World Coal

A.K. Matveev

Professor, Moscow State University, Moscow, USSR

It is a great honour for me to read my paper on coal at this
Symposium. Coal is a wonderful gift of nature, which warmed
primitive man, which has been the principal promoter of
technical progress for hundreds of years, whose role has been
undeservedly belittled for a short period recently, and
which is now expected to rescue mankind during an energy
crisis.

Nature scattered this gift generously all over the world, as
can be seen from the map of coal fields throughout the world,
scale 1:15,000,000, that is shown here. The map shows 2,100
coal deposits of basins of various sizes, fields, coal-bearing
regions and areas. Each object is designated according to
the nomenclature adopted in the country it belongs to.

The map has been compiled by a group of Soviet geologists and
edited by the speaker. It is the only map of its type that
has been published in the world since the "Atlas of world
Coal" issued in 1913.

Great progress has been made in coal geology over this period;
many new basins and deposits have been discovered and the
geological coal reserves have thus been increased from 7-8 to
16-20 trillion tons. This, and the interest recently revived
in coal, have made it necessary to study the present state of

the world coal economy. This report is an attempt at such
an analysis, the result of which can be seen in this map.
This study has made it possible for us to reveal the new
laws governing the distribution of coal deposits, such as
their association with major structural features, for example.

The principles of compilation and the content of the map are
given in the legend. The map shows: I) the age of coal
formations, 2) the degree of reliability of the deposits'
mapped outlines, 3) types of coal (brown or hard coal) and
4) amounts of geological coal reserves within the mapped
outlines; the deposits containing coking coal are marked
by a special sign.

The ages of coal formations are indicated with respect to
geological systems. The coal formation of the Antarctic is
distinguished as an individual system, known as the Bicon
system.

Depending on the degree of reliability, the information is
classified as ascertained and unascertained, and it is shown
as the solid line of a contour and as a broken line, respectively. The areas of black or hard coal are outlined in black
and those of brown coal in red.

According to amount of geological reserves, coal basins and
deposits are classified in five groups: 1) over 500 billion
tons; 2) 200 to 500 billion tons; 3) 500 million to 200
billion tons; 4) less than 500 million tons; 5) deposits of
unascertained reserves.

Of the above mentioned 2,100 coal objects, seven belong to
the basins of the first group known as "giant basins": the
Lena, Tungusska, Taimyr, Kansko-Achinsk, Kuznetsk, Alta-
Amazona and Appalachian basins. The second group includes
four basins: Lower Rhein-Westphalian, Donetsk, Pechora and
Illinois. The third group is comprised of 210 basins and
fields.

The largest number of basins and fields (almost 1700) belong
to the group with reserves of less than 0.5 billion tons.

Some difficulties were experienced in compiling this map
because of the lack of knowledge of the exact position of
some fields and the differences in the stratigraphic nomen-
clatures adopted in different countries. To overcome the
latter difficulty the age of coal formation is given in terms
of one geological system.

The first estimates of world geological reserves of coal
were made in 1913 to be presented at the XII Session of the
International Geological Congress. Calculated in accordance
with the unified requirements of the smallest coal seam
thickness of 0.3 metres and the greatest depth of 1,800
metres, the reserves amounted to 7.3 trillion tons. Later
the reserves were recalculated as new coal discoveries were
made and the existing coal fields were developed. The
calculations were not based on any strict requirements with
respect to the thickness and depth of coal. As a result,
the estimates of world reserves reported in the literature
were totally inconsistent. For this reason, it is advisable
to divide world coal reserves into the following groups: 1)
geological, 2) local, 3) special and 4) relative reserves.

The first group includes coal reserves calculated in accordance with the international requirements of the XII Session
of the Geological Congress, irrespective of the currently
existing means of coal extraction.

The second group combines the geological reserves belonging
to large regions which include several countries. Here the
estimates may differ considerably from international standards.

For instance, in 1968 the coal reserves of the USSR calculated
as part of the reserves in the countries belonging to the
Council of Mutual Economic Assistance, were 6.8 trillion
tons, while in 1957, when calculated according to international requirements, they amounted to 8.7 trillion tons.

Special coal reserves include coals of special applications
such as, for example, gasgenerating coal, anthracite, coking
coal, etc.

The most complete and dependable estimates of coal reserves
in this group are those prepared for each session of the
International Energy Conference, when only workable coals
are included in the calculations. The calculations are made
by almost all countries according to the IEC requirements
which are stricter than those adopted for the calculation of
general geological coal reserves (the greatest depth for
brown coal being 500 metres and for hard coal 1,200 metres).
Naturally, there is some reduction in world coal reserves as
a result.

Relative estimates of coal reserves are made for various
reasons depending on certain circumstances in order to
evaluate the potential usage of coal when the conditions of

the competition with other fuels or sources of energy change, in order to unify zonal standards or revise the existing requirements, etc. Calculations of this type are generally regional, the amounts of coal reserves estimated within the same regions varying greatly, depending upon the combination of circumstances.

Most constant are the general geological reserves which do not depend on fluctuating present-day demands. The reserve estimates of the last two groups are not reliable in most cases.

The general geological coal reserves of the world are at present estimated on a variable basis.

After the first estimates in 1913 the world coal reserves were calculated by various organisations and specialists. The final estimates of the world reserves, made by means of a wide variety of techniques and approaches, and reported in the literature in the early sixties, amounted to 16-18 trillion tons. The subsequent discovery of the Alta-Amazona basin with 2.2 billion tons of coal increased the reserves to 20-21 trillion tons.

Because of insufficient geological knowledge in many regions of the world, and also because of the more strict requirements in some countries where the geological coal reserves had never been calculated before, the given figures of 20-21 trillion tons apparently stand for the official minimal amount reduced by the strict requirements, and do not reflect the actual state of things.

The estimates of 20-21 trillion tons of world coal obtained predominantly from the official national sources appear to be under-estimated if one makes a thorough analysis of the geological evidence available and uses the more reliable figures presented by competent authors. This primarily applies to China where the amount of geological coal reserves is officially given as 1.5 trillion tons which is highly underestimated.

Bearing in mind that in many countries the reserve estimation has been based on more strict requirements, first of all on the greater thickness of the coal seam, it would be right to state that the total world wide estimate of geological coal reserves of 30 trillion tons is quite realistic.

The coal reserves are distributed in the world as follows:

sia 58%, North America 30% and the remaining continents 12%, including 8% in Europe and less than 1% in Africa. It is worth noting that in the USSR, the USA and China, the countries possessing the greatest coal reserves, coal-bearing formations occupy 13-15 percent of the territories. In other countries coal-bearing areas do not exceed 1-3 percent and Africa less than one percent.

The average coal concentration in the coal-bearing areas is estimated to be 2 million tons per square kilometer, and when related to the total area of the continents it is 0.1 million tons, varying from 0.003 to 0.2 m.t/sq. km.

According to national estimates, the distribution of the total geological reserves between nations is such that the following ten countries head the list: (in billion tons) the USSR (6,800), the USA (2,911), China (1,500), Canada (1,234), FRG (287.3), Australia (172), Great Britain (170), India (125), Poland (124), and GDR (over 29). About 51 percent of the coal reserves are concentrated in the socialist countries, 34 percent in the capitalist and 15 percent in the developing countries.

The larger part of the geological coal reserves estimated in the leading countries occur at depths of less than 600 metres. However, in some large basins (Donbass, USSR, Ruhr, FRG, and some basins in Great Britain and Belgium) coal has been worked out to a depth of 600 metres and the remaining resources are concentrated at great depth, the working depth occasionally exceeding 1,000 metres.

Although the general demand for coal is satisfied by the geological reserves available throughout the world, there is shortage of coking coal. The largest reserves of coking coal are found in the USSR (670 billion tons) and in the USA (270 billion tons). They are less abundant in Great Britain, FRG, Poland, Canada, Australia, China, the Republic of South Africa and Rhodesia.

From a practical viewpoint it is interesting to discuss the degree of coal exploration. An analysis shows that only some ,000 billion tons, i.e. about 7.5 percent of the total geological coal reserves have been explored.

As regards the geographical distribution of the coal reserves that have been explored as A+B+C categories (the USSR standards roughly equivalent to the actual and probable reserves), the highest degree of exploration of up to 30 percent is found in Europe where coal basins and fields have been thoroughly explored and developed. In particular, this applies to the FRG, Belgium, the Netherlands and to some extent France, where the degree of exploration is 50 percent and over.

At the same time we find that the coal reserves of North America and Asia have been poorly explored (4.4 and 4.8 percent, respectively). This is explained by the great abundance of coal reserves on these continents, which greatly exceed demand.

The following is the distribution of the coal reserves explored in some of the leading countries (in billion tons): the USSR - 256, the USA - 140, FRG - 130, Great Britain - 127, China - 125, Poland - 15.6.

From a geological point of view it is interesting to consider the rules of coal distribution in the earth. We have attempted to qualify coal distribution according to the geological systems, genetic types and structural features with which coal is associated.

As was established by Academician P.I. Stepanov in 1937, the distribution of coal reserves grows by leaps in a succession towards younger geological formations.

The geological investigations conducted since then have corroborated this statement and, furthermore, revealed significant changes in the contribution of individual geological systems. For instance, the role of the Permian, Jurassic and Cretaceous coal potential has substantially increased as compared with that of the Peleogene-Neogene coal accumulations, although the latter still holds its leading position with respect to the other coal ages. As regards the distribution of the coal explored by eras, the Paleozoic comes first (47%) and is followed by equal percentages of the Mesozoic and Cenozoic. This can be explained first of all by the predominance of hard coal in the Paleozoic formations and also by the fact that the coal-bearing formations of this age were largely formed in the foredeeps where the conditions for coal accumulation were most favourable.

The geographical distribution of coal formations is as follows. Carboniferous coal formations hold the leading position in Europe and the western regions of North America, where 81 percent of the geological and 95 percent of the explored reserves of Carboniferous coal are found. Permian coal formations are predominantly concentrated in Asia and to a lesser extent in Africa and Australia. Triassic coal accumulations are abundant only in Australia where 72 percent of the geological and 83 percent of the explored reserves of Triassic coal are found.

Jurassic coal formations are almost exclusively concentrated in Asia where 99 percent of the geological and 95 percent of the explored reserves of Jurassic coal are found (the Lena, Kansko-Achinsk and Irkutsk basins of the USSR). Cretaceous coal formations are mainly confined to the Pacific belt with 99 percent of the geological and 98 percent of the explored Cretaceous coal.

Paleogene-to-Neogene coal accumulations are found everywhere, but have been little studied in Africa, South America and Asia. As a result, the majority of their estimated reserves (about 60 percent) are concentrated in North America.

Another point that is theoretically and practically important is the dependence of the coal potential of various basins upon their position within the major structural features of the earth.

As a rule, the largest basins, such as the Tungusska, Lena, Appalatian, Alberta and some other basins, are situated in synclines as belts running along the peripheries of continental platforms which extend into marginal platforms the depressions or the foredeeps of folded areas.

Outside of the marginal platform belts mentioned, coal formations generally occur as scattered coal fields of incommensurably smaller dimensions, irrespective of their location in platforms or in adjacent folded areas (in the foredeeps and intermontane troughs). The same applies to the median masses, the folded belts of which are also known as the centres of coal accumulations, although of smaller scale owing to the limited area.

The coal potention of basins is closely related to their association with different types of structural features. Basins situated in marginal platform depressions hold about 70 percent of the world geological coal reserves while those in the foredeeps hold about 20 percent. The remaining reserves are distributed among the basins scattered inside the platforms and median masses, in salt domes and other types of structural features.

In addition to the principal conclusions drawn from examination of the map presented, there are some points of local

significance, such as, for instance, the problem of the
Pacific mobile belt, which deserves further special study.

Apart from the conclusions indicated which are largely of a
statistical character, we can discuss some geological rules
for identifying large coal-bearing areas, which are based on
the similarity or affinity of the origin and on the analogy
of stratigraphic relations. Such areas are known as coal
provinces. Coal provinces are divided into four groups or
types (in decreasing series): 1) megaprovinces, 2) transcontinental provinces, 3) inland or mesoprovinces and 4)
local provinces.

Among the most prominent megaprovinces is the Gondwana
province which includes the south-eastern parts of South
America, the southern parts of Africa and India, the eastern
part of Australia and the whole of the Antarctic.

The coal-bearing formations in all the areas indicated are
remarkably similar in geological sequence, lithology, flora,
the mode of occurrence and coal petrography. This, and the
fact that the coal-bearing formation is ubiquitously overlain
by dolerites, allows one to attribute these areas to one and
the same province or, as in this case, megaprovince.

A group of transcontinental provinces includes: 1) the Andes
province of Cretaceous age extending through South and North
America and 2) the Pacific province of Tertiary age extending
almost parallel to the former.

The inland provinces naturally consist of a much larger
number of members. Inland provinces can be differentiated
by systems and further divided into smaller local or sub-
provinces which will still preserve the similarity to the
main inland province designated.

An example of such an approach is the author's map "Local
provinces of the USSR and some other countries". This map
forms just a fraction of the enormous amount of work that
needs to be done throughout the world in the detailed systema-
tization of world history of coal accumulation.

The new coal basins and deposits that can be discovered in
each province will be different in geological structure and
types of coal, but will be similar to one another if they
are found in the same province.

For instance, the distinctive features characteristic of the new basins and deposits to be discovered in the Gondwana province will be association with rift zones, restriction by steep faults, almost horizontal bedding of coal formations lying on much older rocks inside the grabens, the presence of basalt sheets on top of coal formations and the preservation of coals of an intermediate and high degree of metamorphism. The finding of extensive coal areas is unlikely.

New coal discoveries in mountain areas of the Andes province will be characterised by local distribution, significant faulting responsible for a block structure or chess-pattern alteration of the coal and barren areas and the presence of coals of different degrees of metamorphism owing to extensive intrusive activity. In foothill areas coal occurrences will be less faulted, larger in area and will contain coal of a low degree of metamorphism. Coal may be found at small depths favourable for open-work mining.

In the Pacific Tertiary province the beginning of coal accumulation is generally assumed to be connected with the end of the Lower Cretaceous global transgression, appearance of various new land forms and the formation of new structural features or the rejuvenation of the previously existing ones as fairly flat-lying faulted synclinoria. The coal accumulation, which began in the Maestrichtion time and continued in the Paleogene, migrates westwards towards the Pacific into younger stratigraphic horizons. The largest amounts of coal are likely to be found in the western part of the province. The occurrence of karst rocks in the underlying Lower and Middle Mesozoic formations suggests that Karst-type coal deposits may be found, which are small in area but abundant in brown coal, and also caldera-type deposits with coals of various degrees of metamorphism.

Local provinces of this type also have a number of distinctive features, but these are confined to more limited areas and are more prominent. Such provinces have been distinguished by the author in the USSR and adjacent areas, by D. Minchev in Bulgaria and by other researchers.

The location and mapping of coal provinces requires much work. Yet this work will be highly beneficial for all countries for it will facilitate the prediction of potential coal reserves in them.

For the world as a whole this does, in fact, need a tremendous amount of work. We believe that it can only be done through the combined efforts of geologists from many countries. It would be unscientific to try to find and propose any better alternative. Here we can add that the possibilities of finding new coal areas in the world are far from being exhausted. For instance, the discovery of coal fields is expected in Africa, polyfacial in Nubian sandstones of various ages. Scattered isolated coal areas may be found in the eastern half of Laos, in the eastern and north-eastern regions of the USSR, in North West Canada, etc. Naturally, more detailed and fundamental predictions can be made only by National Geological Surveys armed with all the necessary geological information.

To conclude my fragmentary paper and not take up too much of the audience's time, I would like to emphasize one point, namely that we must put our records of the world coal resources in order. The records should be kept according to international standards common to all the nations. The standards should be such as to allow us to differentiate the world coal reserves according to their application, particularly those of coking coal, and to find out on a comparable basis how individual continents and nations are provided with this material in the light of the further development of technical progress.

Naturally, the accomplishment of this task should be preceded by the work of an authoritative international commission sponsored by UNESCO or by the Committees of our Symposium. The commission should work out the standards and techniques of calculation of the total geological reserves including the requirements of the greatest depth, the smallest thickness and reliability.

Without predetermining in any way the future decisions concerning the critical thickness and depth, I presume that the new technique of reserve estimation should include a broader combination of these parameters. For instance, the number of the horizons in the upper 800-900 metres to be included in the calculation should be larger than was adopted in 1913; the calculations for every horizon should be differentiated and made for two or three critical thicknesses from the smallest used in the total calculation to larger ones which usually characterise the real energy potential of a nation.

In fact the seemingly large amounts of total coal reserves will be greatly reduced by the beginning of the next century as a result of the ever increasing output of coal. Future generations will criticise us for any delay in solving this problem.

REFERENCES

1.   Averrit, P., "Coal resources of the United States", January 1, 1967, Washington, 1969.

2.   Beschinsky, A.A. and Volfberg, D.B., "National reviews of energy resources", Papers to VII International Energy Conference, Moscow, 1968, pp. 1-28.

3.   The coal resources of the world. Canada, 1913.

4.   Harnish, H., "Coal mining in the People's Republic of China, World Coal, October 1974/Premier Iss.

5.   Lardinois, P., "Les reserves Mondiales de combustibles mineraux solides", Annales de Mines, Belgique, 1958, No. 2.

6.   Laverov, N.P. (Editor) "Review of the mineral resources of the capitalist and developing countries (beginning of 1969)".

7.   Matveev, A.K., "The geology of the coal deposits of the USSR", Gosgortekhizdat Publ., 1960.

8.   Matveev, A.K., "Coal deposits outside the USSR". Eurasia (1966), Australia and Oceania (1968), Africa (1969), America and the Antarctics (1974), Nedra Publ.

9.   Matveev, A.K. "The coal resources of the world and their distribution in geological systems", In: The modern problems of combustibles, Nauka Publ., 1973, pp. 115-119.

10.   Stepanov, P.I., "Some rules of the stratigraphic and paleographic distribution of the geological reserves of world coal", Proceedings of XVII International Geological Congress in Moscow, 1937, v. I, 1939.

11.   Terentiev, E.V. and Matveev, A.K. "The dependence of the size and structure of coal basins on structural position", Proceedings of VIII Congress on Carboniferous Stratigraphy and Geology, Nauka Publ., v. I, 1939.

12.   "The world mineral resources with reference to the developed and developing countries", Annual review, edited by N.P. Laverov, All-Union Geological Files, Moscow, 1971.

13.   Zheleznova, N.G. and Matveev, A.K., "World coal resources", Soviet Geology, 1973, No. 1, pp. 76-85.

14. Matveev, A.K. (Editor), "Map of the coal fields of the world", USSR Ministries of the Coal Industry, of Geology, and of Higher and Specialized Education, Moscow, 1972.

This full color map with 15 separate colors consists of eight sheets which, when trimmed and mounted as one full map, measure approximately 106 by 47 inches (2692.4 x 1193.8 millimeters). The projection is Equatorial Conformal, and the scale is 1:15,000,000. Detailed insets of Antarctica, the British Isles, and Central Europe are included. A 76 page booklet alphabetically indexing all coal deposits shown and listing country and coordinates for location of the deposit accompanies the map. The map is available through Miller Freeman Publications, Inc., Book/Map Dept., 500 Howard Street, San Francisco, California 94105, U.S.A.

3

OBJECTIVES AND ORGANISATION OF

COAL EXPLORATION PROJECTS

DELL H. ADAMS

Vice-President Exploration Consolidation Coal Company

As you all know, coal is a form of Fossil Fuel. It was recognised much earlier than oil because it occurred much closer to the surface and is often quite visible and far more accessible. It was widely used for several hundred years, mostly for space heating; but it also became the first basic energy resource of the so called industrial revolution, which is hardly 200 years old.

However, the discovery and development of oil, with its inherent advantages and ease of application, together with the later development of natural gas, resulted in a rather dramatic shift from coal as a basic source of energy. As a result of the historical abundance of oil and gas in the United States, coal now accounts for only about 1/5 of our domestic energy needs, and now we suddenly find ourselves in the position of having a scarcity of oil and gas while having an abundance of coal. It is this very scarcity of oil and gas, not only in the United States, but in the rest of the world, that has prompted renewed interest in coal.

As I am sure many of you are aware, coal is the most abundant fuel resource, both in the United States and in the rest of the world. (See Figure 1). World resources of coal total over 8 trillion tons; this is equivalent to

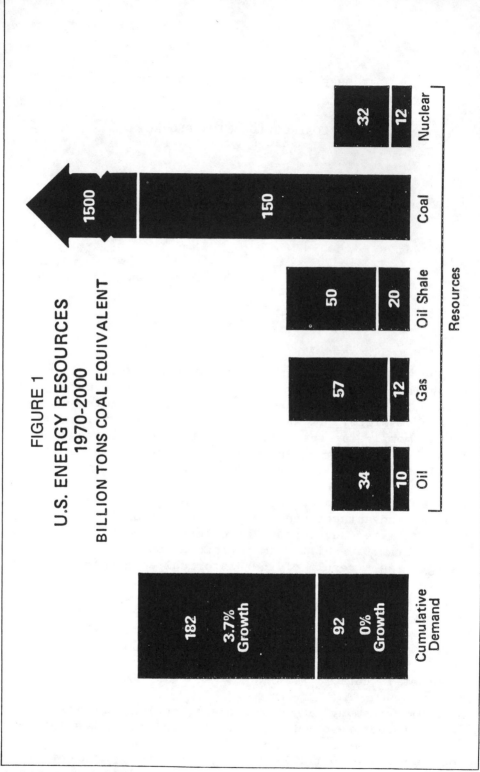

FIGURE 1
U.S. ENERGY RESOURCES
1970-2000
BILLION TONS COAL EQUIVALENT

33 trillion barrels of oil or 10 to 20 times the estimates of ultimately recoverable oil resources in the world.

In the United States, we have recoverable coal resources of about 1.5 trillion tons ... sufficient to meet our domestic needs for thousands of years at the present rate of production. While much of these coal reserves can be produced only at costs much higher than present mining, we estimate that over 150 billion tons (enough to continue current production levels for over 250 years) can be mined at approximately current cost levels. These coals are located in many different geologic provinces and vary from lignites to anthracite, with each province having its own set of unique advantages and disadvantages. Since all coals are not alike, but rather range in both quality and mine-ability, I will point out some of the different types as they occur in the United States. I believe most of these coals are represented by similar types at various locations around the world.

Figure 2. shows the Eastern United States coal field. It is important to note here that not all Eastern coals are alike.

Eastern coals are currently being produced for both steam and metallurgical use and to a much lesser extent certain blocks have been assembled for synthetic conversion use.

First, the Southern West Virginia area produces some of the highest quality metallurgical grade coals in the world and there is simply no other area in the United States from which this type of coal could be substituted. Indeed, there is hardly any source in the Western World that could replace the low volatile coals produced from that area. Metallurgical grade coals are also produced from certain areas of Eastern Kentucky, Central West Virginia, Alabama and South-western Pennsylvania. We anticipate a modest increase in demand for these coals into the foreseeable future.

Secondly, steam coal production is centered in four principal areas. Namely, the famous Pittsburgh seam in South-western Pennsylvania Northern West Virginia and Eastern Ohio, and the high quality coals of Eastern Kentucky. Lesser production can be found in virtually every area of the entire Eastern coal field. It is these coals that will face increasing competitive pressures from Western coals. Eastern steam coal reserves presently require prices in the range of $1.00 per million BTU, but remain competitive in mine mouth or short-haul markets. These prices are well

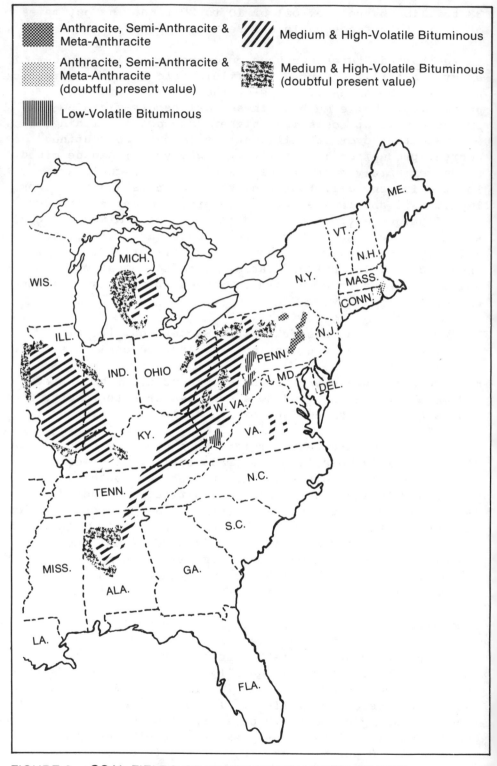

FIGURE 2   COAL FIELDS OF THE EASTERN UNITED STATES
92 COAL EXPLORATION

below the range of $2.00 per million BTU, presently required to import conventional oil and gas supplies from foreign sources. I should note that coal's apparent price advantage is partially offset by higher costs associated with its transportation and utilisation.

President Ford has set a goal of reducing the dependency of the United States on imported foreign crude oil. If this goal is to be accomplished, then coal must most certainly assume a greater role in our total supply of energy. Much of the oil presently being imported is used to fire steam-electric generating stations in the Northeastern States. It is this very type of application that coal lends itself to efficiently. Indeed, many of the plants in the North-east, now using oil, were previously converted from coal in order to take advantage of what was at that time low-cost imported oil.

We believe that present and potential new markets for coal in the Northeast will be supplied with Eastern coals because of their inherent higher heat content and proximity to market. Eastern United States coals have traditionally supplied steam coal markets in the Eastern and Midwestern States. Some of these markets, notably in the Midwestern area, are facing increasing competition from Western coals, for example, the Detroit Edison Co., a long-time user of Eastern coal, has recently entered into a long-term contract to purchase about 180 million tons of Montana coal. This particular Montana coal represents some of the very best of the low-cost Western strip coals, however, and we do not believe that large volumes of Western coal will penetrate Eastern markets beyond the Midwestern States.

Probably never before has the coal industry had such exciting prospects for growth, yet been confronted with as many formidable obstacles. I will comment on some of these and how they must be considered during the exploration phase of development if one is to develop viable projects.

Many problems ... not obvious to newcomers to the industry ... convince us that if the United States is to depend on coal for an increasing share of its total energy, that we must expand our production in virtually every major coal-producing district in the country.

Some of the problems associated with mining and transporting Western coals are:-

- Remote location

- Sparse Population

- Long Transportation

- Relatively Low Quality

- Federal Leasing Policy

The task of relocating and training miners, providing housing and the schools, hospitals and social development will not be an easy one. To move Western coal to its remote markets will require the construction of thousands of railroad cars, and rebuilding and construction of thousands of miles of railroad track and coal slurry and gas pipelines. In addition, most Western coals are of lower heat content than Eastern coals and result in substantially reduced boiler efficiency when utilised in plants designed for higher quality Eastern coals.

Perhaps most important of all however, is the United States Government coal leasing policy ... or lack of a policy, as the case may be ... in the West. Approximately 80% of the coal in the West is Government-owned. There has been a moratorium on Federal coal lease sales since 1971 and, as of this date, there is no firm schedule for the resumption of large scale leasing.

I don't mean to convey the impression that Western coal production will not experience a dramatic and sustained growth rate. My Company has made ... and is continuing to strengthen ... a major commitment in Western coals. My point is simply that the United States coal industry cannot meet the challenge of producing the demand fuel without expanding in all areas.

By now, many of you are probably wondering how exploration can be important if coal resources are truly far greater than demand for the coal. Let me attempt to draw an analogy between coal exploration and oil and gas exploration and attempt to explain why.

In the case of oil and gas exploration, there is nearly always a significant value of discovery. Another way of expressing this would be to say that there is an immediate market for the product at a price which would yield an attractive return on investment. If we now examine coal development with this factor in mind, certainly there is a point in the development stage at which a coal project

could be said to have this same type of value. That point
is reached when a mining project has been engineered
sufficiently to predict with reasonable confidence the cost
of production and when a customer is willing to purchase the
mine production at a price-level that results in an
attractive investment for the mine construction.

As we noted earlier, the abundance of coal reserves is such
that relatively little value can be placed on coal reserves,
per se. Accordingly, the term "Exploration", as commonly
used in the oil and gas industry, is somewhat of a misnomer
when applied to coal.

If we accept the case that coal resources are relatively
abundant and by far exceed the projected demand for the
foreseeable future, then we may conclude that only those
coal resources, equal to their demand, will have any so-
called value of discovery. The one notable exception to
this is that of low-volatile metallurgical coal in which
case demand is actually in excess of supply and is expected
to remain so for some time.

Those phases of work included in coal exploration are:-

- Reconnaissance
- Initial Leasing
- Initial Drilling or Geologic Mapping
- Preliminary Evaluation
- Leasing of Total Project
- Development Drilling
- Feasibility study

Certainly, the first three steps shown on this slide,
namely Reconnaissance, Initial Leasing and Initial Geologic
Mapping have their counterparts in oil or other mineral
exploration.

Probably the best way to identify and select target areas
for coal exploration is to conduct an initial reconnaissance
of those areas and regions known to contain the type coals
of interest. This would normally include a review of
published data and literature and a preliminary visit and
review of the areas under consideration. Reconnaissance
would also determine if areas of the region were still
available for property acquisition. If an area was found to

have potential, then the next phase should be some type of land or property acquisition through leasing or optioning. This provides a firm control base from which to conduct more detailed studies and work programmes; the first of which would normally be geologic mapping or initial drilling. Certainly, some knowledge of structure and probable location of the coal seams is desirous and necessary before planning and committing large expenditures for drilling and additional land control. Field geologic mapping, together with a few selected drill holes, can often provide a rather clear picture of the type deposit being studied. Location of seam outcrops or marker beds can often be made, as well as the existence of faults or other disturbances that would affect the deposit. This phase of work becomes increasingly important as the costs of property acquisition increase and environmental constraints limit the amount of drilling and trenching work that can be performed. However, it is at this point where I believe the similarity between coal exploration and oil or other mineral exploration ends. We estimate that less than 5% of the United States coal resources will be consumed within the next 40 years. Thus, it can be reasoned that only the 5% best or most favourable of the United States coal reserves will have any real value of discovery for the next several years. Bearing this in mind, the objective of coal exploration is no longer one of simply finding coal, but rather one of finding and developing blocks that are within the best 5% of the total coal resources.

Since there are many types of grades of coal, as I noted earlier, and further assuming that one desires to develop a reserve for a specific application, such as metallurgical coke, the target area within which to aim an exploration programme becomes further restricted.

To further define these higher rank projects, preliminary mining economics must be prepared early in the project development and the project under consideration must be ranked with other known deposits to determine its potential competitive rank. If the preliminary project evaluation is favourable, the next logical step is acquisition of all property within the project boundaries, either by lease or purchase.

Probably the most important and difficult phase of coal exploration is total consolidation of property ownership. Figure 3 depicts a hypothetical project whereby alternate tracts of land are under control and the intervening

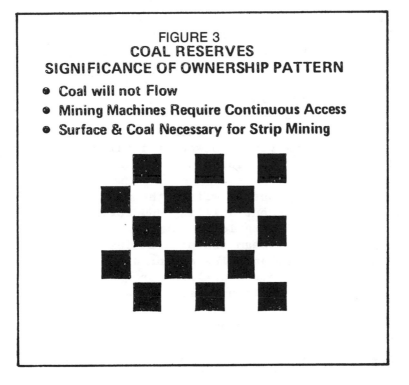

FIGURE 3
COAL RESERVES
SIGNIFICANCE OF OWNERSHIP PATTERN

- Coal will not Flow
- Mining Machines Require Continuous Access
- Surface & Coal Necessary for Strip Mining

parcels are outstanding ownership. Such an ownership pattern would not prevent development of an oil and gas reserve. One would drill his leased properties and subsequently pump the oil from underneath all the project. Unlike oil, however, coal will not flow from one location to another. While such an ownership pattern might actually contain more than enough coal tonnage to support a project, it more than likely contains absolutely no coal that can be commercially produced with the existing ownership pattern. Since coal will not flow to machines used to mine the coal, it follows that mining machines must move to the coal. This machine-movement requires a continuous access or, to state it a different way, this requires a "blocked-up" or contiguous ownership pattern. Strip mining projects are further complicated since both coal and the overlying surface are required in contiguous fashion for those projects. The problem of establishing contiguous ownership of a desirable coal deposit, more often than not, means a lag-time of several years between initial discovery and actual commencement of mining operations.

As acquisition of property within the project progresses, close centre development drilling and coring must be performed to permit detailed measurement of the reserve, analysis of the coal seam and provide insight to the problems that can be anticipated during the actual mining process. The development drilling programme is needed not only to define recoverable reserves, but equally important, if not more so, to define the quality of the coal and how well the quality can be beneficiated in order to determine what type markets it can compete in. Other information from the development drilling includes knowledge of the geologic structure, type of overburden or roof rock material, as well as any unusual conditions that may exist, such as faults or cut-outs in the coal seam. This knowledge becomes invaluable in the process of designing a mine plan and selecting equipment. Recent developments in testing equipment now permit us to measure the rate of methane gas emission from a coal core. This data will be extremely useful in designing mine ventilation requirements.

To summarise, project information obtained from coal boring includes:-

- Extent of Recoverable Reserve

- Coal Quality

- Type Benefication Needed

- Geologic Structure

- Unusual Conditions

- Type Overburden or Roof Rock

- Anticipated Rate of Methane Gas Emission

Once development of a project is complete, a detailed
reserve evaluation and a mining feasibility study must be
prepared. If this study indicates that the project can be
operated and the production marketed at a price which yields
an acceptable return on investment, then at this point the
coal can be assumed to have a value of discovery. For the
most part, the coal-bearing geologic provinces have been
mapped geologically and the existing coal seams are known.
The problem then is to identify those localised blocks that
have some type of competitive advantage over other blocks.
This competitive advantage could be in the form of greater-
than-average seam thickness which would reflect itself in a
reduced mining cost. It could be in the form of better-
than-average quality or it could be in the form of
strategic location which would give it a transportation
advantage. In the case of steam coals each of the above
factors remains an important consideration and will for so
long as available resources exceed demand.

With these objectives in mind, I would like to consider
some of the complexities involved in the considerations of
what determines a viable coal project.

Let us now look at the Eastern coal field and the advantages
it has and the constraints it faces to the development of
new production. The quality advantages are:-

- Low Volatile Matter Content

- High Fixed Carbon Content

- High Fluidity

- High Coke Strength

- High Coke Yield

As we noted earlier, many Eastern coals, especially those
of metallurgical grade, supply specialised markets for which
there is no alternate source of supply at this time. Coking
coals from the East, not only provide the mainstay of supply
for the United States steel industry, but are an important
mix to the blast furnace fuels of Europe, Japan and South

America. Other nations, such as Canada and Australia, are important exporters of coking coals, but coke made from those coals simply does not perform as well as coke made from blends with higher quality coals such as those produced in Southern Appalachia. Indeed, blends of coking coals containing a relatively small percentage of these more ideal coals often permits the production of a superior coke with an increase in the total percent yield of coke from its feedstock coals. Similarly, should those generating stations in the Northeast that once converted from coal to imported oil be required to re-convert to coal, they will require a high BTU coal of the type they were originally designed for or else suffer a severe loss of plant efficiency.

The higher heat content of Eastern coals, about 50-60% greater than most Western coals, and lower than average ash content, partially offset their higher cost of mining. This factor, together with the more favourable proximity to market for Eastern coals permit them to compete on a cost-per-delivered BTU basis in mine mouth and short-haul markets.

On the other hand, Eastern coals face some serious problems and constraints. The Federal Health and Safety Law, which was passed in 1969, has had a very serious impact on productivity. Parts of this law have contributed little or nothing to the actual safety of the mine worker, but have had a tremendous adverse effect on productivity. We are hopeful that technological advances in mining systems will help restore part of this loss.

Since most Eastern coal is mined by underground methods, the problem of recruiting and training large numbers of new people to work in underground mines becomes significant. The fact that coal miners are now one of the highest paid work forces in the United States should provide some relief to the manpower problem.

Perhaps, most critical of all the problems facing Eastern coals today, however, is that of sulphur dioxide emission restrictions. The presently-existing timetable for enacting environmental protection agency sulphur emission standards is such that 70% of the steam coal currently being produced East of the Mississippi could not be used after July 1, 1975. Considering that the United States has an almost critical present supply of domestic energy, there is simply no alternative source of energy available within that time span. The answer is obvious ... Utilities

FACTORS TO CONSIDER IN USA COAL EXPLORATION AND DEVELOPMENT

EASTERN U.S. COALS PRINCIPAL TYPES UTILIZATION

- Metallurgical
- Steam-Electric Generation
- Planned Conversion Plant Feedstock

PROJECT INFORMATION OBTAINED FROM COAL CORING

- Extent of Recoverable Reserve
- Coal Quality
- Type Benefication Needed
- Geologic Structure
- Unusual Conditions
- Type Overburden or Roof Rock
- Anticipated Rate of Methane Gas Emission

WESTERN U.S. COAL DEVELOPMENT CONSTRAINTS

- Remote Location
- Sparse Population
- Long Transportation
- Relatively Low Quality
- Federal Leasing Policy

EASTERN U.S. METALLURGICAL COALS QUALITY ADVANTAGES

- Low Volatile Matter Content
- High Fixed Carbon Content
- High Fluidity
- High Coke Strength
- High Coke Yield

COMPARISON OF OIL AND COAL VALUES

- OIL
  - Strong demand - Short supply
  - Value of discovery

- COAL
  - Resources greater than demand
  - Only small part has value of discovery

EASTERN COAL EXPANSION

ADVANTAGES
- Demand for Specialized Coals
- High Quality
- Proximity to Market

CONSTRAINTS
- Safety/Productivity
- Trained Manpower
- SO_2 Emission Restrictions
- Capital Requirements

COAL EXPLORATION

- Reconnaissance
- Initial Leasing
- Initial Drilling or Geologic Mapping
- Preliminary Evaluation
- Leasing of Total Project
- Development Drilling
- Feasibility Study

EXPLORATION PROGRAM ESSENTIAL INPUT

- Geology (Discovery)
- Mining Engineering (Mineability)
- Geographic Location (Transportation)
- Quality (Marketing)

presently burning high sulphur coal must be granted
variances to the emission standards to provide time for
technology to solve the problem of scrubbing sulphur dioxide
gases from stack emissions. We believe that such a system
can and will be developed to do the job efficiently.

The other constraint I want to comment on is that of the
huge capital investment required to construct and develop
a new coal mine. We now estimate the cost of a new 1.6
million ton-per-year underground steam coal mine to be
about 55 million dollars. In order to attract the required
capital for further development, the industry must have an
assured outlet for its coal at a price which justifies the
investment.

I might summarise by saying that the discovery and development of a viable coal project involves several variables
that impact on the economic value of the project. Any coal
exploration programme that is to be successful must involve
the necessary staff support functions into the programme
at their critical points in the project development. The
uniqueness of the resource is such that factors such as
mineability, geographic location and quality, more so than
discovery as such, determine the success or failure of the
exploration programme.

In conclusion, I believe that even though coal faces some
very difficult problems, that its use and growth is an
essential part of the world's energy requirements. Given
reasonable utilisation standards and sound energy policies
I believe that an appreciable growth in production from
coal can be forthcoming.

I think I might best summarise the charge given to coal
explorers, thusly:-

CHARGE FULL SPEED AHEAD
BUT!!
DO IT CAREFULLY

Section Two Discussion

D.C. ION: The high figures usually quoted for coal resources encourages a complacency in some quarters which inhibits the great effort needed to add to the true working inventory of the industry. Figures such as those produced by Professor Matveev of 30 teratonnes for 'geological resources' confuses the layman, for he does not understand the difference between 'geological' and 'explored', or between 'resource' and 'reserve', when really only the 'proved' fraction is of practical importance. The normal bar-chart of energy resources, with coal dwarfing all others, put the big emphasis on the least known, whereas it is the best known which should be stressed. Citing as Proved Reserves, comparable with those of other resources, the 97 gigatonnes of Identified, Demonstrated, Measured plus Indicated, Economic coals of the USA, (USGS 1975), rather than the 150 gigatonnes used by Mr. Adams, world Proved Coal Reserves build up to 344 gigatonnes, rather than the 1,100 gigatonnes quoted by Mr. Collins. When this figure for coal and the other usually quoted coal resource figures, together with the comparable figures for the other fossil fuels and uranium, all in coal equivalents, are plotted on a semi-log grid, (see 'New Scientist', 29 April, 1976, p.222), the highly speculative Resource Bases are shown and can still be compared, but the truly important Proved Reserves are given the prominence they warrant, because it is from this working inventory that the coals will be produced for the foreseeable future.

H.E. COLLINS: I can agree with my friend Dan Ion on one statement which he has made. Namely, that the reserves, or perhaps resources is a better word to use, of coal bandied about gives a figure so great that people tend to be rather wasteful in its exploitation. I fully accept that, but he contradicts himself if I may say so. He says this figure is far too great and then he reduces it on paper to a figure more commensurate with the resources of oil, but in my view merely reducing it on paper won't alter the situation one iota. The fact remains, that the resources of coal in the world are far, far greater than oil, and the net result is that we have got to intensify exploitation and, I think we have got to mine those coals less extravagantly than we are doing. I fully agree with that and it applies particularly to the United States and to South Africa, where the percentage recovery of the mineable resources is exceedingly low, but I don't agree with his writing down, as he has done, in the presentation which he has given.

DC. ION: Mr. Adams, you used the figure of 150 gigatonnes for the recoverable reserves of the United States.

D.H. ADAMS: No Sir, I said that we estimate that approximately 150 million tons could be mined at approximately cost levels. These figures of course do not agree with the figures which I was using which I said were a year old, but were differentiated between: 50 gigatonnes measured and 47 which were merely indicated.

D.C. ION: Well perhaps we don't have any conflict.

R.S. QUINTON: (Conzinc Rio Tinto of Australia). Mr. Collins, whilst it has been indicated that a coal to liquid hydrocarbon conversion process is a long term prospect in economic terms, I wonder if you would like to suggest a possible time scale in years?

H.E. COLLINS: Well, as you know, the use of coal as a feedstock for the production of crude oil has already been operative on a commercial scale in South Africa for some 10 or 15 years, but of course that is an isolated case, and it has proved commercially viable for two reasons, one: South Africa has virtually no indigenous oil, secondly: landed imported oil has a very high transportation cost before it gets to the industrial conurbations. Now, with the upsurge in the price of oil, obviously it is going to be much easier to achieve commercial viability in the liquifaction and gasification of coal. The new project which I mentioned, is already in the course of construction and will be producing before the end of this decade. It is my view that the production of synthetic crude oil and substitute natural gas, will not come into its own much before the turn of the century.

P.J. FEMIA: (U.S. General Accounting Office). You said the percentage of coal was 28% of the total energy in 1972. What I would like know is, what percentage will it be in the year 2000?

H.E. COLLINS: If my sort of star gazing comes off, then coal is likely to be taking about 20% at the turn of the century, but that is a mere guess.

P.J. FEMIA: May I add to that, the United States are presently using 18%. The Bureau of Mines says it will also be 18% in 2000.

P.K. GHOSH: Production of coal in India is not 80 million tonnes as mentioned in the paper. Current production i.e. for the year 1975/76 is of the order of 103 million tonnes including 3 million tonnes of lignite - rest being hard coal.

During the last few years, after the nationalisation of the coking coal mines in India in 1971 followed by nationalisation of the rest of the coal properties, i.e. non-coking coal mines, in January 1973, Government has been laying special stress on sustained increase in the production of coal in planned manner to cater to the growing demand of the country. With the energy crisis that engulfed the world in 1973, importance of the solid fuel resources has been brought in to focus once again and all efforts are being made to use coal as an alternate fuel to replace oil and oil-products to the maximum. Government has also formed a high power committee on Synthetic Oil to report on the feasibility of adoption of oil-from-coal conversion technology in India. The report of the committee is shortly to be submitted to the Government for its consideration. The production of coal during the last few years has been as follows:

		(Production in m. tonnes)	
72/73	73/74	74/75	75/76
80.1	81.5	91.4	102.4
(including 2.8 lignite)	(including 3.3 lignite)	(including 2.9 lignite)	(including 3.0 lignite)

Production target as drawn up for future years is given below. With greater availability of coal vis-a-vis demand, the production targets are now under consideration for revision:

76/77	77/78	78/79	80/81	83/84
108.0 (plus 4.5 lignite)	123.0 (Plus 4.00 lignite)	135.0 (plus 4.5 lignite)	145.7 (plus 6.5 lignite)	182.2 N.A.

With the enhanced production of coal, India today has indeed a surplus of non-coking coal which, with agreements, could possibly be exported on a long term basis, if need be. Indian coal no doubt, by and large, has a high ash content compared to traditional sources of this commodity, but its sulphur content is invariably very low being within 0.5%. Prevention of environmental degradation being an important consideration in the generation and utilisation of energy, it is believed that the coal from India with low sulphur content will find a sustained demand in the export market. Currently, only coking coal in India is beneficiated. Non-coking coal can also be conveniently washed to bring down its ash content.

For shipment of a large quantity of coal, infra-structures have been and/or are being developed in the Haldia Port (Calcutta) close to one of the most important coal fields of the country, i.e. Raniganj. This port having a present draft of 10.4 m. (34 ft.) increasing to 12.8 m. (42 ft.) by 1982 can handle large colliers. The port is equipped with modern mechanical loading facilities at the rate of 3,000 tonnes per hour i.e. 3.5 million tonnes per year and is expected to be ready by early 1977. By 1982, it would be in a position to handle 8 million tonnes annually. The present port of Calcutta with a draft of 6.4 to 8.2 m. (21-27 ft.) can deal with 2.5 to 3 million tonnes per year as in the past. Facilities for export from the nearby coal fields have also been developed at the Paradip Port in Orissa. As a matter of fact, export to some of the E.E.C. countries has started from this port.

G.B. FETTWEIS: (Montan University, Leoben, Austria) I would like to make some remarks on the very important suggestion Prof. Matveev has made. He suggests a coordinated international calculation of the coal resources and reserves. In this respect, I would like to refer to an older proposal of Netsikert and others of the USA to the kind of classification the Canadian Department of Energy and Resources has used since last year, and to the new classification of

Geothermal Energy of the U.S. Geological Survey. Referring
to these, I would propose to divide the known or inferred
coal deposits in three classes instead of the two of resources
and reserves. That means that the McKelvey-Diagram of the
U.S. Geological Survey and the U.S. Bureau of Mines with two
classes, resources and reserves may be extended with a third
class, that is the resource base. Thereby would be included:

- the reserves, the coal which is well known today and
  which is economically mineable today, as it is understood already in the McKelvey-Diagram

- the resources, the whole known and also the inferred
  coal, so far as the knowledge is not hypothetical and
  speculative, which probably can be mined economically
  in a foreseeable future of, let me say, 1 or 2 generations

- the resource base, which includes all other coal which
  is geologically known and surmised but which

  (a) probably cannot be mined in a foreseeable future
      for instance because of a thickness of less than
      0.6 m. or of a depth, more than 1,500 m. or which,

  (b) is not known well enough to give any assessment in
      the question of mineability. This probably will
      include a great part of today's so-called geological
      resources.

I think the same kind of classification can be used in the
case of all energy resources and minerals.

Furthermore, I would propose the introduction of an international classification system which can bring together the
different kinds of national coal classification systems
regarding feasibility and certainty in the Eastern and
Western countries. For better international understanding
it may be good to introduce and adopt in the English speaking
world a classification with the letters, A, B, C1, C2, D1,
D2, as is already used not only in the Eastern European
countries, but also more and more in the German and French
speaking states.

I will give these proposals in detail in a book on World
Coal Resources, a comparative analysis of their recording
and assessment, which will be published soon by the German
publisher Gluckauf, Verlag Essen, in which I continue a
paper on similar questions to the International Institute of

Applied Systems Analysis, Saxenburg, near Vienna last year.
In any case I believe that an international coordinated kind
of coal resources classification and assessment can help to
give a better view on the abundance of coal in the world and
the contributions coal can give to the world economy.   Up
to now there are very strongly different views on the definition of coal resources in the minds of different people.

D.H. ADAMS: I might point out that my company is doing
something quite similar to what you suggest. We are doing
this in accord with standards adopted by the United States
Geological Survey. We are now categorising reserves into
measured, indicated, inferred, hypothetical and speculative
categories with a firm definition as to the degree of confidence that can be assigned to each of those categories. I
think this very issue causes much of the frustration in
analysing the discrepancy among the various numbers that are
bandied about around the world today. Various people quote
different numbers regarding reserves, but maybe they are
referring to reserves that are assigned to very different
confidence levels. We have adopted, internally within our
company, standards that do conform to the U.S.G.S. I think
you are merely suggesting that perhaps we should adopt world
wide standards which would be fine.

A.F. LESSARD: Is that practice in your company similar to
the other major coal companies in the States?

D.H. ADAMS: I am not sure.

A.M. CLARKE: (Chairman of the International Energy Agency,
World Coal Reserves and Resources Databank Study Group): We
found in examining the various systems of looking at coal
reserves, that people had produced telephone directory-thick
systems for establishing coal reserves and they were suited
to two things. They were suited firstly, to the type of
geology in the coal fields of their country, (for instance
the Germans gave far greater attention to how to cope with
descriptions of tectonic disturbance), or secondly, they
related to the degree of maturity of their coal industry.
The Americans and Canadians had some excellent systems for
dealing with strict mineable reserves because this is what
they were really interested in. If you have any idea of how

long it takes to get a national system of reserve classification accepted and also how long it takes to make the slightest change whatsoever in it, you will realise why the I.E.A. adopted a system rather like a monetary exchange. What we are going to do is to work out how many tons of Italian reserves are equivalent to how many tons of U.S. reserves and so forth. The way we do this is quite simple and I can talk about it to anybody who is interested.

L.E. SCHLATTER: (Petroconsultants SA, Switzerland). My remark is prompted by the paper of Prof. Matveev, which is of rather a general nature. All the lecturers of this afternoon have compared energies in the way of coal equivalent but when we look at reserve figures or at surveys and resource statistics of purely coal, then we realise that everything is given in tons of coal, completely neglecting the fact that the heat value or the quality of the coal varies according to rank and ash content. That means we have a wide range of quality for example in heating value between 1,000 and 8,000 kilo-calories, per kilo. So my suggestion would be that statistics in future relating to coal, should also calculate the coal as standard coal equivalent.

H.E. COLLINS: I would just like to say this, that all the figures I have quoted for coal are in hard coal equivalent to a standard.

D.H. ADAMS: And I might add that the numbers I quoted were in terms of equivalent 11,000 BTU per pound coal.

Surveyors examine a test core from New Zealand's Huntly field.
Photo courtesy World Coal

SECTION 3

4

APPLICATION OF WIRELINE LOGGING TECHNIQUES

TO COAL EXPLORATION

D.R. REEVES

Managing Director
BPB Industries (Instruments) Limited

INTRODUCTION

Wireline Logging Techniques have become an important
aspect of Coal Exploration Programmes. The main reasons
for this are the growth of coal exploration due to the
steep rise in oil prices and the increase in conventional
exploration costs due to the rapid inflation over the
last few years.

The object of this Paper is to review the uses of these
techniques and to consider the advances made since this
subject was previously reviewed by the Author in 1970 (1).
Some likely developments for the next five years are
suggested.

THE MEASUREMENTS

The technique is based on a series of electro-physical
measurements made within a borehole and displayed for
examination on a continuous depth synchronised chart (the
Log). The following measurements are usually made -
Natural Radioactivity, Rock Density, Hydrogen (Neutron)
Density, Resistivity, Temperature and Caliper, and
sometimes Velocity and Strata Dip.

None of these measurements can be considered to have been introduced during the last five years as they have been used for many years by the Oil Industry, but most have been improved upon recently and the many thousands of holes logged have resulted in improved understanding of their meaning.

Natural Radioactivity

The Gamma Ray Log has maintained the position as the prime indicator of lithology. The log is particularly useful in Coal Measure sequences, not only for lithology and correlation but often for identifying marine bands. A typical Gamma Ray Log showing all the usual features is shown in Figure 1.

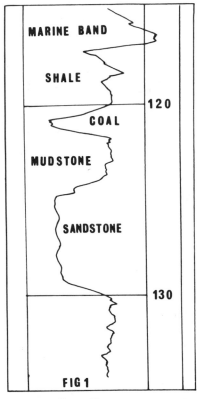

Gamma Ray Response over Typical Coal Measures Sequence

There have been no technical advances concerning coal applications but in the oil business spectral natural gamma logs are being used which can, under favourable conditions, correlate the spectra of rocks, thus allowing, for example, a sandstone or shale to be as specifically identifiable as a marine band.

Density

The value of rock density measurements is the basis of the coal logging technique in view of the unique low density of coal compared with other rocks. Positive identification and measurement of thickness and structure of coal seams is usually possible by this log alone, and, in combination with a caliper log, a certainty.

Recently, the resolution of density tools has been increased with the introduction of the Bed Resolution Density Tool (BRD) which is particularly useful in small diameter drillholes.

Figure 2 shows the improvement in resolution compared
with the LSD (Long Spacing Density) and HRD (High
Resolution Density); however, this improvement is at the
expense of horizontal penetration, thus this tool has
limitations in badly caved boreholes.

Further improvements have been made with on-site
calibration techniques, permitting a standard density
measuring unit to be developed.

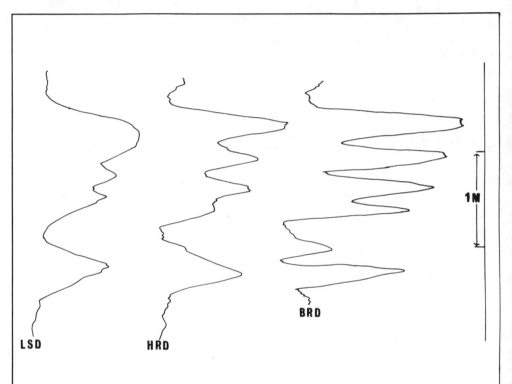

FIG 2

Comparison of Long Spacing Density (LSD), High Resolution
Density (HRD) and a Bed Resolution Density (BRD), showing
improvements in resolution.

Neutron

Measurement of hydrogen distribution is useful for
lithology interpretation and determination of formation
porosity. Apart from improved calibration techniques
this log has not been developed further in recent years,
but its importance is likely to increase in the future as
a means of analysing coal seams. Figure 3 shows the
typical response of a conventional neutron log over a
Coal Measure sequence.

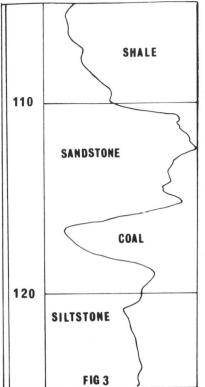

Response of Neutron Log
over Coal Measures Sequence

Caliper

Caliper logs continue to increase in importance. Apart from being essential for correcting density measurements, Caliper logs show considerable detail which can allow seam thickness and partings to be measured with great accuracy. Logs are regularly run on extended vertical scales and horizontal scales which actually magnify the hole section. Tool development has continued and measurement accuracies of $\pm$ 1/10 inches are now achieved. Caliper measurements are also useful for determining incompetent strata.

Figure 4 shows a typical Caliper log over a coal seam.

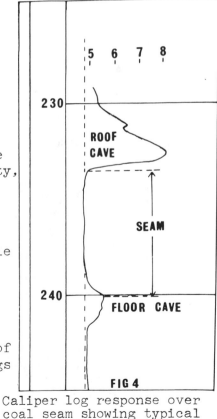

Combined Display of Logs

In many parts of the world the display of Gamma Ray, L.S. Density, Neutron and Caliper Logs on one record is standard procedure for what is termed the general log (usually on a scale of 100 or 200-1). Figure 5 shows an example of this combination.

Detailed logs over seams are re-run separately and often as a combination of Gamma Ray and LSD as one log, and Caliper and one of the higher resolution density logs (HRD or BRD) as a second combination (see Figures 6 and 7)

Caliper log response over coal seam showing typical roof and floor caving

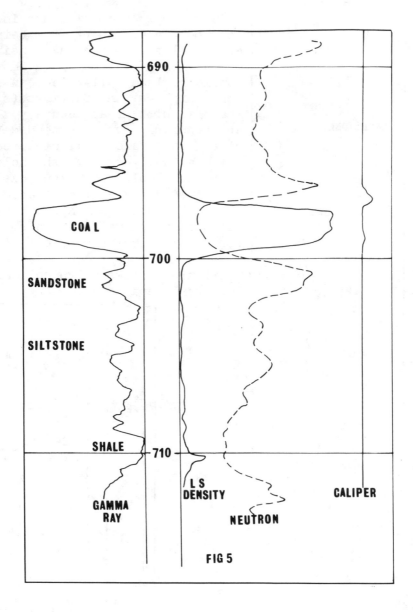

COAL

SANDSTONE

SILTSTONE

SHALE

GAMMA
RAY

L S
DENSITY

NEUTRON

CALIPER

FIG 5

Four-log presentation on 100-1 scale suitable for
identifying lithology, coal and for correlation

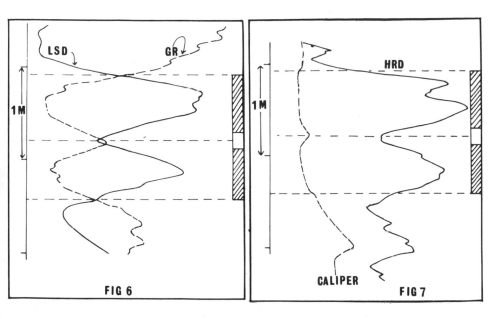

Presentation of Gamma Ray and L.S. Density on detailed scale 1-20 for seam thickness and structure interpretation.

Presentation of Caliper and H.R. Density on detailed scale 1-20 for seam thickness, dirt parting and caving examination.

Resistivity

Simple apparent Resistivity Logs are still run in favourable conditions, but the inability to distinguish between coal and sandstone and lack of calibrated results are severe limitations.

Of more importance are the more recent developments of Focussed Resistivity tools which can measure the true resistivity of the coal seam as well as having sharp resolution. The measurement of true formation resistivity will have important implications in coal seam analysis. A conductive fluid within the borehole is required for this measurement and it cannot be used in cased holes.

Technical developments now permit this tool to be used in fresh water as well as drilling muds.

Figure 8 shows an example of this log.

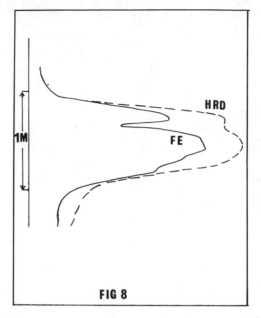

FIG 8

Comparison between Focussed Electric Log (FE) and
High Resolution Density (HRD) showing resolution
over 3-centimetres dirt parting.

Temperature

Continuous temperature logs are often run in potential
new coal fields where deep and large mines are planned
and ventilation studies are undertaken. Measurements can
sometimes reveal water inflows and indicate coal seams.

Sonic Velocity

The measurement of rock velocity can be diagnostic for
coal and is valuable for interpretation of seismic surveys.
Development of this technique has been hampered by the
non-availability of small sized tools; however smaller
tools are now becoming available and this measurement
will become increasingly important.

Strata Dip

The production of a continuous dip record is very
valuable. Unfortunately, this is technically a difficult
and expensive log and can only be used in large diameter
boreholes (6 inches minimum), and at present a simple low
cost method appears unlikely.

EQUIPMENT

Coal exploration calls for light, rugged equipment
which can be mounted either in cross-country vehicles or
underslung on a helicopter. Operating conditions vary
enormously, covering most types of topography and climate.

Fortunately, developments in micro-electronics over the
last decade have permitted miniaturisation and also given
increased reliability, with the result that complex systems
can be developed into quite small units. Figure 9 shows
a skid unit capable of running to 6,000 feet Gamma Ray,
Density, Resistivity, Caliper, Neutron, Temperature and
Hole Position logs, digitising the results and recording
on tape. Figure 10 shows a small cross-country vehicle
in which the unit shown in Figure 9 can be mounted.

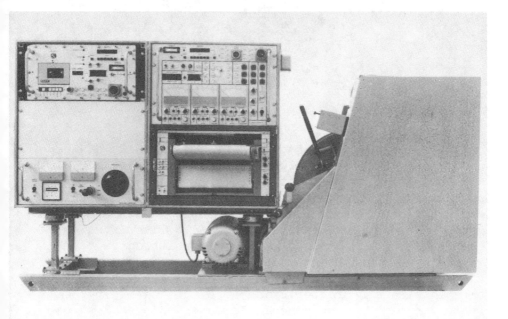

FIG 9

Logging Skid

Light Cross-country Vehicle   **FIG 10**

APPLICATIONS

The use of Wireline Logs can be conveniently divided
into two categories: Geological Control and Petrophysical
Analysis. Geological Control refers to the day-to-day use
of the logs by the geologist, helping both to control the
exploration programme and to make rapid assessment of the
data. Petrophysical Analysis refers to later studies of
the results which are processed by sophisticated techniques
involving computers and other data-processing devices from
which information such as ash content, moisture, rank and
certain strata characteristics important in mining
(strength), may be obtained. To date, the greatest use of
logs has been for geological control, and this application
has been widely accepted by Exploration Companies. Petrophysical analysis is just starting to be applied and the
required data handling equipment is not always available.
However, with the interest in coal exploration now being
shown by major International Oil Companies, who have both
the data handling capabilities and the petrophysical
expertise, rapid developments of this method are now certain

Geological Control

The main elements of geological control are as follows:

Coal Identification. Obviously the positive
identification of coal is essential and the combination of
Gamma Ray, Density and Caliper Logs achieves this, except
in the rare and difficult cases where there is substantial
caving within the coal seam.

In the limit, in a really extensive cave a coal seam
cannot be identified by the logs but fortunately this
occurrence is very rare.

The identification of coal as opposed to sandstone and
shale can be seen in Figure 5.

Lithology. The Gamma Ray is usually sufficient to
give reasonable lithological identification. This is
essentially based upon differentiations in the coal, shale
and sandstone and the various gradations between these
units. It is often easy to identify some special lithologies such as Basalt or Dolerite, and the use of the
Neutron Log in some cases can be useful. One of the most
difficult problems with simple tools is to identify the
difference between limestone and sandstone.

Correlation. As logs respond quite well to lithology
changes, correlation is usually easier than by using hand
specimens, except where special stratagraphic markers,
such as marine bands, are present. Often, however, even
in a standard and uninteresting geological sequence, the
logs will show a unique pattern which can facilitate
correlation which might otherwise have been difficult.
In certain openhole drilling operations the logs can often
pick out marker horizons, thus permitting seam and
formation identifications. Figure 1 shows the Clay-Cross
marine band which is a well-known stratigraphic marker
within the Coal Measures in the U.K.

Seam Thickness. Overall seam boundaries and seam
thicknesses can be interpreted to accuracies of about
$\pm \frac{1}{2}$ inch. There can occasionally be problems where there
is caving at the roof and floor of the seam but in this
case the use of the caliper log, as well as the conventional
density logs, can usually solve the problem.

Figure 7 shows a caved seam where the caliper log is used to interpret the density response at the base of the seam, and the thickness can be determined from either the density or caliper log.

Seam Structure and Dirt Partings. Many seams show a regular "fingerprint" on the logs which enables very rapid identification, again permitting the geologist to know precisely the horizon being proved by the borehole. The seam's "fingerprint" is due to the variations in coal quality or banding of dirt partings. (See Figures 5, 6 and 7). At present, dirt partings cannot be resolved to the same accuracy as seam boundaries, and partings smaller than 8 centimetres require a fairly high interpretation factor before they can be understood.

Core and Sample Control. Detailed logs of coal seams are very useful when examined in conjunction with the core to verify the core sequence. Often, parts of cores are missing and the log enables the missing area to be identified. The continuous detail log can be used for identifying the natural breaks in quality or structure of the seam prior to division for chemical analysis, thus enhancing the value of the chemical analysis.

The control benefits of adding logging to a drilling programme appear considerable, and should result in better and faster evaluation with costs in some cases being reduced by integrating the logging and coring programmes. As an example, the growth of the use of logs by the NCB over the last ten years may be cited. Originally, logs were run in holes that were cored from the surface to verify seam thickness. Gradually, as experience in understanding the lithological response of the logs grew, coring in measures above the Coal Measures was abandoned and this information was obtained from logs, thus permitting some economy. This technique was improved sufficiently for coring to be abandoned in parts of the Coal Measure sequence, spot cores for important seams only being taken.

Ultimately, and in special circumstances, the NCB now drill holes to 1,200 metres depth without any coring at all in order to take a "quick look" in a new area. This has clearly given further economy, but more important in this case, an increase in speed. Although the uncored

holes rely almost entirely upon logs, it is anticipated
that if they indicate favourable coal seam developments,
drill hole patterns involving spot coring of the important
seams will be initiated later.

There are many examples of logging programmes being
extended in shallow hole exploration prospects which are
unable, by conventional coring techniques, to provide the
required information within a given time. There are also
cases of unlogged programmes being repeated because of
poor recoveries in some areas which have prevented an
adequate evaluation being performed. There is also an
interesting case of logging being used when, in a shallow
hole programme, because of an unexpected structure, the
drilling equipment was unable to take cores in the deeper
part of the prospect but could manage to drill open holes.
The increase in flexibility by incorporating logging in a
programme has proved to be important.

There are many ways in which logging can be applied to
any programme to improve geological control, but essentially
it simplifies to a decision as to how much coring and
manpower, and hence cost and time, can be saved if logging
is used, and to what extent do the vital parameters of the
exploration programme need to be checked? These assessments must be made by the geologist against the background
of the normal economic constraints.

Petrophysical Analysis

By carefully examining responses of a family of logs in
great detail, it is possible to extract information
concerning the constituents of a coal seam. The method,
which is outlined below, is tedious and requires the use
of a computer system with logs presented in digitised form,
but the information derived on ash, carbon, moisture, rank
and volatile content is important enough to warrant the
extra sophistication.

The method depends on accepting a general model for a
coal seam, and making a series of measurements of different
physical properties of the coal. By expressing the results
as equations it is possible to obtain a solution which
quantifies the various constituents of the coal seam model.
Models of various complexity can be arranged and it is
instructive to start with a very simple model:

Assume - Coal + Ash = Coal Seam = 100%

then, if we have one good measurement, say Density, we get

$$x \, Coal_{(d)} + y \, Ash_{(d)} = Seam_{(d)}$$

where x = per cent Coal

     y = per cent Ash

and (d) represents the density of the material concerned.

Since $seam_{(d)}$ is measured and x + y = 1, we can solve for x and y.

This is the basis of ash content determination directly from the density log and it works very well when such a simple model is valid.

For the sake of illustration, let us consider a much more complicated model, ignoring intermediate models which may well be applicable in many cases.

The new model is

Carbon + Volatile Matter + Mineral Matter + Moisture + Ash
$$= 100\%$$

This is much more like a full analysis, but to make matters easier, let us assume

Mineral Matter + Ash = 1.2 Ash

So, the model becomes, using this substitution and convenient lettering:

$$xC + yVm + zM + 1.2qA = 1.$$

where x, y, z and q are the respective proportions.

To solve this equation, we need four different measurements, such as density(d), resistivity(r), neutron(n) and sonic(s).

So, for Density

$$xC_{(d)} + yVm_{(d)} + zM_{(d)} + 1.2qA_{(d)} = Seam_{(d)}$$

Similar equations can be written for other logs, and thus x, y, z and q can be solved.

Solving these equations for every inch of a seam is much too laborious by hand but can easily be handled by a computer - hence the need for digitised field results.

Sometimes, however, important simplifications can be made where some logs respond more significantly to one parameter, in which case a "cross-plot" technique may be used. For example, if the density log responds more to ash than moisture and coal, and a resistivity log responds mainly to carbon, then cross-plotting the density log against the resistivity log is effectively plotting carbon against ash from which an elementary indication of rank can be estimated as shown in Figure 11.

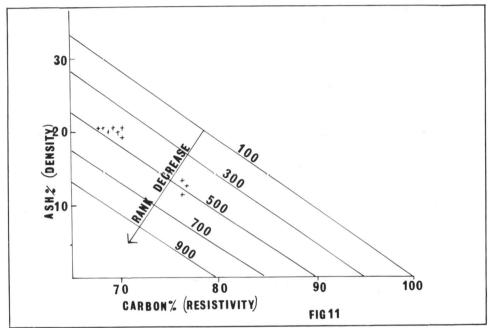

Cross-plot representation of Density v Resistivity to indicate rank changes (diagrammatic).

To date, except for ash content, the results of petro-physical analysis are too few to be cited with confidence although there is considerable optimism. Figure 12 shows why such optimism is being generated, by showing a sequence of seams interrupted by an intrusion. The four logs run over this sequence (Resistivity, Neutron, Gamma Ray and Density) show just the sort of responses which would be expected to seams subjected to progressive metamorphism (rank increase) and it is worth noticing how the logs react differently to the changes.

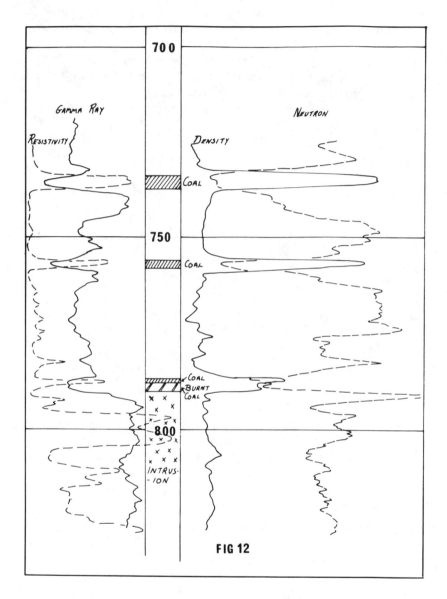

700

GAMMA RAY

RESISTIVITY

Coal

750

Coal

Coal
BURNT
Coal

x x
x x
x x
x x
800 x
x x x
x x x
INTRUS-
-ION

NEUTRON

DENSITY

FIG 12

Responses of principle logs over coal seams progressively
metamorphosed by intrusion (note base of seam adjacent
to intrusion is carbonised).

FUTURE DEVELOPMENTS

If the current demand for coal logging services is sustained, then the next five years should see significant advancements. Forecasts are always dangerous but it would be disappointing if most of the items mentioned below are not progressed.

New Measurements

Practical and economic slimline tools such as Sonic and Formation Resistivity are now about to be introduced and will become important. These tools will greatly improve the range of petrophysical analysis and hopefully serious investigations into the Sonic tool's use for rock strength measurements will begin.

Digitisation

On-board digitisation will become standard, resulting in wider use of detailed interpretation via the computer.

Resolution of Thin Beds

Two techniques are likely to improve thin bed and dirt parting resolution. Firstly, special resistivity techniques, some already under test, have indicated that beds as thin as two centimeters can be fully resolved. Secondly with the advent of digitisation, the possibility exists for the computer to synthesise a recorded log from a store of rock densities and bed thicknesses, thus interpreting many of the repeatable "wiggles" on a density log which at present cannot be understood.

Tool Combination

Small tools capable of a combination of up to four logs in one run, which can be operated by one man, are now quite feasible and their introduction will speed up operation.

Elemental Analysis

Attention is now being given to developing techniques for direct determination of sulphur and sodium. Theoretically, there are techniques which can achieve the answer, but calculations indicate that such a technique is working at its lower limit. However, given some luck and some more advances in technology, five years could see the development of a viable technique.

ACKNOWLEDGEMENTS

I would like to thank the NCB for their unrestricted permission to publish parts of logs from their current exploration programme.

REFERENCES

(1) In-situ Analysis of Coal by Borehole Logging Techniques by D.R. Reeves

CIM Transactions: Vol. LXXIV, pp. 61-69, 1971

(2) The Radiation Density Log applied to the resolution of thin beds in coal measures by J.R. Samworth

Third European Symposium Transactions London, October 1974

5

RECENT DEVELOPMENTS IN COAL PETROPHYSICS *

B.A. Lavers and L.J.M. Smits

Shell Internationale Petroleum Company
The Hague – The Netherlands

ABSTRACT

In 1973 a major programme of coal exploration was
initiated by Shell Companies operating in various parts
of the world. While it was recognised that Coal Petrophysics was still a fairly new and inexact science, a
policy decision was made at the outset to log every
bore hole.

Requirements are given for a suitable package of
wireline instrumentation.

Qualitative and quantitative use and interpretation
of the logs are discussed. These include control on
coring, correlation, determination of ash content,
calorific value, rank, identification of and character-
ising of overburden lithologies for mining engineering
purposes.

* Reprinted with permission from the Transactions of
the Seventeenth Annual Logging Symposium of the
Society of Professional Well Log Analysts.
Denver, Colorado, USA – 9-12 June 1976.

An integrated computer system is described to handle coal core analysis and petrophysical evaluation. This is based on a data bank of coal core measurements 'COGEO', on an interpretation system 'COPLA' and a data bank of interpreted values 'COPET'. A number of application subroutines generate Pseudo-Washability curves and vertical selection for mining feasibility studies.

1. INTRODUCTION

In 1973, as one of several diversification projects, a major programme of coal exploration was initiated by Shell Companies operating in various parts of the world. This programme was primarily aimed at assessment of surface mineable coal deposits.

The traditional tools in the coal industry for such a prospect evaluation consisted mainly of drilling a large number of shallow core holes. Drill cuttings and cores were described by geologists. In specialised laboratories coal quality parameters were measured such as seam thickness, ash content, calorific value, rank, moisture content, volatiles and sulphur. These data were used for mapping, volumetric estimates and economic evaluation of the deposits discovered.

In the oil industry basically similar economic appraisals are made by drilling most times a few wells, often less extensively cored. "Quality" parameters such as net sand, porosity, hydrocarbon saturation, type of hydrocarbons are obtained for a major part from logs, and a large number of sophisticated logging tools and evaluation techniques have been developed over the last three or four decades.

Having a well established experience of logging, although it was recognised that coal petrophysics is still a fairly new and inexact science, it was felt that logging in coal would lead to much improved evaluations. Therefore, a policy decision was made at the onset to log every bore hole.

To obtain maximum results the logging tools should be specifically designed for coal logging. That is to say, their range of measurement should be specifically geared to the physical properties of coal. However, the lithology of the overburden and base layers should also be properly detectable. Geologists and petrophysical engineers should be adequately trained for qualitative and quantitative interpretation of coal logs.

The huge amount of data, from cores as well as from logs, generated in a short time span can only adequately be managed with an integrated computer system.

A number of these objectives are further discussed in this report.

2. LOGGING TOOLS

Routine logging for coal requires a suitable package of wireline instrumentation, purpose designed to do the necessary job.

Core holes are generally drilled with light equipment and have a diameter of 3-4 inches, therefore the logging tools should have an outer diameter of 2 inches or less.

Operations generally require that the logging equipment is light weight and can be transported in vehicles of limited size. In some cases it was necessary that it could be transported by man-power along small bush tracks.

The tools should have a good vertical resolution. A maximum resolution of 15 centimetres seems satisfactory.

The measurements should at least include Formation Density, Hydrogen Index (neutron logs), natural Gamma Rays, Resistivity and hole Caliper.

The pressure and temperature ratings of the tools can be very modest.

It is imperative that the job is done at low cost. An extremely large bore-hole footage is to be logged and the price list of the logging service company should not be out of proportion to the overall cost of an exploration campaign.

The following tools are routinely run in our holes:

Gamma - Gamma density logs:

Long - spacing density log   (LSD), spacing 48 cm
High - resolution density log   (HRD), spacing 24 cm
Bed - resolution density log   (BRD), spacing 15 cm

Natural Gamma Ray

Neutron - Neutron or Neutron-Gamma Ray

Caliper (3-arm)

Resistivity. So far a single electrode reference log has been run. Currently a calibrated focussed resistivity log is being introduced.

The holes are generally logged from top to bottom with a depth scale of 1:200 and over the coal seams with a depth scale of 1:20.

The complete logging service is, generally on a contract basis, provided by a British logging company, with which close liaison is maintained, regarding mobilisatio of logging units, tool performance, calibration procedures, research and development.

3. OPERATIONS

The operations have already become quite extensive and there are now twelve units working for Shell in South Africa, Botswana, Swasiland (5 altogether), Indonesia (5), Australia (1), and Canada (1). In 1975, durin a short campaign one unit operated in Sarawak.

These units are built in small trucks. Several of these units have been built to fit into two 700 pounds packages for transport by light helicopter.

The above level of activity is supported by petrophysical engineers assigned to each of the major areas. Their basic experience has hitherto been in hydrocarbon petrophysical engineering, but prior to their new assign ments, they have been trained for some time in SIPM's coal petrophysics section.

4. QUALITATIVE INTERPRETATION

The logs are in the first instance used as qualitati

geological tools for coal seam identification, characterisation and correlation. In this respect full advantage could be made of Shell's experience with hydrocarbon logs, and this knowledge was directly available and fully applied in coal geology from the start.

In Figure 1 a schematic response of the logging tools is given for various lithologies. It is seen that identification of coal seams is generally not too difficult, in view of the very low density of coal (1.3-1.8 grammes per cubic centimetre). Therefore, the suite of density logs are best for coal identification.

The density logs also show generally the most variation with differences in coal properties, which makes these logs also most suitable for individual seam characterisation. In specific cases, depending on coal composition, also other logs, like the Gamma Ray log or the resistivity log, give useful additional help for correlation. In fact, logs provide a continuous and much more detailed picture of coal seams than can be achieved with core descriptions by a geologist. This is exemplified in Figure 2. This holds in many cases also for core analysis data, where, consecutive core sections of 1-2 metres are taken, crushed and thoroughly mixed prior to measurement.

Therefore, logs are considered superior to core description and analysis for field-wide seam correlation, and, together with a quantitative evaluation, for economical appraisal of a prospect. An example of correlation of a complicated set of coal seams on the basis of logs is given in Figure 3.

Two more advantages of logging should be mentioned at this stage. Logs proved an unexpected bonus for operational control of driller's depth, coring depth and interval cored. The latter is of importance because generally in drillers contracts a minimum of (say) 95% core recovery should be guaranteed against the penalty of free redrilling. In fact, drillers performance in these respects have improved considerably upon the introduction of logging. Also, as readily visualised in Figure 2, core sampling for analysis is greatly facilitated when the cores are compared with the logs. Then reasonably homogeneous pieces can be selected, consistent over the whole field, and admixtures of clay partings or shoulders can be avoided.

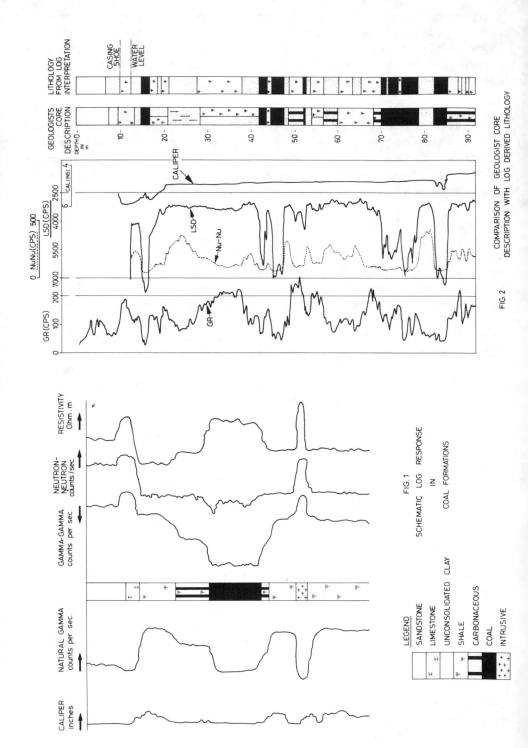

CALIPER
inches

NATURAL GAMMA
counts per sec.

GAMMA-GAMMA
counts per sec.

NEUTRON-
NEUTRON
counts/sec.

RESISTIVITY
Ohm.m

FIG. 1

SCHEMATIC   LOG   RESPONSE
IN
COAL   FORMATIONS.

LEGEND

SANDSTONE
LIMESTONE
UNCONSOLIDATED CLAY
SHALE
CARBONACEOUS
COAL
INTRUSIVE

LITHOLOGY
FROM LOG
INTERPRETATION

GEOLOGIST'S
CORE
DESCRIPTION

CASING
SHOE

WATER
LEVEL

DEPTH
IN
m

10 -
20 -
30 -
40 -
50 -
60 -
70 -
80 -
90 -

CAL.(INS.)

CALIPER

NuNu(CPS)

LSD(CPS)

LSD

Nu-Nu

GR(CPS)

GR

COMPARISON OF GEOLOGIST CORE
DESCRIPTION WITH LOG DERIVED LITHOLOGY

FIG. 2

134 COAL EXPLORATION

VERTICAL   SCALE 1 : 1000
HORIZONTAL SCALE NONE (AVG HOLE DISTANCE = 1 km

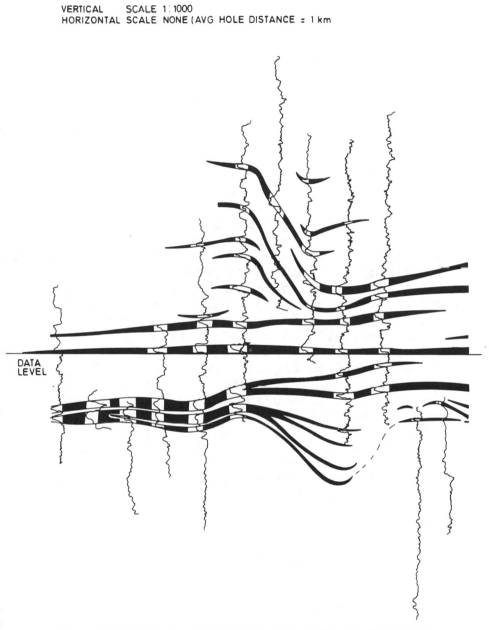

DATA
LEVEL

FIG. 3   EXAMPLE OF LOG CORRELATION IN COMPLICATED COAL FORMATIONS
(LOGS ARE LINED UP ACCORDING TO THE DATA LEVEL INDICATED)

We feel that the above qualitative benefits would already provide sound justification for logging all bore-holes.

5. QUANTITATIVE INTERPRETATION

The coal properties which should be determined from logs are the quality parameters determining their economical value. These include precise seam thickness, ash content, calorific value, rank, moisture content, volatiles and sulphur.

With the exception of precise seam thickness neither of these properties are directly or uniquely related to the responses of logging tools. This is in part due to the fact that routine measurements made on coal samples do not always reflect in situ conditions. For instance, ash content of coal, as used in the coal industry is the weight percentage of residue after complete incineration of a coal sample. During this process, in general, the ash constituents, which may vary from clay to carbonates and pyrite, are subject to strong chemical alterations. The moisture content is measured on air-dry samples rather than on fully saturated samples. One would argue that, in principle, measurement of mineral content in a virgin sample and total moisture content would be more meaningful for logging. On this basis Bond et al (1), and Kowalski (2), have described quantitative evaluation methods. On the other hand, present ash and moisture data obtainable in bulk with simple and cheap routine determinations are employed throughout the coal industry and are considered to serve the purpose. Reeves (3) has implicitly used these latter data to derive calibration curves correlating ash content and log response.

Rather than, at the present stage, seeking more fundamental correlations between, for instance, in situ mineral content and laboratory-determined ash content, we have, for practical reasons, chosen to establish indirect correlations between routine laboratory data and log deflections in our present exploration campaigns.

Because of the high density contrast between coal seams and adjacent beds, boundaries between the two can accurately be determined with the BRD tool. Upon passing such a boundary, the tool response will gradually and linearly change, over an interval equal to the source/

detector spacing. The mid-point of this change indicates
the boundary with an accuracy of a few centimetres. This
is true for seams, or vice versa, for shale partings,
equal or thicker than the tool spacing, 15 centimetres.
For intervals thinner than this tool spacing, no accurate
seam thickness can of course be determined.

It has been found in all our areas that ash content
is best reflected in density log variations. This is
specially true in southern Africa and Australia, where
ash content is the overriding coal quality parameter.

In principle the factors influencing ash determination as a function of density are:

1. Type of mineral matter associated with the coal. This
   can be clay, carbonates, quartz, pyrite.

2. Moisture content. This moisture can be associated
   with the mineral matter and with the pure coal. In
   pure coal the amount of moisture is generally associated with rank (degree of coalification((4).

3. Degree of coalification.

4. Maceral composition, the main groups being inertinite,
   vitrinite and exinite.

Therefore, one can not expect a world-wide unique
correlation between density log deflection and ash content.

This might not even be the case in a geological province, as is demonstrated in Figure 4. Here bulk density
is plotted versus ash content for a number of samples
from Natal and South East Transvaal. The enveloping
straight lines intersect at about the points where the
"pure ash" minerals clay/quartz are found, which are indeed the ash minerals in the above samples. Several other
pure mineral points are plotted in Figure 4. Here in-
situ grain densities (as the logs would measure them) are
plotted versus residual ash, after complete incineration,
as a weight percentage of the original, in-situ grain
weight. It is seen that, depending on type of mineral,
there can be large variations in correlations between
bulk density and ash content.

Coal ranks, coal and ash type variations in seams

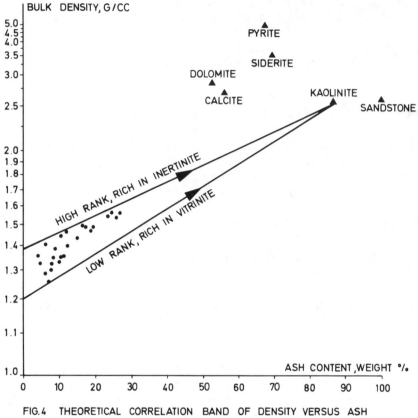

FIG.4 THEORETICAL CORRELATION BAND OF DENSITY VERSUS ASH
CONTENT FOR COAL TYPES WITH RANK BETWEEN HIGH VOLATILE
BITUMINOUS COAL AND ANTHRACITE, MACERAL COMPOSITION OF
INERTINITE / VITRINITE BETWEEN 20/80 AND 80/20 % AND WITH
MINERAL CONSTITUENTS MAINLY KAOLINITE.

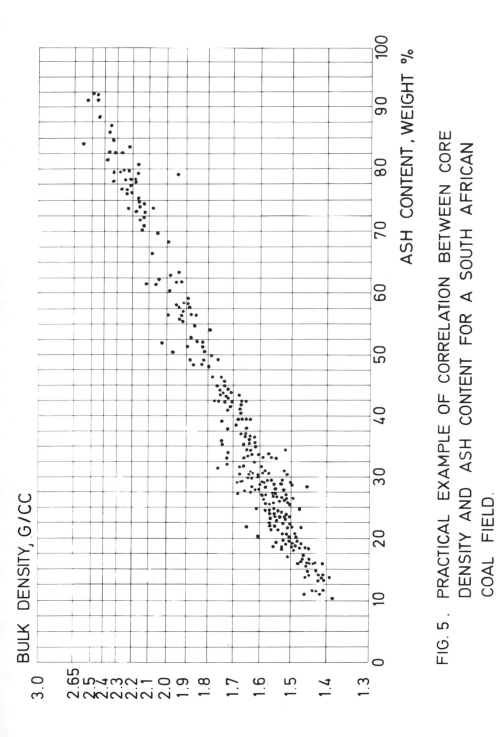

BULK DENSITY, G/CC

ASH CONTENT, WEIGHT %

FIG. 5. PRACTICAL EXAMPLE OF CORRELATION BETWEEN CORE
DENSITY AND ASH CONTENT FOR A SOUTH AFRICAN
COAL FIELD.

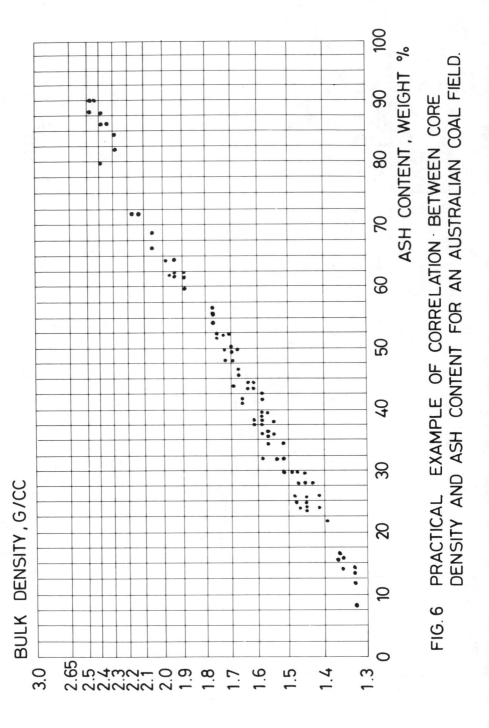

FIG. 6 PRACTICAL EXAMPLE OF CORRELATION · BETWEEN CORE
DENSITY AND ASH CONTENT FOR AN AUSTRALIAN COAL FIELD.

explored until now, in southern Africa and in Australia have not caused serious calibration problems. Examples of these correlations are given in Figures 5, 6 and 7. In Figure 5 a field-wide correlation is given of core ash content versus core bulk density for a field in southern Africa. In Figure 7, for the same field, a correlation is given of core ash content and HRD log deflection. In Figure 6 a field-wide correlation is given of core ash content versus core bulk density for an Australian field.

These and similar relations have already led to the decision in a number of campaigns to reduce drastically the coring programmes to, say, one core hole in five holes.

It is noted that, in general, the HRD could be correlated with appreciably less scatter than the LSD. This is felt to be due to the fact that the HRD spacing of 24 centimetres is better adapted to observed variations in seam quality than the much longer spaced LSD. In above cases, where we could obtain a good correlation of bulk density and ash content, it is generally also possible to obtain a good correlation between bulk density and calorific value. Hence this latter property can then be derived using the HRD log. Figure 8 shows such a correlation between core-determined calorific values and HRD log deflection.

In Sumatra, Indonesia, where ash content is usually low, rank and resin content are the more important economic parameters. This interpretation is much more difficult and not all problems have been solved. However, progress is being made with the estimation from the logs of volatiles, ash and calorific value, related to the above properties.

In Figure 11 a cross-plot is given of neutron logs versus calorific value (which on ash-free basis can be considered as a rank parameter) in a coal field which is locally upgraded due to igneous intrusions. In Figure 9 the response of a single-electrode electrical log is given, showing markedly higher resistivity in resin-rich bands. Volatiles are related to resin content, and there appears to be a relationship between electrical resistivity and volatiles, as exemplified in Figure 10. This relation cannot be used field-wide, since the single-electrode resistivity log is not calibrated. The focussed

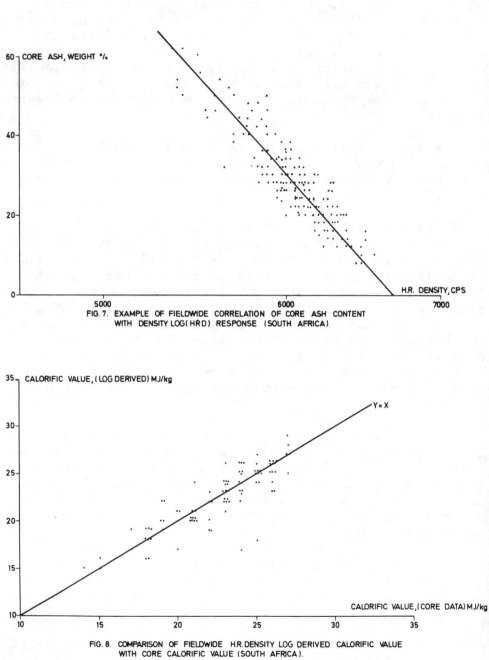

FIG. 7. EXAMPLE OF FIELDWIDE CORRELATION OF CORE ASH CONTENT
WITH DENSITY LOG(HRD) RESPONSE (SOUTH AFRICA).

FIG. 8. COMPARISON OF FIELDWIDE H.R.DENSITY LOG DERIVED CALORIFIC VALUE
WITH CORE CALORIFIC VALUE (SOUTH AFRICA).

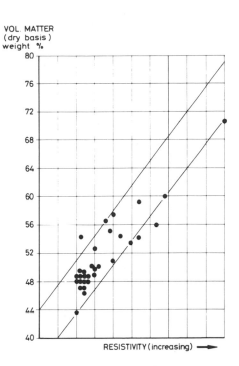

FIG.10 CORRELATION OF SINGLE ELECTRODE RESISTIVITY
WITH VOLATILE MATTER IN A COAL ZONE INCLU-
DING RESIN RICH BANDS (SUMATRA)

LIGNITOUS	BITUMINOUS	CARBONACEOUS	

ANTHRACITIC

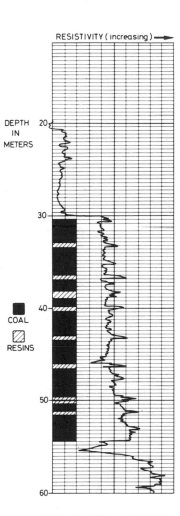

FIG.9 SINGLE ELECTRODE RESISTIVITY LOG
INDICATING THIN RESIN RICH BANDS.
(SUMATRA)

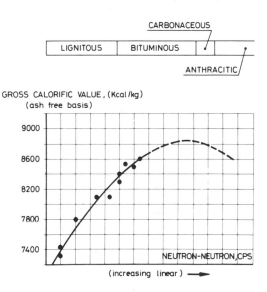

FIG.11 CORRELATION OF NEUTRON-NEUTRON LOG WITH
CALORIFIC VALUE IN AN UPGRADED (INTRUDED)
COAL FIELD (SUMATRA)

DEVELOPMENTS IN COAL PETROPHYSICS 143

resistivity log, which is being incorporated in the log suite, will be of much more value for this purpose.

Cross-plots of LSD and Gamma Ray give information on coal type, as exemplified in Figure 12, where cannel coal can easily be distinguished.

Neutron, Density and Gamma Ray logs are generally run over the total depth of the hole and cross-plots suffice to delineate overburden, parting and floor-rock lithology. This is illustrated in Figure 13 for the Neutron-density combination and in Figure 14 for the Neutron-Gamma Ray combination.

6. DATA MANAGEMENT

An integrated computer system has been developed to handle coal core analyses and petrophysical evaluations. This system is based on a data bank of coal core measurements COGEO, and interpretation system COPLA and a data bank COPET, on which the results of the interpretations from COPLA are stored. The system has been adapted from the SIPM digital log processors and interpretation programmes developed over the years for oil and gas logging. Figure 15 shows a schematic diagram of this system.

Up to now, digitisation of all coal logs has been carried out by a specialised contractor, using the paper logs. Very recently, digitisation at the well site became possible. The signals of the various tools are recorded on cassettes. They are converted, in the logging company's centre, to a 9-track tape in a SIPM format, suitable for further processing on our computer system.

The log data, originally recorded in equal time increments, are transformed to readings in equal depth increments and stored on a permanent master file. A subsequent series of programmes blocks or squares the logs (see Figure 16). This squaring procedure restores sharp boundaries between layers of different properties, which were rounded off during logging due to the finite distance between source and detector. The squared logs are also stored on a permanent master file.

Finally, a relevant set of the logs and core data from the COGEO file are multiplexed on a work file,

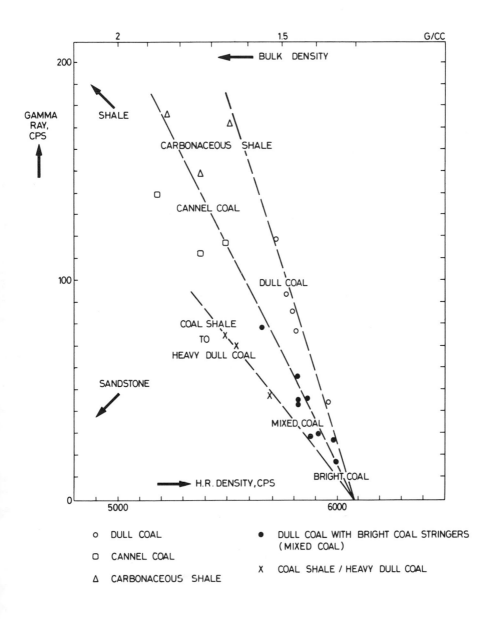

FIG. 12   COAL TYPE FROM GAMMA-DENSITY CROSSPLOTS (SOUTH AFRICA)

DEVELOPMENTS IN COAL PETROPHYSICS 145

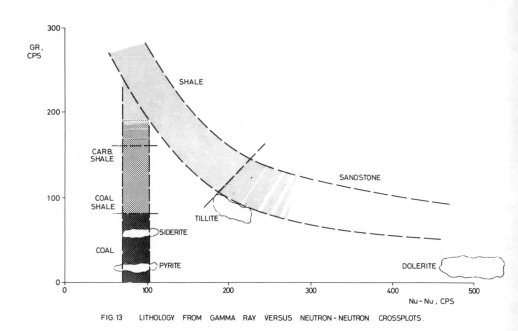

FIG. 13   LITHOLOGY FROM GAMMA RAY VERSUS NEUTRON - NEUTRON CROSSPLOTS.

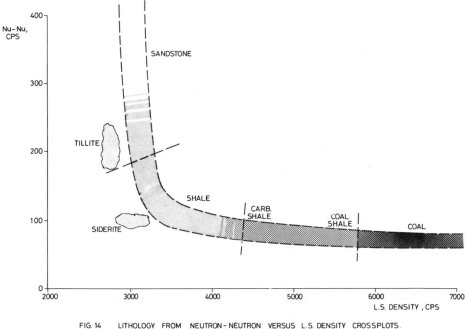

FIG. 14   LITHOLOGY FROM NEUTRON - NEUTRON VERSUS L.S. DENSITY CROSSPLOTS.

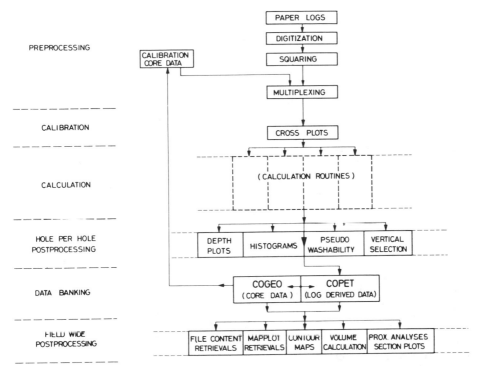

PREPROCESSING

CALIBRATION

CALCULATION

HOLE PER HOLE
POSTPROCESSING

DATA BANKING

FIELD WIDE
POSTPROCESSING

FIG. 15   SCHEMATIC OF COAL PETROPHYSICAL COMPUTER SYSTEM WITH APPLICATION PROGRAMS.

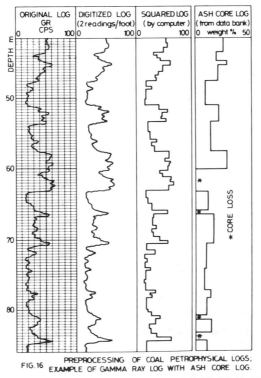

FIG. 16   PREPROCESSING OF COAL PETROPHYSICAL LOGS;
EXAMPLE OF GAMMA RAY LOG WITH ASH CORE LOG.

DEVELOPMENTS IN COAL PETROPHYSICS 147

which serves as the basis for petrophysical interpretation.

As mentioned before, much if not all interpretational effort in coal petrophysics, in the present state of the art, is empirical. Consequently, a first step in the interpretation chain is to establish correlations of core data with relevant log data through cross-plotting. Once these correlations are established, per hole and for each single layer in a seam, the desired property is calculated, e.g. ash content, calorific value, volatiles. Post processing routines are then used to generate depth plots (Figure 16), histograms, pseudo-washability curves, weighted average properties for vertical intervals and vertical selection for mining feasibility studies. Figures 17 and 18 show examples of COPLA output. Figure 17 shows a tabulation of proximate data per specific gravity class separately and cumulatively. This figure also shows a plot of cumulative mass versus specific gravity. It indicates that if the coal is washed to a specific gravity of 1.65 grammes per cubic centimetre the yield will be 55 percent. Figure 17 also shows a plot of cumulative ash content versus specific gravity which shows that the ash content of the floats then will be 26 percent. Figure 18a) demonstrates the vertical selection of the mineable and saleable coal in a hole on the basis of a specified minimum separately mineable dirt thickness and a specified minimum economically mineable coal thickness. The density at which the coal will be washed and the minimum economic yield are also specified. Figure 18b) shows a depth plot of core ash and of HRD derived ash. Four columns indicate, for four different sets of input specifications, various mining possibilities. The first of these four is indicated in Figure 18a. It is seen that the detailed picture of the HRD derived ash content is obviously a better description of coal distribution than that indicated in the core ash column.

The results of the calculations made in COPLA are stored in a petrophysical data bank COPET. Together with the geological data bank, COGEO containing, next to core analysis data, a fairly complete geological review of each hole, provides an integrated data bank from which a number of convenient retrievals can be made with appropriate programmes. For example single value map plots, contour maps and seam volume calculations can be retrieved.

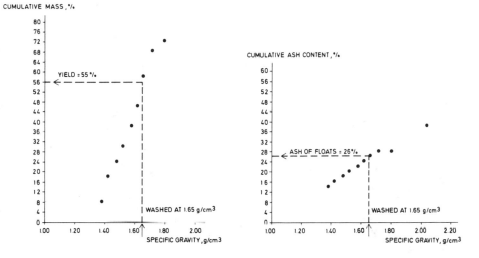

SUMMARY OF COAL PROPERTIES FROM LOG INTERPRETATION

INTERVAL (S) 32.09-46.26 SEAM 7

SPECIFIC GRAVITY CLASS	AVERAGE s.g. IN CLASS	TOTAL METER IN CLASS	MASS FRACTION PER CENT	ASH CONTENT PER CENT	MOISTURE PER CENT	CUMULATIVE AV.	S.G. METER	CUMULATIVE MASS %	ASH %	WEIGHTED MOISTURE %	AVERAGES VOL. MAT. %	CAL.VALUE MJ/KG
1.35 - 1.40	1.38	1.47	8.6	14.8	3.7	1.38	1.47	8.6	14.8	3.7	31.1	25.81
1.40 - 1.45	1.42	1.62	9.7	17.5	3.6	1.40	3.09	18.3	16.3	3.7	30.7	25.31
1.45 - 1.50	1.49	0.89	5.6	22.4	3.3	1.42	3.98	23.9	17.7	3.6	30.4	24.81
1.50 - 1.55	1.53	1.04	6.7	25.6	3.1	1.44	5.02	30.6	19.4	3.5	30.0	24.20
1.55 - 1.60	1.58	1.11	7.4	29.6	2.9	1.47	6.13	38.0	21.4	3.4	29.6	23.51
1.60 - 1.65	1.62	1.18	8.1	32.6	2.8	1.49	7.31	46.0	23.4	3.3	29.3	22.82
1.65 - 1.70	1.67	1.76	12.4	35.9	2.6	1.53	9.07	58.4	26.0	3.1	28.8	21.17
1.70 - 1.75	1.73	1.32	9.6	40.6	2.3	1.55	10.39	68.0	28.1	3.0	28.6	21.17
1.75 - 1.80	1.79	0.45	3.4	45.3	2.1	1.56	10.84	71.4	28.9	3.0	28.4	20.88
1.80 +	2.03	3.33	28.6	64.1	1.0	1.67	14.17	100.0	39.0	2.4	27.5	17.37

FIG. 17 EXAMPLE APPLICATION PROGRAM " PSEUDO WASHABILITY " :
PLOTS AND LISTINGS OF LOG DERIVED DATA SORTED IN
A FORMAT SIMILAR TO A WASHABILITY TEST

VERTICAL SELECTION AND CALCULATED OVERBURDEN RATIOS FROM LOG ANALYSIS

FOR INTERVAL 27.75 – 102.22 M

VERTICAL SELECTION IS BASED ON A MINIUM SEPARATELY MINEABLE DIRT PARTING OF 2.00 M THICKNESS AND A MINIMUM ECONOMICALLY MINEABLE COAL THICKNESS OF 5.00 M.

DIRT IS GENERALLY DEFINED AS MATERIAL HAVING A DENSITY GREATER THAN 1.65 G/CC BUT IT CAN ALSO INCLUDE SOME THIN COAL LAYERS.

OVERBURDEN RATIO IS DEFINED AS THE RATIO OF THE OVERBURDEN PLUS DIRT PARTINGS TO THE TOTAL MINEABLE COAL.
YIELD IS BASED ON COAL BEING WASHED AT 1.65 G/CC. MINIMUM ECONOMIC YIELD IS ESTIMATED TO BE 20%.

| | MINING ZONES | | | | | | | | | | | | | OVERBURDEN RATIO | |
Nº	TOP M	BOTTOM M	THICKNESS M	COAL OR DIRT	AV. S.G. G/CC	YIELD AT 1.65 %	AV. S.G. FLOATS G/CC	ASH CONTENT %	MOIS- TURE %	VOL. MATTER %	CAL. VALUE MJ/KG	AV. S.G. SINKS G/CC	COAL IN DIRT %	VOL/VOL	VOL/WT CU M/TONNE
1	32.09 –	45.53	13.44	COAL	1.67	48.7	1.49	23.4	3.3	29.3	22.82	1.87	–	2.4	1.4
2	45.53 –	47.63	2.10	DIRT	2.17								0.0		
3	47.63 –	60.98	13.35	COAL	1.97	19.7	1.55	27.7	3.0	28.4	21.31	2.11	–	1.3	0.7
4	60.98 –	80.27	19.29	DIRT	2.06								7.3		
5	80.27 –	100.88	20.61	COAL	1.83	25.4	1.54	26.9	3.1	28.6	21.57	1.96	–	1.1	0.6
6	100.88 –	102.22	1.34	DIRT	2.46								0.0		

SUMMARY

| MINEABLE COAL | | | | | | SALEABLE COAL (WASHED AT 1.65 G/CC) | | | | | | |
TOP COAL	BOTTOM COAL	TOTAL THICKNESS	THICKNESS	SPECIFIC GRAVITY	TONNAGE /SQ. MTR.	YIELD %	SPECIFIC GRAVITY	ASH CONTENT	MOIS- TURE	V.M. %	C.V MJ/KG	TONNAGE /SQ. MTR.
32.09 –	100.88	68.79	47.40	1.82	86.45	29.7	1.52	25.6	3.1	28.8	22.05	25.67

FIG. 18 a

EXAMPLE OF APPLICATION PROGRAM "VERTICAL SELECTION" RUN WITH THE SET OF CONDITIONS AS MENTIONED IN THE COMPUTER MESSAGES.

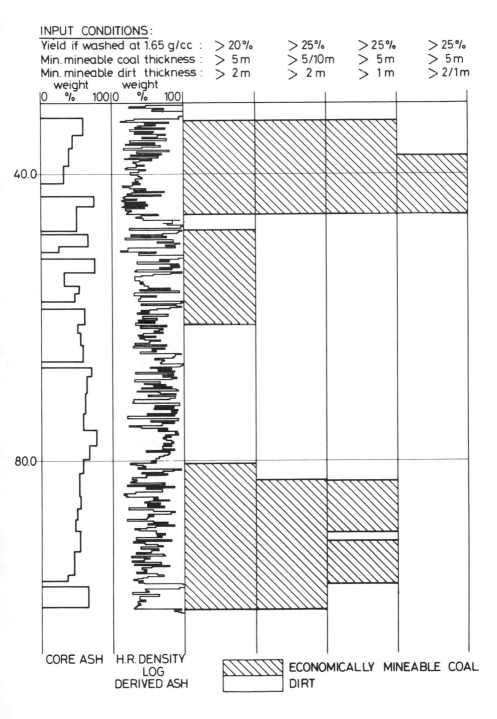

INPUT CONDITIONS:

Yield if washed at 1.65 g/cc :	> 20%	> 25%	> 25%	> 25%
Min. mineable coal thickness :	> 5 m	> 5/10m	> 5 m	> 5 m
Min. mineable dirt thickness :	> 2 m	> 2 m	> 1 m	> 2/1m

CORE ASH

H.R. DENSITY LOG DERIVED ASH

ECONOMICALLY MINEABLE COAL

DIRT

FIG. 18 b   MINEABLE COAL AS DEFINED BY COMPUTER PROGRAM "VERTICAL SELECTION" BASED ON DIFF. SETS OF COND.'S

ACKNOWLEDGEMENTS

The authors wish to acknowledge the contributions of the petrophysicists whose work has been condensed in the above. They are:

R. Kloosterman, F. Driedonks, M.J. Russell, H.M. Howells, A.V. Allington, M. Peeters, R. Blom.

They are also indebted to Shell Coal South Africa, Shell Australia and Shell Mijnbouw N.V. Indonesia for permission to use their data, and to Shell International Petroleum Maatschappij for permission to publish this paper.

REFERENCES

1. Bond, L.O., Alger, R.P., and Schmidt, A.W., "Well Log applications in Coal Mining and Rock Mechanics". AIME Transactions, vo. 250, December 1971, pp. 355-362.

2. Kowalski, J.J., and Holter, M.E., "Coal Analysis from well logs". SPE AIME Paper No. SPE 5503, 50th Annual Fall Meeting of SPE of AIME, Dallas, Texas, 26 September-1 October 1975.

3. Reeves, D.R., "In-situ Analysis of coal by bore-hole logging techniques". The Canadian Mining and Metallurgical Bulletin, February 1971, pp. 1-9.

4. Krevelen, D.W., van, "Coal: -typology, -chemistry, -physics, -constitution". Elsevier Publishing Co., Amsterdam, New York, 1961.

Section Three Discussion

F.A. HADSELL: (Colorado School of Mines): What we call deconvolution or inverse filtering techniques have been thoroughly developed in seismic exploration for oil over the years and the first question is, how many of these techniques do you feel may be applicable to the processing of these well logs to improve our correlation. The second question concerns the development of our state-space filtering techniques in the aerospace industry where we mix a variety of kinds of data to better define what we call the so called state of the system. That is, we have a space craft and we can have a gut feel for what its state is at a moment and we try to estimate that on the basis of various measurements. I wondered if you gentlemen felt there was any application of this technology, the state-space technology, or Kellman photoing technology, of aerospace.

L.J.M. SMITS: To answer the first question, whether you are using deconvolution techniques for having better resolutions even than the spacing would allow: I would say that we are working at it but we have not yet finalised our research in this particular aspect, but we are well aware of its possibilities. I don't know whether I understood your second question correctly, could you repeat that.

F.A. HADSELL: I think it developed in the aerospace programme, the terminology they use is the state space filtering or Kellman Bushey filtering approach. In this approach they define a state generally with seven or eight measurements. So we might have the state of our coal seam defined with the thickness an ash content and maybe a calorific value. These are the things we want to know. We have another thing called a measurement vector which might be seven other measurements. Now the people in Aero-space have developed a rather neat optimum filtering estimating technique where they mix these seven varieties of observations to obtain this state as defined by three components of a vector. Maybe this doesn't ring a bell.

D.R. REEVES: Yes, I think I understand what you mean. I think the answer is 'yes', but there is a long way to go. I think what you are saying is, is there another way of expressing what the geologists and what the exploration engineer want to know of coal properties in terms of a slightly different concept. I would say yes, we are looking at this to see if we can find some expression acceptable, and then I think the sort of analytical techniques you are mentioning, may well be applicable. I would say however, that no analytic technique can improve really on the information the tool can provide and I feel that we have further to go in making the actual information better, before we get too much farther with our analytical techniques.

DR. A. ZIOLKOWSKI: (National Coal Board): I was slightly bothered by the squaring process that you use to make your logs look sharper. The log is a smooth curve compared with a squared curve and, if you think of the log as a time series, the squaring process is introducing frequencies which are higher than you can possibly sample with a sampling rate. Can you give me a theoretical justification of how you are introducing the higher frequencies, making the log look sharper than the information that you obtained in the first place?

L.J.M. SMITS: Because of the finite distance (spacing) between source and detector of our logging tools sharp boundaries between two layers having different physical properties will be reproduced as gradual transition from one value to the other. The values read on the curves in this transition zone are meaningless in terms of actual physical properties of the two adjacent beds. To avoid such values being used in interpretations we have, over the years, developed a computerised squaring or blocking procedure. We start from a physical model that earth strata are built up of beds of finite thickness having distinct properties. Depending on tool characteristics such as spacing and statistical variations in response, critical slopes, thicknesses and variations in response are assigned to a curve, which determine the boundaries and running average value for successive layers.

The result of this squaring should satisfy visual comparison of incremental and squared logs interpreted by an experienced engineer, who compares all the logs in a particular bore hole.

Application of more sophisticated signal theories, based on the same physical model is under study in our organisation, as we have said earlier in reply to Dr. Hadsell.

P.K. GHOSH: While congratulating the author, I like to mention that application of wireline logging techniques, though now much refined and sophisticated have been in operation in coal exploration work in India since 1956. Since then in all coal drilling operations, electrical logging (self-potential and resistivity methods) of the holes have been carried out. This logging, especially the resistivity curves representing individual seams and/or beds, has been very useful in the correlation of the seams which has posed a problem in respect of some of the areas - seams being very close-spaced with almost the same thickness etc. In addition to this method, temperature logging of the boreholes is also being regularly done. It is thus felt that in conditions prevaling in Indian coal measures where the borehole coal cores are invariably analysed, such a sophisticated method of logging at a considerable expense may not always be called for. I would like to ask how you compute the true thickness of a coal seam in steeply dipping strata.

D.R. REEVES: Well, you can't. You are quite right. Steeply dipping beds will present quite a problem. I would point out to you though, in some open pit work, it is in fact quite common to use the apparent thickness because this in fact determines the amount of coal you will get out of the pit operation. I see no way of doing this until some type of dip measuring device is available, other than of course drilling several boreholes. It may be possible to use the kickoff techniques of the oil industry. Some line tools are available which can precisely define where the hole is underground, therefore if you can kickoff say, 150 feet, and drill three holes, you can clearly define the dip that way, but I would agree with you that it would be very nice to be able to measure the dip directly. It is going to be extremely difficult and costly to do this in slim holes.

C.W. BALL: (Canex Placer Ltd.): I was wondering if there is any possibility of reducing the weight in the near future, so that we could make the equipment mobile. The same thing applies I know, to diamond drilling equipment. You can't always break it down, but if we had units that were broken down to say 250 pounds each, they would be more acceptable to us. Is this too difficult for the near future?

D.R. REEVES: Not at all, in fact in Malaysia and Sarawak, which we showed you a picture of, units were hand portered and I believe, the maximum weight of the individual unit was about 250 pounts. The helicopter units I showed you, were specifically designed for high intensity helicopter use where it was a question of picking up the unit in one go and putting it down and then moving it very quickly, at the same time as moving drilling rigs. So, with proper planning it can be split down to about 250 pounds weight, certainly they could be split down further if that was required.

SECTION 4

6

SEISMIC SURVEYING AND MINE PLANNING:

THEIR RELATIONSHIPS AND APPLICATION

A. M. Clarke
Head of Mining Geotechnology and Chief Geologist
(National Coal Board)

THE BASIC QUESTIONS - SHOULD YOU HAVE A SEISMIC SURVEY? IF SO, WHEN AND HOW?

The coalmining industry has managed to maintain itself, more or less to its own satisfaction for over 2,000 years, by siting its new mines through lateral extrapolation of its own workings and surface geological prospecting, supplemented over the last 300 years by drilling. Except in stripmining, mines last so long that the industry does not often need new mines. Its competitor in the bulk fuel market, oil, has found it necessary to foster the development of reflection seismic surveying to a highly sophisticated massive international contracting industry, now larger than the coal drilling industry. Periodically, the more adventurous, or sometimes the more gullible management in coalmining industry, did sample the odd few miles of reflection seismic surveying. But, until recently, the results always made coalmining management return to surface diamond cored drilling whenever it has been necessary either to prove the coal ahead of operations at an existing mine, or (on rare occasions) to site a new mine.

How the Mining Engineer missed the need for seismic surveying

It is submitted that at least over the last couple of decades, the two highly sophisticated and technologically developed industries, coalmining and exploration seismic surveying, have suffered a communications failure. The seismic explorationists have not understood that generally the coalmining industry can only use seismics to find local disturbances to the continuity of production in the first seam they intend to work, rather than explore for it. Unlike oil, coalfields are singularly easy to find. The coalmining engineer has not known how to specify his needs in seismic surveying terms to the seismic contractor, or, at least, he has not been able to recognise situations amenable to seismic techniques and specify his needs (in a manner which will produce the results he requires from the seismic contractor).

The implications of concentration of mining production in deep coalmining

In their struggle to prevent closure of as many of their mines as possible in the era of cheap oil from the 1960s to the early 1970s, the coalmining engineers developed methods which profoundly altered the preliminary exploration requirements for new deep mines, without becoming aware of having done so. This lack of awareness of exploration consequences in the case of new mines was because few, if any, new deep mines (except in rare, virtually undisturbed coalfields) were planned or opened during this era of expansion of oil's share of a growing energy market and contraction of the coal industry. The development and adoption of fully mechanised mining methods in the 1960s has brought about a situation in which there are more production eggs in fewer mining baskets. In the UK, concentration of mining effort, measured as production per square mile of coal worked per year has not yet reached the state of affairs in which all the eggs in each mine are in one production basket, but it has doubled in output per square mile (Fig. 1). It was when the planning implications of this new situation in the case of new mines in geologically disturbed areas of coal were recognised,that 'coalmining seismics' had to be born.

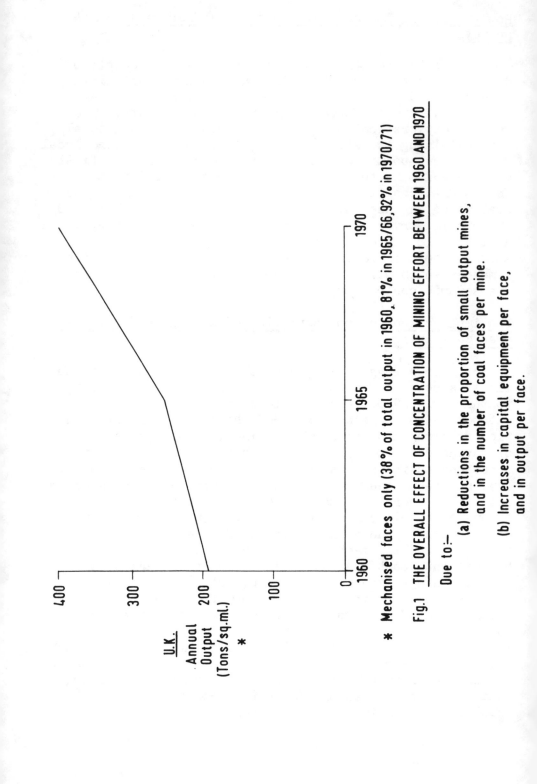

* Mechanised faces only (38% of total output in 1960, 81% in 1965/66, 92% in 1970/71)

Fig.1   THE OVERALL EFFECT OF CONCENTRATION OF MINING EFFORT BETWEEN 1960 AND 1970

Due to:-

(a) Reductions in the proportion of small output mines, and in the number of coal faces per mine.

(b) Increases in capital equipment per face, and in output per face.

U.K.
Annual
Output
(Tons/sq.ml.)
*

400
300
200
100
0

1960   1965   1970

How the seismic surveying contractor missed the mining opportunity

The amount of recoverable energy per acre of coalfields is often of the same order as that of oilfields. But unlike oilfields, because of interaction of vertical ground pressures induced during mining, only one of the multiplicity of seams at any point can be produced simultaneously in a coalfield. The coalmining engineer cannot 'perforate' and make 'completions' to produce all 'pay zones' simultaneously, which is why his surface access points take so much longer to exhaust than those in oilfields.

Prior to full mechanisation, and especially in the earlier era of manual bord and pillar working and pillar extraction (extracting the coal by forming a more or less square grid of tunnels in the seam) individual mines were still producing up to a million tons of coal per year. But this output came from the advance of many individual places (up to several hundred) on a wide front all round the shafts, in the seams being worked.

The inherent exploratory content of this kind of production system was often very high, and it was also impossible to avoid carrying plenty of 'spare pitroom' (the amount of logistically serviced access to coal in the mine which was developed, but not immediately used for production, to which any production unit whose advance was suddenly stopped by meeting an unexpected fault could be transferred). Thus apart from a few widely spaced bore-holes or a geological extrapolation to show that, in general, the seams continued beyond the 'takes' (logistic limits, and/or concession limits) of neighbouring earlier mines, each hundred or two hundred year life mine needed little or no surface exploration - and rarely received any.

Planning was largely a matter of periodic laying out the development and phasing of a succession of production sites on a mine plan wherever conditions were currently the most productive. When production was unexpectedly stopped or disturbed locally, the system grew round the ends of the disturbances or, if it could not, the system was re-developed on the other side of the geological disturbance to the continuity of production in the seam before its 'tail' had caught up. Furthermore, in normally faulted areas, discovery of fault positions in one seam meant that the positions of such faults could be projected upwards and downwards into all of the other seams to be worked later. Thus in all seams but the first to be worked in any area, the production units could be laid out and phased in a manner which avoids interruption by faulting. Deepening the shaft, with occasional underground 'staple pits' or inter-seam drilling, revealed the choice of 'next best seam' to be worked. It was only when concentration of mining operations took place that the exploratory needs of new mines in disturbed areas was radically changed.

The state of seismic surveying prior to 1973

This lack of need for surface exploratory data in advance of production was little understood by the seismic surveying industry, accustomed as it was to a situation in which its art was a vital contribution to reducing the amount of risk capital outlayed in the highest risk sector of the oil industry - that concerned with exploratory drilling. They did not understand why the exploration results in the form then produced were largely unnecessary in the coalmining industry, and hence 'too expensive'. Nor did they understand that, over the long life of a field, because the coal industry was nothing like as 'nose-ended' in its capital outlay requirements as the oil industry's, the greatest outlay of risk capital lay in the maintenance of the recovery process, not the maintenance of the discovery process. (An illustration of this situation in British coalfields and oilfields is given in Figs. 2 and 3).

Thus in coalmining, except in the rare case of new mines, non-productive exploration (drilling and seismic survey) has to be planned integrally with the exploratory content of the production process. It must be understood by the seismic surveyor that every advancing mine opening can be considered as having both an exploration and a production content, the amount of each which it contains depending on its placing and phasing in relation to its neighbours.

Finally, without realising it, the seismic surveying industry had, in developing the techniques to reveal in miniature the full geological cross-section of the ground over a depth of 10 kilometres below a seismic profile line, degraded the quality obtainable in the uppermost kilometre of the section, and then grown accustomed to this state of affairs. The oil industry was not much interested in this 'superficial' part of the cross-section, as generally

Size and Setting of Target

		OIL	COAL
Typical size of field	(sq. miles)	c. 25	c. 700
Typical depth range of producing horizons	(ft.)	9000 ± 1000	c. 2000 ± 2000
Typical number of producing horizons		1 (up to 2)	c. 20 (down to 1)
Typical thickness of producing horizons	(ft.)	c. 450 ± 150	c. 5 ± 3
Typical recoverable reserves/unit area of reservoir	(bbl/acre - ft.)	c. 300 ± 100	c. 7500
Number of vertically superimposed horizons simultaneously in production		1 (up to 2)	≥ 1
Direction of flow or mineral under pressure gradient		Into production facilities	No flow
Direction or advance of installed production facilities		No advance	Into mineral
Typical number of natural disturbances to continuity of planned daily production (per sq. mile under production)		0	3/5
Minimum size of disturbance to continuity of production	(ft.)	> Pay thickness under production (= 450 ± 150)	> Pay thickness under production (= 5 ± 3)
Proportion of total annual output likely to be lost, due to 'unexpected' natural sub-surface disturbances to continuity of production		0	c. 10%
Annual production (mid-1980s)	(m. tons o.e.)	300 ± 80	70 ± 10

Fig. 2. The contrasting characteristics of UK oil- and coalfields

Setting and performance of production units

		OIL	COAL
Typical spacing of production platforms or shaft pairs per field	(mls.)	2 - 3	(modern) 4 - 6
Typical number of production platforms or shaft pairs per field		c. 1 - 2	100 - 200
Typical number of wells/platform or production faces per shaft pair		c.40 - 60	(modern) c. 3
Typical rate of production per production well or coalface (bbl/producing day)		2/4000	3000+/-
(bbl/annual day)		2/4000	1890+/-
Average rate of output per platform or shaft pair (m. tons/year (o.e.) est. for mid-1980s)		6.6	0.8
Typical life of production platform or shaft pair	(years)	10/20	80 +/- 40
Typical life of producing offshore oil/coalfield	(years)	10/20	100/200

Costs of Production units (est. for mid-late 1970s)

		OIL	COAL
Typical initial detailing and development costs per platform or shaft pair:-			
– Seismic (@ 300 $/ml. oil and 6000 $/ml. coal)	$m.	0.03/5	0.02/3
– Platform and drilling or equipping, sinking and developing, etc.	$m.	500/700	50/80
– Transport (shorepipe or to railhead)	$m./ml.	2/2.5	?
Est. Maintenance of production (capital costs) per platform/mine	$m./year	?	0.9/1.5

Fig. 3. The contrasting techno-economic factors affecting U.K. oil- and coalfield production

petrostatic pressure at these 'shallow' depths was not
enough to raise oil in commercial quantities even if any
was present in this part of the section.

Seismic surveying was geared to finding, within very
large (thousands of square miles) basins, changes of dip
at depth, or more especially areas of 'closure' (of
countour lines) around dipping formations (thick enough to
contain porous and permeable sealed potential reservoir
sandstones or limestones) and big enough in size (several
square miles) to be 'commercial'. Seismic data acquisition
and processing was therefore directed to enhacing continuity
of events on the seismic profiles from which dips could be
read.The industry was unaware of the changes in coal mining.

The coalmining equivalents of 'is it prospective?' - 'do we have a structure?'

In coalmining, the questions (so common in the oil
industry) of 'does the regional geology of the basin
indicate that it might be worth exploring for a coalfield?'
- and, if yes, 'has the presence of a suitable target
structure for exploration been shown?' - rarely arise.
The initially developed coalfields in any country have
nearly always outcropped at the surface and, being an almost
unmissable and distinctive geological target, their presence
and location have usually been noted during the early history of coalmining, or over the last hundred years, during
primary survey of the country's surface geology. The
corresponding fundamental questions, directed towards
maximising the profitability of coalmining, are:-

(A) Where is the most, or next most, accessible area
of coal in the coalfield or mine? This question becomes
serious when production from stripmineable coal at outcrop
cannot wholly satisfy demand - and it should be remembered
that surface mineable coal in any particular locality is
commonly recovered at as little as half and one-third the
cost per ton mined of the corresponding deep mineable coal,
but commonly only forms a few percent of the total recoverable
reserves in a coalfield.

(B) Where is the thickest/least disturbed area of coal
seam, within an accessible area of coalfield or mine (for
any given threshold of minimum quality)? This question is
always valid provided sufficient coal exists (in the area
of the coalfield where the answer occurs) to recoup the cost

of making access to it. The seriousness of the question is
due to the fact that, at any period in time in a producing
coalfield, the costs of mining coal are largely costs per
unit area worked, whilst the proceeds, for a given quality,
are proceeds per unit volume worked. (In the case of
surface mineable, open-cast mineable or stripmineable coal,
it is necessary to add 'at a given dip'. In surface mineable coal, the more important control factor is the coal
to overburden thickness ratio, rather than coal thickness.)
Furthermore, in deep mineable coal, variations both around
the mean thickness of seams and in their mineability due to
roof and floor variation, are generally independent of mean
thickness. Above a threshold minimum, the scale (or areal
intensity of) disturbance to continuity of coal seams in
relation to the scale of mining operations, can also be of
importance in determining the productivity of mining
operations. An illustration of the effect on productivity
of seam thickness and continuity in the UK is given in
Fig. 3 of Clarke (1), reproduced here as Fig. 4 .

It is only by boreholes (or neighbouring mine workings)
that the thickest seam can be distinguished and a vertically continuous sample for analysis of the quality of the
seam by subsections can be obtained. Hence the importance
of direct, rather than indirect, methods of exploration in
coalmining. However, it will be shown that coalmining
type seismic reflection surveying, in geophysical situations
suited to it, can be cheaper than the corresponding amount
of structure drilling, to investigate degree of disturbance
in seeking the thickest/least disturbed area of accessible
coal seam.

As far as seismic surveying is concerned, one answer to
the basic question of 'should you have a seismic survey?'
is no, you should not, if you have sufficient evidence to
show that the intensity of local disturbance is too low to
affect choice of area for future mining operations.

When to employ seismic surveying in the four phases of
coal exploration

The four phases of exploration in coalmining arise in
answer to each of four simple and sequential questions.
It will be seen that questions 1 to 3, although fundamental,
rarely arise, and that question 4, the one which arises
most frequently in coalmining, demands the existence of a
plan to mine the coal. Whether or not you should include

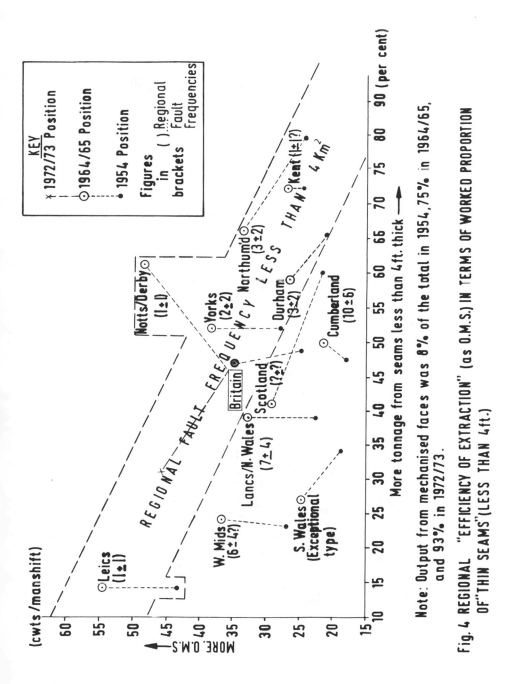

Note: Output from mechanised faces was 8% of the total in 1954, 75% in 1964/65, and 93% in 1972/73.

Fig. 4 REGIONAL "EFFICIENCY OF EXTRACTION" (as O.M.S.) IN TERMS OF WORKED PROPORTION OF "THIN SEAMS" (LESS THAN 4ft.)

seismic surveying in the exploration requirements of each
phase is obvious, except in the fourth, the most commonly
occurring phase. The answer in any case is that seismic
surveying should be used when it is likely to be cheaper
than the alternatives.

COAL EXPLORATION PHASE I - IS THERE ANY COMMERCIALLY MINEABLE COAL AT ALL?

This is only asked of a locality in which previous
coalmining has given little or no indication of the likelihood of mineable coal being present within accessible
depths. It is restricted to the cases of separate,
individually isolated coalfields, totally concealed from
the surface, such as the Kent and the Oxfordshire fields
in the UK; in coalfields which are economically separated
although geologically continuous beyond a zone of regional
impoverishment in all seams, marking the economic boundary
of a coalfield, as in the giant new Belvoir-East
Leicestershire field in the UK; of extensions of coalfields
which lie beyond a previously assumed structural boundary,
as in the new Selby extension of the Yorkshire coalfield
in the UK described by Rees (2).

Action in answer to the Phase I question - One borehole in the locality

Depending on why the Phase I locality is prospective,
e.g., on the basis of long range interpolation between
neighbouring geological indications of a possible field,
structural analysis of the regional geology, or where
gravity survey indicates possibility of a concealed coal
basin - if there is doubt as to precisely where the thickest
potential coalbearing section within 1 kilometre of the
surface (the geothermal gradient dependent limit to depth
of coal working) then a preliminary cross-spread of almost
normal (except shallow) 'oil company' type seismic surveying
might be indicated, in order to find or confirm the best
locality for the one Phase I borehole.

Outcome of action in answer to the Phase I borehole = a yes or no

If the answer is yes, the discovery hole also gives an
indication of the depth to the most likely 'target seam(s)'

in the sequence (the seam(s) upon which the cost of initial
access would have to be justified) together with the richness
of supporting seams in the sequence, and the overburden
thickness and nature (= costliness of access, if the overburden is water bearing or otherwise difficult to sink or
drift through from the surface). Hence, from these
considerations, the one borehole, if it gives a 'yes' answer,
also gives an indication of the minimum area in the newly
discovered field which would contain sufficient reserves to
justify making access.

COAL EXPLORATION PHASE II - IS THERE ENOUGH COAL?

(or, Is the commercially mineable coal (whose existence is
known) present over more than the minimum area necessary to
justify making new access to it?)

This question is only asked of a locality which is so
'unknown' from neighbouring mining that genuine doubt
exists as to whether the coal is sufficiently extensive.
Otherwise exploration action starts in answer to the
Phase III question - and confirms the reasonable assumption,
rather than discovers, that the coal is sufficiently
extensive ... etc.

Action in answer to the Phase II question = 3/4 boreholes, at the corners of the 'minimum area'

The three or four boreholes must be boldly stepped out
to the boundaries of a minimum area containing the
reserves necessary to justify making new access to the
area (generally between 10 - 30 square miles for a new deep
mine in the UK). If not already performed at Phase I, in a
structurally unknown area, at least one of these holes would
be sonically logged, and shot for uphole seismic interval
velocity, preparatory to Phase III, as the cost of this is
not likely to exceed about ten percent of the total cost of
the hole. If general indications at Phase I of the marginal
richness of the area justifies it, but the direction of the
principal tectonic 'grain' of the field is totally unknown
from regional evidence, a small 'diamond spread' of coalmining type seismic profile lines around the Phase I hole,
linked in to the hole by one line might be justified to site
the Phase II holes in an appropriately oriented rectangle.
Otherwise action in answer to Phase II B (to find the best

of a number of accessible takes) precedes this. The idea
of not going straight into Phase III, after a successful
Phase I hole, a temptation to the pure coalfield geologist
(who has no mining engineer to ask, or experience of his
own to guide him as to minimum area) is to avoid ending up
with a certain amount of 'redundant' drilling or, alternatively,   to avoid premature commitment to a particular
access site and/or too small a scale of mining operation
on the part of the hard pressed and enthusiastic mining
engineer often sitting on his tail. (He fails to realise
that a 'no' from any one of the holes at Phase II can save
redundant drilling from the point of view of the mining
undertaking.)

Outcome of the action in answer to the 3/4 Phase II boreholes = a yes, a no, or maybe

If the answer is no, proceed to the next Phase I
locality.

If the answer is yes, and there is still room for more
new minimum sized new colliery takes, the pattern of 3/4
holes 'at the corners' is repeated as a 'Phase II B' in
the most promising direction, until the location of the
overall 'best minimum area' can be located. In doing this,
it is necessary to take into account other resource
requirements such as the vicinity of suitable existing (or
potential) routes for surface transport systems. (There
is, of course, an ultimate limit to Phase II B provided by
exploration licences, or by the view of present mining
capacity in the light of future market demand giving the
the total sum available for finding and developing new
access, which might limit Phase II B to the best that can
be found in a smaller area than a whole new concealed
coalfield.) If the answer is 'maybe', then usually the
action at Phase II has found indications of interruption
to the continuity of mining on a considerable scale (or
indications of the possibility of such phenomena as
volcanic necks, igneous sills, areas of limestone or
oxygen burning, or the incrop of seams to unconformity) in
one of the 3/4 initial Phase II holes, or in any of the
Phase II B holes. If this is the case, then a suitably
open grid of seismic lines, on a scale suited to detection
of the presence, but not the detail of these phenomena, is
indicated - provided sonic logs indicate that they are
likely to offer sufficiently contrasting seismic
impedence against the normal coal bearing strata to be
detectable.)

COAL EXPLORATION PHASE III - WILL A KNOWN BEST MINIMUM AREA OF COMMERCIALLY MINEABLE COAL BE NATURALLY PRODUCTIVE ENOUGH TO JUSTIFY MAKING ACCESS?

(or, Is the intensity of geological disturbance to the continuity of the target seam(s) low enough to give an annual rate of production high enough to justify choosing thise as a new mine site?)

The Phase III question only arises when there is no neighbouring mining evidence on which to assume that the answer is more likely to be yes than no, in which case making a mining plan can be justified. It should not be confused with very similar questions which arise at Phase IV after at least some plan to mine an area of coal has been made. At the beginning of Phase III, there is rarely enough detail available to choose the best site for initial access, let alone to make a mining plan. Also, at the beginning of Phase III, matters are still in the hands of the geologically oriented coal exploration professionals, rather than in the hands of the production oriented mining engineer, who may be aware that a little data can give an attractive picture for raising the risk capital necessary for him to pursue his art, and that more data can sometimes make reality seem less attractive. Commonly, if new mine sites are involved, the mining capacity/future demand situation will have led to more than one site having been run to the beginning of Phase III. The really critical choice, of which site is the best, in which the really expensive part of the whole operation must be committed, i.e., mine sinking, equipping and development, follows later.

Action in answer to the Phase III question = a grid of sedimentary framework boreholes, and a grid of seismic lines or structure boreholes, infilling the best area

The efficiency of exploration for deep mines depends, apart from luck or inspiration at the cheaper Phases I and II stages, very greatly on the grid spacing at Phase III.

Grid spacing of sedimentary framework boreholes

Needless to say, the coal seams are part of the sedimentary framework. The object of the grid of these boreholes is to identify (in the target seam(s) and the nearest alternative seams) any lines along which seam splitting takes place; boundary lines of regional seam impoverishment and thinning; together with evidence for the presence or absence and general intensity, but not the detailed locations of major individual washout channels in the seams, belts of any channel sandstone with undulating base within the proximate roof shales of such seams; also evidence of the presence or absence and general intensity of any non-linear and non-sinuous 'patchy' areas of disturbance to the continuity of production or proceeds per ton, such as cannel-coal, limestone or other 'burning' within the seam. The grid spacing of such boreholes depends on (a) A geological factor. The extent to which the general spatial relationships between envelopes of each kind of facies, which together make up the sedimentary framework of the region, is known. For instance, the course of a major distributory channel sandstone, maybe recognised, without actually penetrating it more than once, from recognition of the presence of its laterally associated rock facies in other holes; and (b) A mining factor. The extent to which failure to recognise the presence of phenomena occurring on a given scale would interfere with average annual productivity from most likely mining system to be employed (familiarity with scaling factors implicit in different mining systems in general is through familiarity with the Geosimplan techniques described under Phase IV below).

In the UK, application of these two criteria in the various coalfields generally gives rise to a grid spacing of between 1 and 3 kilometres between boreholes.

Grid spacing of seismic lines or structure boreholes

The object of the grid of coalmining type seismic lines is to identify the orientation pattern and intensity of major faults. From this, in conjunction with the sedimentary framework borehole grid, and the statistical frequency distribution of maximum throw of all faults detected seismically, it is possible to infer the intensity and orientation pattern of the faults down to 'zero throw', i.e., the generally more numerous minor faults which are not detectable seismically from the surface, see Krey (3). The final spacing of the grid, which is shot iteratively, starting with about 2 mile spacing, is determined by the strike length of faults at the lower tail of a statistical plot of frequency distribution of the strike length against

maximum throw of the smallest faults detectable. This is
built up from an examination of the plan interpretation of
the initial seismic cross-sections in the grid. (The
final spacing between grid lines being sufficiently close
to prevent ambiguity in reading both 'ends'of fault of smallest detectable throw.) Rapid processing and interpretation
is a pre-requisite of the 'feed-back' involved in a coalmining type seismic survey. Normal oil company practice
will not do as lag times have to be in days or, at the most,
weeks, not months,between data acquisition and its interpretation.

In UK coalfields, except Scotland and South Wales,
where it is smaller, this grid spacing generally works out
at about the same 1 - 3 kilometre spacing as is the range
for sedimentary framework boreholes.

Structure boreholes

Depending on (a) the depth to the first identifiable
stratigraphic 'marker horizon' in the productive coal
sequence (from which levels of the seams below can be
inferred), and (b) the extent to which it is statistically
determined in the later phases (and judged in the early
phases) that the marker horizon gradient can vary due to
local 'anomalies' in the regional dips across unfaulted
distances between holes - it is possible to estimate the
spacing of a grid and the total footage involved in drilling
structure holes which would distinguish between small faults
and local variations in dip to the same degree as a seismic
survey grid. It is possible to estimate the overall cost
of the structure boreholes, estimate the number of rigs
required to complete the job to the same degree within the
same time and compare them with the 'yardstick' costs of
a coalmining type seismic grid, shot at about 10 - 15
line miles per month (if the highest resolution is required).

Unless there is evidence that the regional gradients are
going to be sensibly flat, in which case all but the smallest
differences in levels between neighbouring boreholes must be
due to the presence of faults (or local monoclines) coalmining seismic lines usually work out to be several orders
of magnitude cheaper to do the same structure borehole
sampling job at Phase III.

Surface mineable reserves and seismics

This obvious advantage of coalmining type seismics in
the matter of disturbance to continuity of production by
faulting is, of course, untrue at the shallow depths which
prevail in the case of strip- or surface mineable reserves.
In such cases, the cost of even 10/20 metre spaced bores
with, say, not more than 1 to 40 coal to overburden thickness ratios, is still only a few percent of the total mining
costs, so that in general the case for coalmining type
seismic methods as a partial substitute for drilling will
be correspondingly weakened when the calculations are made.
There are exceptional circumstances, however, e.g., when
the scale of disturbance is similar to or larger than the
scale of mine units at Phase II (see, for instance,
Hasbrouck and Hadsell (4)), or when the thickness of coal
is very great, giving abnormally great depth to the deeper
part of the grid of boreholes at the same coal to overburden
ratio as in the case of reserves in normal (thinner) seams.

Actual grid spacings in UK Phase III exploration for new mines and exploration costs as a proportion of the total

Given the 10/30 square mile area of new deep mine
'takes' obtaining in the UK, the grid spacings given above,
it will be seen that Phase III exploration per prospect
usually works out at between a dozen and three dozen boreholes for minimum readability of target seam and host strata
variation, and between 20 line miles (in rich fields with
abundant vertically stacked coal reserves) and 100 line
miles of coalmining type seismic line.

At the 2,000 to 3,000 ft. depth range of the average
new deep mining prospects in the UK, this can give total
exploration costs of up to about 10% of the total access
and development costs of a new mine, most of which fall at
Phase III, as Phase IV is only required to cover the first
5 or so years of full production. However, production
delays and shortfalls on annual output over as little as
the first three or so years of planned full output (due
to inadequate exploration) can lose (and have lost in the
past) sums equivalent to the total access and development
costs without interest charges or discounting. Currently
new deep mine costs for 2/3 million tons per year output
are running in the high tens of millions of pounds, for
revenues of the order of £15 - £20 per ton.

Outcome of the action in answer to the Phase III grid of
sedimentary framework boreholes and seismic lines =
sufficient evidence to choose the best alternative as the
target seam and to make an initial plan - and,hence, a
yes/no answer

The outcome of Phase III exploration means that the
mining undertaking has an initial phased mine layout plan
and supporting estimates of annual expenditure and revenue
from planned output and proceeds per ton from the target
seam. It also has knowledge that the prospect is not only
of sufficient extent to justify making access, if sufficiently productive, but also it now knows whether or not
it is likely to be sufficiently productive to justify the
access, equipping and development costs. In the latter
stages of Phase III, the coalmining geotechnologists fall
back into their accustomed role as part of a substantial
mining management,engineering and logistic team associated
with the prospect,with a bias towards the planning side of
activities.

COAL EXPLORATION PHASE IV - IS THE RATE OF YIELD OF THE COAL PER UNIT OF MINING EFFORT APPLIED TO IT AT A MAXIMUM?

(or, Assuming the costs of mining the coal in undisturbed
conditions are at a minimum (implying a best fit of layout
plan and its considerable engineering content to the state
of knowledge of the ground) - is the sum of the costs of
(a) exploration (productive and non-productive), (b)
insurance (for the unexplored residue of disturbance) and
(c) geological disturbance - at the minimum possible for
the ground to be mined?)

The Phase IV question (of which exploration is a part)
is the most common one in day-to-day deep mining operations.
New coal mines are 'rare birds' in the life of a substantial
coalmining undertaking operating in mature coalfields.
The question assumes that carrying 'associated' exploration
(not necessarily by drilling and seismic methods) is just
as much a part of the total cost of mining, as is carrying
a traditional level of spare mining capacity. The spare
mining capacity (or insurance) is in the form of (a) spare
face-shifts (for instance, 6 shifts/day out of a possible
9, as 2+2+2 shifts, or 4+1+1 shifts, on a 3 face, or a 3
district mechanised bord and pillar mining operation.(the
highest number of shifts being deployed on whatever is the

most productive face or bord and pillar leg of the mine at
the time), and/or (b) spare (meaning developed but not
immediately used access to new faces) pitroom and/or (c)
coal stocks, or coal in transit, to iron out fluctuations
in production. The Phase IV question also caters for the
situation in which the long run average annual costs of
meeting unexpected local geological disturbance to continuity
of production, can involve disturbances of a frequency and/
or nature which makes them more costly to explore for, or to
carry insurance against, than to allow for, in the annual
budget estimate for the mine. This is an unheard of or
unbelievable situation to the seismic surveying industry,
accustomed as it is to the situation obtaining in the oil
industry. In general, there is both a 'traditional' amount
of non-productive (as opposed to productive) exploration,
i.e., none or almost none per quarter or year at each
average existing mine, and a traditional amount of spare
capacity carried.

The productive exploration content of mine layouts

With regard to the pure geometry of a phased mine
layout plan, layouts with the highest productive content,
and lowest exploratory content (a developer face or heading
advancing into the middle of a block of coal with close-
packed stepwise advancing 'flanker' faces or, alternatively,
mechanised bord and pillar districts on either side) have
the least logistic (overhead) costs associated with them.
Layouts with the highest exploratory content (dispersed
L-shaped or U-shaped development heading or retreat
districts, generating plenty of 'pitroom' in advance of
strict need for phased replacement faces in the production
plan) have both the highest logistic costs, and offer the
least productive employment for at least some of the labour
force and capital equipment otherwise deployed on 'coal-
getting'. (Wages constitute around half of the total costs
of fully mechanised deep mining.) In most existing mines, a
best traditional 'compromise' layout between the two
extremes, suited to the average annual amount of geological
disturbance present in the field, has usually been found
intuitively over the years, without much formal analysis.

Traditional planning techniques

Traditionally, the plan for a coal mine is usually
updated at monthly , quarterly or annual intervals. The
layout aspects of planning at a mature coal mine usually

involve projecting,etc.,such fault and sedimentological
disturbances are known by extrapolation or interpolation
from neighbouring adjacent (plus, in the case of faulting,
sub- or superjacent) workings; revising a phased layout of
the production units and their associated access and
development tunnels; calculating the quarterly, etc.,
production outcome of the phased layout from average 'tons
per cut' and 'cuts per day' on the phased production units
in the plan, plus the output from the advance of any
development headings. Electrical,mechanical, strata
control and ventilation, engineering;and the transport
system for men, materials and coal; plus the provisioning
of all these and total manpower deployment, falls out of
the revised layout plans.

    The key point, at least in disturbed conditions is that,
because the location of all disturbances to the continuity
of seam cannot be known in advance, the calculated outcome
of the plan is always better than long experience would
lead the planner and his senior management to expect.
The point to remember is that it is always necessary, in
disturbed conditions, to put 'a suitable relaxation' on
the outcome of the calculations from the plan in order to
get 'a realistic estimate' of the outcome of mining, for
budgetary purposes.

Inadequacies of traditional planning techniques when:-

(a)Plans embodying a change in the serial state of
   knowledge of unchanged 'average geological conditions'
   (layout changing the traditional production :
   development ratio);

(b)removing a logistic bottleneck (changed phasing due to
   increased rate of advance in a production unit);

(c)mining is entering changed, or new geological conditions.

    For all new mines, and from time to time in existing
mines, or whenever new exploration techniques become
available, the traditionally evolved balances, which are
accepted as ensuring that the rate of yield of coal per
unit of mining effort applied is about the best that can
be achieved, are no longer valid. It is the relaxation
part of the traditional system of 'plan-for-what-you-know'
and (because you don't know everything) then 'relax-for-
what-you-don't' (know, but have learned from bitter

experience to be a reasonable expectation) which cannot cope in new circumstances of the type described. This is because the outcome of the plan depends on a succesion of complex interactions between the mine, as a definite complex shape, in its growth, by fits and starts in different advancing places, into the fixed but only partially known geometry of the coal seams.

The Geosimplan technique

The Geosimplan method was evolved for cases in which traditional planning methods are inadequate, now that in mining,modern fully mechanised systems do not automatically carry 'more than enough spare capacity' as a by-product of the system. The method was evolved in order to estimate and demonstrate the effect on overall costs of not changing traditional layout or phasing to something with a higher productive exploration content, and/or not building in a certain amount of non-productive exploration. The management of a mine or a coalmining undertaking was always painfully aware of the immediate costs of 'doing so'but they had no convincing means of estimating the costs of 'not doing so'. It was through this method that the necessity for developing something that would do what coalmining geophysics now does was first recognised.

In essence, the Geosimplan approach is now recognised to be more a management gaming technique, than the simulation technique it was thought to be when first conceived. (Hence the name.) It involved simulating the growth of a mine, as it would eventuate on average, in practice. The phased layout plan for the mine is placed on a light table, with an underlying sheet of plain white paper; underlying this 'a typical or representative configuration of the geological disturbances in the mine'. Provided the 'repeat unit area' of the configuration (the minimum area which, no matter where it is placed within neighbouring fully worked ground, would always include a sample of both best and the worst disturbance conditions) is several times smaller than the area now being planned, the configuration is sometimes lifted 'en bloc' from a neighbouring worked area of the mine,with the disturbances at its periphery adjusted to fit the extrapolation of any disturbances proved in workings neighbouring the plan area. Realistic rules, as to rates of advance of development and production units, in both undisturbed and various degrees of disturbed ground, are agreed. The average annual costs of running the mine are known.

With the light table switched off, a suitable first increment of execution of the plan, say, one month's or one year's work, is coloured in. The light table is then switched on, to see whether any 'unexpected' disturbances have been encountered in mining during that period. If one or more has, then only the portion of the plan up to the point of first encounter, together with the portion of the disturbance encountered, is inked in. The light table is then switched off, so that the rest of the configuration cannot be seen. Irrespective of the 'actual' continuation of the disturbance (as shown on the configuration of disturbances beneath the white paper) a typical local extrapolation of the portion of the disturbance encountered is then made. A typical management decision is then made as to how the plan would be changed in practice (without knowing what lies ahead), in response to this disturbance to the continuity of production. The outcome of the plan up to this point, in terms of mine production (average per week or per month) is recorded.

Then the next increment of execution of the plan is coloured in. The check against having encountered any disturbances, and the 'replan' decision are made, and the effect on production is recorded with all units advancing according to the agreed rules. Constant repetition of this execute an increment, switch on, switch off, decide, execute sequence, over the period embraced by the plan, gives an 'annual average outcome' of the plan in terms of proceeds. The annual costs are known. (It can be checked against traditional estimates of the outcome of the plan and relaxed for the average effect of unexpected disturbances.) The method, including techniques for coping with 'repeat unit areas' which are large compared with the area being planned, has been described by Clarke (5-6), Rees (7), Hawes (8), the NCB (9) and Sloman (10).

Identification of when, where and to what extent seismic surveying should be used

The Geosimplan method had to be evolved in order to test, and later demonstrate, whether a brand new mine (Parkside) with a brief history of appalling annual losses, could be saved, or whether the losses should be cut and the mine closed (see discussion p.43, Clarke, 1). It has been used for testing the balance of exploration, insurance and disturbance costs and the outcome of plans, both with and without various phased programmes of boreholes built into the annual mining costs, and 'discovering' certain items in a statistically homogenised series of two or three representative configurations; and with and without changes in the production : development ratios, and hence changes in the productive exploration content and average pitroom carried in mine layouts, in some hundreds of case histories since its inception in 1966. It can be seen that, if additional rules for the extent to which faults, or other disturbances, detectable on a seismic profile can be agreed, then the effect on mining profitability of various configurations and intensities of seismic survey lines can be examined as well. When this has been done, it has sometimes shown that, when the plan is changed more by intervening minor disturbances (below the threshold of seismic survey detection) than by the periodic major disturbances which would be detected, the application of seismic surveying would be unprofitable, although it would prove the position of much or all of the major faulting. In other cases, the ratio of survey cost to the return on seismic survey, in terms of revenue from additional output and saving on abortive development (Sloman, op. cit.), has been shown to vary from 'very large' down to two or three fold. Particularly sensitive points in the plan can also be identified (as areas repeatedly sensitive in several Geosimplan 'runs'; which usually occur when and where the mine 'bottlenecks' more than usual in the spare capacity planned to be carried). These indicate the areas in which exceptionally intensive seismic survey efforts are worthwhile, particularly if an example of 'somebody's law' occurs, and unexpectedly bad seismic 'weathering' conditions occur in just that area. These necessitate either expensive delays for experimentation, to obtain good seismic records at target seam level, or acceptance of 'a hole' in the final plan showing faulting in that particular area or zone where seismic conditions are 'bad'.

THE SUPPLEMENTARY QUESTIONS - The evolution of what should
                              be specified to the seismic
                              contractor

Having defined the operating reserves of the NCB undertaking as a whole in a very strict manner, which enabled the rate of depletion (by both annual rate extraction and annual rate of fall in percentage recovery of accessed coal) to be monitored annually (see Shaw (11), Fig. 1) and thus shown that if the long term trends

continued, the Board would be short of capacity, even on
a falling market, within the lead time for creation of new
mines, the NCB started systematic exploration for new mine
sites in 1970/71.

The then latest (oil-type) manifestation of the seismic
reflection method was tested at two sites - one in a
relatively undisturbed coalfield at Hatfield and the other
in a relatively disturbed coalfield near Lea Hall mine, in
1971. The results are shown on Figs. 5 and 6.

It is noteworthy that the individual seismic traces are
fifty feet apart and the aligned sequences of wavelets on
the traces (the 'events') are from 100 to 300 ft. thick.
From knowledge of the average thickness of beds of individual strata, and as demonstrated from coalfield sonic logs
by van Riel (12), these events seen on the record must
therefore be internal reverberations from certain groups
of beds which happen to have got into phase, and not
reflections from individual beds. Beds, as groups providing
an event on a seismogram, would have a lateral persistence
greater than is common for beds as individuals in the
deltaic coal measures of the UK. No sonic logs were run
in boreholes on/near to seismic lines to correlate seismic
events with particular groups of beds, but it was thought
that the beds, as groups, might well be as laterally persistent as those seen in the seismic records obtained in
1971/72.

At that time, not a grid but a succession of single
profiles, employing oil company seismic methods with
slightly shorter spreads and offsets, were found to be
'adequate'. That is, they could detect the position of
extrapolated major (over 200 - 300 ft. throw) faults in
the more disturbed coalfields, any unknown (and therefore
unexpected) change in which would seriously affect the
continuity of production in the cases investigated. Some
25 miles of seismic line were executed in 1972, in the
disturbed coalfields of the West Midlands. However, the
most attractive looking potential places for new mine
sites lay in the less disturbed coalfields of the East
Midlands and Yorkshire. It was recognised that the absence
of structural data, in the new concentrated mining circumstances now obtaining, would be very costly, wherever
faults of over 200 - 300 ft. throw are rare, e.g., in the
Selby prospect in Yorkshire (see Rees, 2, for a description
of the seismic surveying to Phase III, and the change of
mine layout orientation which followed it at Selby).

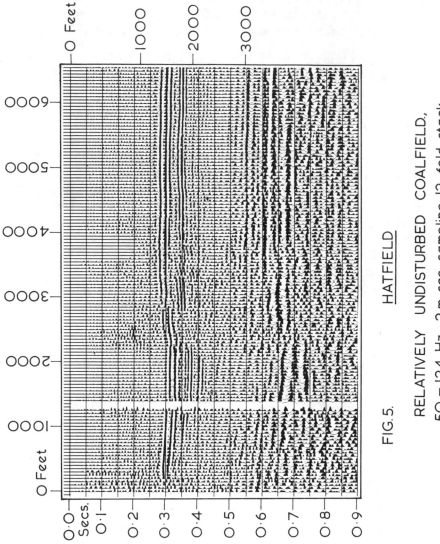

FIG.5.   HATFIELD

RELATIVELY UNDISTURBED COALFIELD,
50 – 124 Hz., 2 m. sec. sampling, 12 –fold stack.

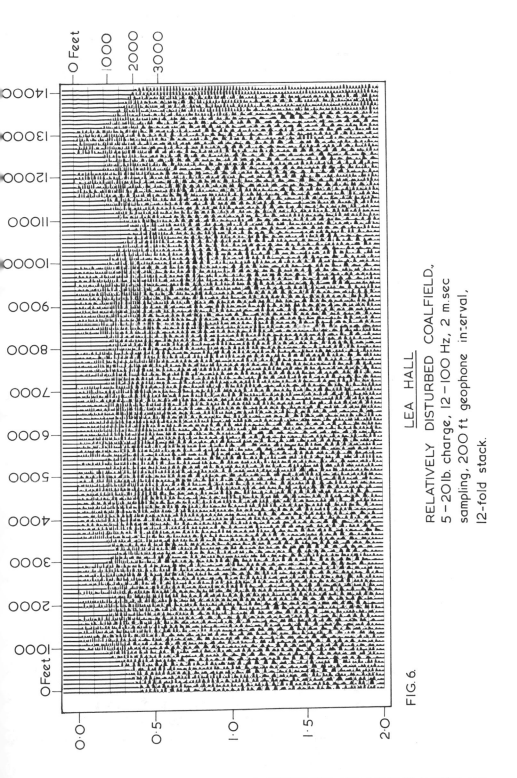

LEA HALL

RELATIVELY DISTURBED COALFIELD.

5–20lb. charge, 12–100 Hz, 2 m.sec
sampling, 200 ft geophone interval,
12-fold stack.

FIG. 6.

SEISMIC SURVEYING AND MINE PLANNING 183

Higher frequency high resolution experiments

After discussing the situation, on taking advice both from the Geophysical Research Department at British Petroleum (P. J. Stanley, personal communication, 13) and from the Department of Applied Mathematics and Theoretical Physics at Cambridge (J. H. Hudson, personal communication, 14), an experiment on an £8,000 budget was devised in 1973. Its purpose was to resolve thinner groups of beds, or even thick individual beds at an abandoned airfield site in the centre of the Selby field known as Acaster Malbis. It was hoped that if continuity of thinner events could be seen, then, in high quality ground conditions, faults with a displacement down to about one-third of the event thickness could be seen.

The same line was shot at 25 to 50 ft. station spacing, with a 5 lb. charge at 50 ft. depth and a 1 lb. charge at 10 ft. depth, with 12 fold stacking, also played out at 6 fold, recorded at 1 millisecond intervals, with $\frac{1}{2}$ to $3\frac{1}{2}$ station offsets and damped geophones (to give better higher frequency response). The results were played back at various additional simulated station intervals, by omitting alternative, etc., traces where necessary, and filtered over various frequency ranges at both 1 millisecond and (by resampling the field tapes) at 2 and 4 millisecond intervals.

The best combination of field parameters for this site was found to include, contrary to some theoretical considerations, 1 millisecond sampling and, in line with predictions, a frequency range of 50 - 160 Hz., 1 lb. charges at 10 ft. depth on 12 fold stack. The results of this particular combination are shown on Fig. 7. It is noteworthy that, although the events look superficially similar to those in Figs. 5 and 6, being printed on a larger (equal vertical and horizontal scale in the coal measures) the events are here only 60 - 80 ft. instead of 100 to 300 ft. thick. As there are more events per 1,000 ft., it is apparent that thinner, laterally less persistent groups of beds are being resolved. However, because of the absence of a distinctive individual event, or of distinctive seismic character, over the whole of vertical strips of the coal measures part of the seismogram, although the position of faults of about one-third the thickness of the events could be resolved under good conditions, the displacement of events on each side of a large fault could not always be matched. The throw of larger faults would have to be

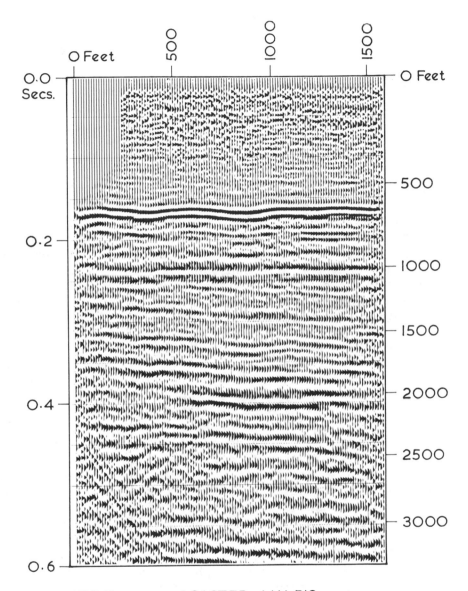

FIG. 7　　ACASTER MALBIS

1 lb. charge, 10 foot deep holes, 50–160 Hz.,
1 m. sec. sampling.

confirmed by carrying levels from a borehole out along
seismic lines to the margins within each fault block.
Later, it was found that,with experience, interpreters
said they could see enough 'character' to correlate across
faults, and a few of their many predictions have been
checked by suitably placed boreholes and found to be
correct.

Owing to the decrease in lateral persistence of the half
dozen or so events at around the level of the target seam
(the level at which the smallest resolvable faults are
required), on the seismic cross-section, it was found that
25 ft. station spacing (giving $12\frac{1}{2}$ ft. between traces on
the seismogram produced far better results than 50 ft.
spacing,and much detail was lost at 100 ft. spacing).
The coal industry commonly pays for exploration drilling
on a 'footage-drilled' and 'coal core loss penalty' basis.
Seismic contractors charge largely on a time rather than
a mileage basis. Shortening the station interval therefore
increases the cost per mile considerably. The answer
appeared to be to select the widest spacing compatible with
desired continuity of 'the average' event at or around the
target horizon.

Having found a system which resolved enough of the
faults in the less disturbed coalfields (and learned the
lessons that it is well worthwhile conducting more elaborate
preliminary tests and experiments than is customary in oil-
type seismics and having a geophysicist who understood
coalmining needs) production shooting, mainly at Phase III,
was started in the East Pennine coalfields. Grid spacing
was tailored to the degree of disturbance in new mine
prospects and, at Phase IV, to needs at some of the
existing mines. A similar evolution took place in processing, in this case, with the advice of Resources &
Engineering Management International Inc. who, in the
absence of any other independent geophysicists, maintained
a technical audit on these operations, see Daly (15) -
until the NCB took on its own geophysicists. The problem,
in this case, was that 'a good record' in the opinion of
the processors, was one on which continuity of dipping
events was most clearly readable. Automatic static
corrections and deconvolution techniques produced 'better'
records because they smoothed out minor displacements
between adjacent traces and suppressed 'noise' due to the
superimposed diffraction patterns associated with minor
faulting. Our 'signal' was often, by long tradition,
'noise' to the seismic processor.

Future developments

To-date, nearly a thousand miles of what can now be considered coalmining seismic data has been shot, largely at 25 to 30 ft. station spacing, giving the time equivalent of, say, thirty or so thousand miles shot at oil company station spacing. The build-up of annual mileage and cost is shown on Fig. 8. During the current year, it is hoped that, in the technical field, problems of inadequate static correction (the current limit on frequency and hence on resolution) will be overcome. On the interpretation side, it is hoped that the total folios of results from multiple faulting in configurations systematically generated by seismic modelling will have been created for interpreters' reference before the end of the year. It is hoped that the total time taken from call-off (of a seismic crew under contract) for work at a particular mine, right through to addition of the latest line data to the structural inter- pretation plan and the modification of a Geosimplan run at that mine, can either be shortened, or can be dovetailed with parallel work at other mines, in such a way as to direct the Party Chief in the field to modify the data acquisition plans given to him according to unfolding results in the light of mining needs.

Seismic exploration of a number of areas, each the size of a whole new mine 'take', as part of a mining undertaking's annual seismic exploration programme, is a rare event. A multiplicity of small prospects, in such of the mines as will shortly be entering disturbed, structurally unproved ground, with first seam workings, is the more general case. Before the coalmining seismic surveying system can meet these needs with the maximum efficiency, the time taken to complete the feed-back loop between Party Chief in the field and Mine Planner at the mine, referred to above, must be shortened (or offset) so that it can take place on nearly a mile by mile basis. Finally, all mining engineers will have to become as familiar with the strengths and limitations of coalmining seismic exploration as they are now with exploration by boreholes and suitable production layouts underground. This will only happen by word of mouth between them, although it can and will be speeded up by presentation of convincing case histories (when current exploration commitments allow time for their preparation). Coalmining seismics in their complementary surface and channel wave form, see Krey (3), when both are fully developed and accepted by the industry, will

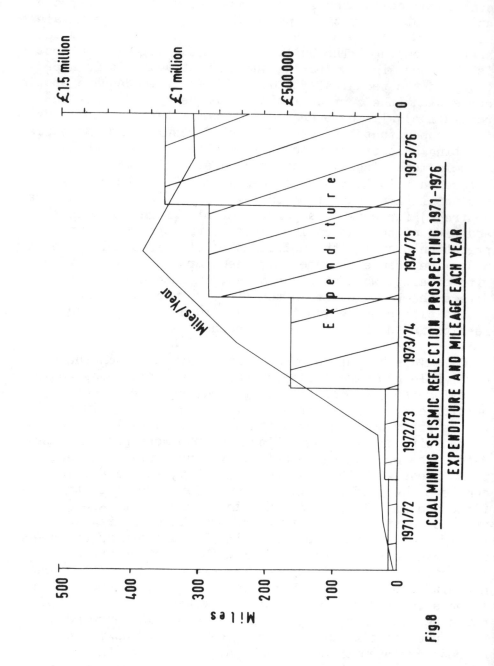

COALMINING SEISMIC REFLECTION PROSPECTING 1971–1976

EXPENDITURE AND MILEAGE EACH YEAR

Fig.8

revolutionise the economics of mining in the more disturbed coalfields of the world.

CONCLUSIONS

1. In spite of both a need for their products in the coalmining industry and the existence of the necessary basic seismic surveying technology over the last two decades, it is only the last three years that seismic surveying has been applied on a continuous rather than a 'one-off' basis in coalmining.

2. It appears that the best way of bridging the communication gap between the two industries, as far as regular needs at existing mines are concerned, is through integration of seismic exploration into mining via the planning function.

3. The mining industry has yet to be convinced that (apart from the phases of exploration associated with the (rarely occurring) search for new mine sites (Phase I: Is there any coal? II: Is there sufficient area of it to justify access? III: Is the new area sufficiently productive to justify access?)) coalmining type seismic exploration should be treated as one of the normal mining techniques reviewed in the routine process of maximising efficiency of production.

4. In achieving routine acceptance of the seismic surveying method, once drilling (or previous mining) has identified the choice between any alternatives of the next most accessible/productive area of coal, any seismic costs have to be treated as complementary to (a) the costs of any coal drilling, (b) the costs of such exploration as is inherent within the coal producing systems of the mine, (c) the costly spare capacity (insurance level) carried within the coal producing systems and (d) the equally costly consequences of meeting unexplored for, uninsured for, geological disturbances to continuity of production.

5. Experience in the National Coal Board indicates that, except in the case of widespread occurrence of simple large throw fault disturbance, the best forum for interaction between the seismic surveyor and the mining engineer is offered by the Geosimplan method (for

testing the robustness and outcome of changes in the balance of the elements in the plan given as 4(a) to (d) above.)

6. Coalmining seismic surveying is still in its infancy. Although 'the new arrival' offers great promise, much remains to be done if it is to grow to full stature. This includes improvement to the technical efficiency at all stages from data acquisition to final profile and,above all, to the development of coalmining seismics as a totally heuristic exploration system for existing mines (each seismic survey extinguishing itself in the course of balancing its findings in relation to the local needs of the mine).

REFERENCES

1. Clarke, A. M., 'Pay-off v. Risk Assessments in Planning and Mining Geology', Mining Engineer, 73, October, 1966, pp 29-46.

2. Rees, P. B., 'New High Output Deep Mines in Great Britain, with specific Reference to the Selby Project', Proceedings IX World Mining Congress, Dusseldorf, May, 1976, in press.

3. Krey, T.C., 'In-seam Exploration Techniques, proceedings International Coal Exploration Symposium, London, May, 1976, in litt

4. Hasbrouck, W. P., and Hadsell, F. A., 'Geophysical exploration techniques applied to Western United States coal deposits', ibid.

5. Clarke, A. M., 'Simulation techniques in the geological assessment of engineering projects' Proceedings, Geological Society, London, 1637, March, 1967, pp 14-18

6. Clarke, A. M., 'The effects of geological pattern on the geometry of mine layouts', Proceedings 2nd International Mine Surveying Conference, Budapest, IV, June, 1972, pp 1-24

7. Rees, P. B., 'Establishing the reliability of deep coal reserves', Proceedings 6th International Mining Congress, Madrid, June, 1970, pp 403-427

8. Hawes, D. M., 'Operational gaming in the planning of the geologically troubled colliery' in 'Operational Research 72', M. Ross, Edtr., North Holland/American Elsevier, 1973.

9. Anon., 'Report and accounts 1969-70' National Coal Board, I, July, 1970, Her Majesty's Stationery Office, London, p.9

10. Sloman, M., 'How operational gaming can help with the introduction of a new technology' Operational Research Quarterly, Pergamon, England, in litt. 1976

11. Shaw, K., 'Development and adaptation of drilling equipment to coal exploration' proceedings International Coal Exploration Symposium, London, May, 1976, in litt.

12 .van Riel, W. J., 'Synthetic seismograms applied to the investigation of a coal basin', Geophysical Prospecting, XIII, 1965, pp 105-121

13. Stanley, P. J. (Personal communication) 1973

14. Hudson, J. H. (Personal communication) 1973

15. Daly, T., 'Surface seismic exploration for coal deposits' proceedings International Coal Exploration Symposium, London, May, 1976, in litt.

The views expressed and any errors in this paper are solely due to the author.

7

SEISMIC METHODS FOR

THE DELINEATION OF

COAL DEPOSITS

T. E. DALY
Director of Exploration Services

R. F. HAGEMANN
Senior Geophysicist

Resources Engineering and Management International
Denver, Colorado, U.S.A.

Once the presence of coal has been determined, the problem then becomes a matter of determining if the coal is economically mineable. That of course depends upon a number of factors, but this paper will only concern itself with geological factors and a method of more precisely defining the geology in the area of interest.

In the past, once the presence of coal deposits had been determined, it was customary to construct geological maps and cross sections based on any surface and subsurface information available to determine as accurately as possible the attitude of the coal seam. In areas of steep dip, faulting or thick overburden, an accurate geological map is very difficult to construct; and it is often necessary to supplement the available geological information by the drilling of bore holes. Unfortunately, this tends to be rather slow and expensive, particularly in areas of thick overburden; so one must balance the utility of the additional information against the time and cost involved

in acquiring the information. Any method that could provide
additional geological information more rapidly and at less
cost would certainly be of interest to a mining company.
A method that would enable the geologist to project from the
known to the unknown with greater accuracy would be extremely
useful. Not only would it enable the economists to determine
more precisely the potential worth of the coal deposit in
the area of interest, it would also be extremely useful to
the mine planners in locating shafts, drifts, roads, etc.

The petroleum industry has been using various
geophysical methods for a number of years as an aid in the
projection of geological information. However, the petroleum
industry usually requires considerably less detailed
geological knowledge than the mining industry, so one must
examine the various geophysical methods used by the
petroleum industry to determine which, if any, of the methods
are applicable to the mining industry.

Geophysics is the study of the physical properties
of the earth, so listing some of the physical properties
that are often measured, and determining to what extent
coal possesses these physical properties is of interest.

Figure No. 1 shows some of the commonly measured
physical properties and the physical characteristics of
coal. It will be noted that coal possesses very unusual
physical characteristics and should, therefore, respond
to a number of geophysical tools. This is indeed the case.
Theoretically, a gravity survey (measuring changes in
density) a magnetometer survey (measuring changes in the
earth's magnetic field), an electrical resistivity survey
(measuring changes in resistivity), a radio-active survey
(measuring changes in radio-activity) or a seismic survey
(a function of velocity changes) could all provide some
geological information concerning the coal seam. Unfortunately, the resolving power of most of the above methods
is not very great, and the utility of the method is directly
proportional to its resolving power. Of the five geophysical methods mentioned above, only the seismic survey
approaches the resolving power required by the coal industry.
Therefore, this paper will be confined to a discussion of
the seismic reflection survey and its use in coal exploration and exploitation. The purpose of this paper is to:

Define in a general way the uses of seismic
data in coal exploration and exploitation

RESISTIVITY	Medium to High	1×10^2 to 5×10^5 ohm-cm
RADIOACTIVITY	Very Low Gamma Ray	
	Good Moderator Neutrons	
MAGNETIC SUSCEPTIBILITY	Very Low	Less than 2×10^6 cgs units
DENSITY		Lignite 0.7 to 1.5 gm/cm^3
		Bituminous 1.2 to 1.5 gm/cm^3
		Anthracite 1.4 to 1.8 gm/cm^3
SEISMIC WAVE VELOCITY	Low	6000 to 9000 feet per second

PHYSICAL PROPERTIES AND CHARACTERISTICS OF COAL

FIGURE 1

- Discuss techniques and problems in seismic data acquisition, processing and interpretation as related to coal mining
- Outline limitations of the present "state of the art" seismic methods and discuss the possibilities of overcoming these limitations.

In order to define the uses of seismic data in coal exploration and exploitation, one must have at least a general understanding of what seismic data is, how the method works and why it works.

Figure 2 shows in a diagramatic way how the seismic reflection method works. An acoustic signal (frequently produced by an explosion) is introduced at the point S.P. and radiates through the earth. The velocity at which the signal travels depends upon the media through which it travels. Typical velocities of an acoustic wave in air are about 1129 feet per second and in water, 4800 feet per second. The velocity of the wave in a hard, dense limestone may be 18,000 feet per second, while the velocity of coal may vary from 6000 to 9000 feet per second depending upon the type of coal. Since the velocity of the wave is a function of the media through which the wave is travelling, one intuitively expects something to happen to this wave when there is a large change in the physical characteristics of the traversed media. Intuition is correct and what happens is controlled by one of the basic laws of physics called Snell's Law. In the general case when a ray radiating from the source, shot point S.P., reaches a boundary marking a change in media, it will produce a reflected ray and a refracted ray. The incident ray from the shot point makes an angle alpha 1 with the normal at the media interface while the refracted ray makes an angle alpha 2 with this normal. The relationship of these two angles is expressed by Snell's Law; that is:

$$\frac{\text{sine alpha 1}}{\text{sine alpha 2}} = \frac{V_1}{V_2}$$

or

$$\text{sine alpha 1} = \frac{V_1 \text{ sine alpha 2}}{V_2}$$

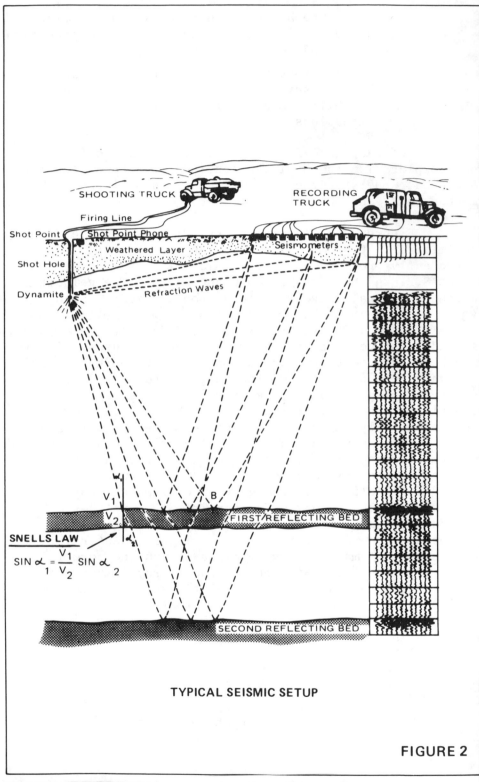

TYPICAL SEISMIC SETUP

SHOOTING TRUCK

RECORDING TRUCK

Firing Line

Shot Point

Shot Point Phone

Seismometers

Weathered Layer

Shot Hole

Dynamite

Refraction Waves

V_1

B

V_2

FIRST REFLECTING BED

SNELLS LAW

$$\text{SIN} \, \alpha_1 = \frac{V_1}{V_2} \, \text{SIN} \, \alpha_2$$

SECOND REFLECTING BED

FIGURE 2

where V_1 is the acoustic velocity in the first layer and V_2 is the acoustic velocity of the second layer. Obviously the angle alpha 2 is limited to 90°, as the sine cannot be greater than one. This places a limit upon the incident angle alpha 1, called the critical angle. All incident rays impinging upon the interface greater than this angle are said to be totally reflected. It is obvious then that the reflection and refraction of a ray path is a function of the distance between the source of the energy and the receiver as well as a function of the relative velocities of the two layers. It follows that as the depth to the reflecting-refracting interface decreases, the distance between the energy source and the receivers must be decreased to obtain a reflected ray. Therefore, the geometry of the "spread," the colloquial expression for the configuration of energy source and receptors, is a function of the depth of the zone of interest, as well as the velocities.

When the change in velocities at a boundary is large, a strong reflection is generated, and this reflection is detected by the receivers, converted to an electrical signal, and recorded. The instant the signal was generated is also recorded, and by recording the time it took for the signal to reach the reflecting point and return,it is easy to determine the depth of the reflecting point,providing the velocity of the traversed media is known. Since coal has a low velocity relative to most sedimentary rocks, it tends to be a good reflector - providing the seam is thick enough. The exact thickness required to generate a reflection is dependent on several factors, but a thickness of approximately 7 feet or more is required to generate a good reflection.

Assuming that a reflection is generated at or near the level of interest, some of the geological phenomena that may be examined with the seismograph shall be reviewed. One may determine the strike and dip of the seam. It is possible to identify major faulting. One should be able to identify "washout" areas. One should be able, to a limited extent, to predict changes in seam thickness and seam splitting. All of these geological factors are important to the mining engineer and mine planners, but the factor that creates really major problems for the engineer is the presence of faults. Even very small faults can cause problems, but faults exceeding the thickness of the seam can cause severe problems for the mining engineer.

If these faults can be anticipated before they are encountered at the coal face, it would be extremely useful information for the mining engineer.

This brings one back to the resolving power of the seismic reflection method. As mentioned earlier, the depth to the reflecting point is determined by measuring the time it took for the acoustic wave to travel from the energy source to the reflecting point and back to the receiver. One can see from Figure 2 that the depth from the surface to the first reflection would be approximately equal to the time it took the wave to travel from S.P. to point B. However, since the reflecting point in the subsurface will be located midway between the S.P. and the receiver, if the beds are horizontal, it is obvious that the time contains a certain amount of horizontal travel time and is therefore longer than the true vertical time. The observed times are corrected for this angularity to vertical time, and these corrections are called "dynamic corrections."

If the elevation of the energy source and the receiver is not the same, one must also make allowances for this change in elevation in order to arrive at a vertical time. If the reflecting bed is dipping as well, the reflecting point is not located at the mid point between the energy source and the receiver, but is actually located in an up dip direction. In Figure 3 when the bed A-A is horizontal the reflecting point will be at point "B" midway between the shot point and geophone. When the bed is dipping $A^1 - A^1$ the reflecting point will be located at B^1.

Up to this point only factors that change the recorded time have been discussed and the correction of these times to a vertical time. In order to correct this vertical time to depth, one must know the velocity of the media in which the wave has travelled. As already discussed, the velocity of different types of media vary. In actual practice, the media traversed may be made up of a large number of different types of sedimentary material generating a particular average velocity, but if the composition of the sedimentary layers changes, the average velocity will change as well.

Supposing that the average velocity is 10,000 feet per second at a particular point, and that there is a

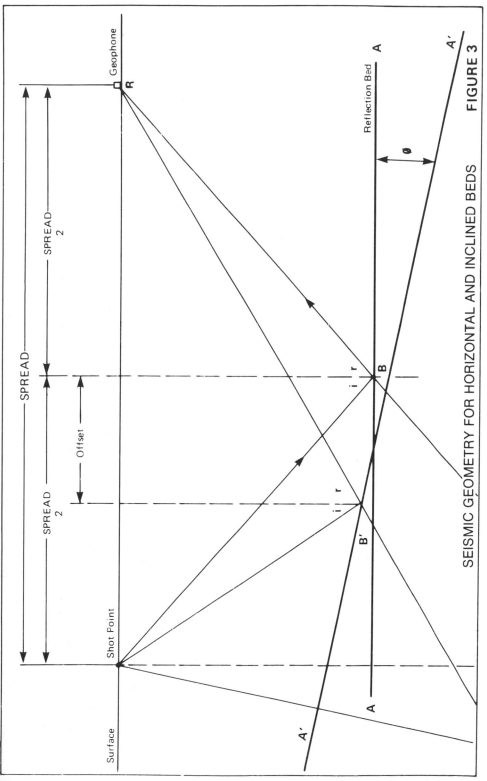

SEISMIC GEOMETRY FOR HORIZONTAL AND INCLINED BEDS

FIGURE 3

horizontal coal seam at a depth of 1000 feet, it is now
possible to determine the time it should take for an acoustic
wave to travel from S.P. to a reflector and back to a
receiver by the equation (see Figure 4) TR = 2Z ÷ 10,000
feet per second, or TR = 2000 ÷ 10,000 = 0.2 second.
Normally the times recorded on a seismic record section can
be estimated to one thousandth of a second, so in this
particular case, one would expect to see a reflection from
the coal seam at a time of 0.200 second on the seismic
record.

Now let us examine what happens if one moves 100
feet down the line of seismic profile with the coal seam
still horizontal, but having crossed a 10 foot fault
between the two points. Assuming that the throw of the
fault is down in the direction moved, and that the depth of
the coal seam at the new position is now 1010 feet,
(Figure 4), what time would now be recorded on the seismic
record? The equation would remain the same that is:
TR = 2 x 1010 ÷ 10,000 = 0.202 second. This assumes, of
course, that the average velocity between the two points
has not changed. If, on the other hand, the geology
between the two points has changed and the average velocity
at point 1 was 10,000 feet per second but the average
velocity at point 2 is 10,100 feet per second, the time
recorded would be TR = 2 x 1010 ÷ 10,100 feet per second
= 0.200 second or exactly the same time as point 1 even
though the two points are separated by a 10 foot fault.

If the coal seam is no longer horizontal but dips
at the rate of 10 feet per 100 feet toward point 2, the time
recorded at point 2 would be 0.202 second or exactly the
same as the time if the beds were horizontal but cut by a
10 foot fault.

From the above, it is obvious that a simple
examination of the differences in recorded times between
two points will not enable one to determine the cause of
that difference, and that a lack of time difference
between two points does not necessarily indicate a lack of
geological change. How then does one determine when or if
geological changes are taking place? By making a series of
closely spaced recordings, one tends to limit the number
of changes that will occur between two points, but a
determination of what changes may be taking place between
any two points requires seismic data of good quality.
When a geophysicist is asked to define "quality," he
usually finds it difficult to give a simple answer. He may

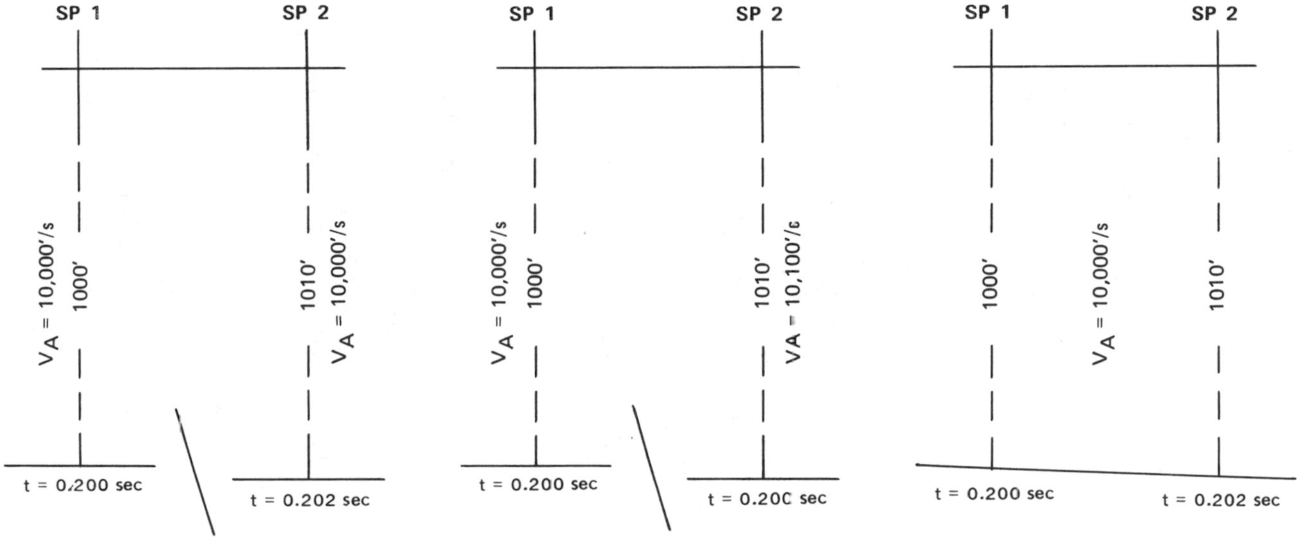

FIGURE 4

CALCULATION OF REFLECTION TIMES

$TR = 2Z (Z) \div V_A$

TR = Reflection Time
Z = Depth
V_A = Average Velocity

SEISMIC METHODS FOR DEPOSIT DELINEATION 201

say it is data showing a good "signal to noise" ratio, good
continuity, high amplitude reflections, lack of multiple
reflections, no static correction problems, good dynamic
corrections, correct scaling, proper filtering, good
amplitude control, etc. etc. etc. In short, he lapses into
the jargon peculiar to the industry without being able to
define precisely what is meant by good quality data. For
example, what does good "signal to noise" ratio mean? Signal
in this case means the reflected energy that is coming from
the point or points of interest. Noise, on the other hand,
is defined by a geophysicist as any <u>unwanted</u> signal. The
act of setting off a charge of dynamite not only generates
the acoustic wave that produces the wanted reflection
(signal), it also produces many other waves that are not
wanted (noise).

In effect, good quality data can best be defined
as data that accurately depicts the geology of the subsurface. However, geophysicists actually tend to classify
data according to the ease of interpretation without regard
as to whether or not the interpretation accurately depicts
the subsurface conditions. In other words, the classification of data quality tends to be a highly subjective
opinion of the interpreter. For example, most interpreters
would tend to classify the seismic data shown in Figure 5
as good data, and the seismic data shown in Figure 6 as
fair to poor data. However, it is entirely possible that
that data in Figure 6 depicts the subsurface conditions
just as accurately as the data in Figure 5.

Assuming that good quality data is recorded that
depicts subsurface conditions accurately, one may then
reach certain conclusions concerning what geological
changes may be taking place between two points. For
instance, in Figure 7 one would assume that the change
in time between points 115 and 85 at the level of the "C"
Horizon is due to dip toward point 85, because the time at
this point is .016 second greater than it is at point 115.
How does the interpreter know there is not a fault between
these two points that would account for the time difference?
Primarily, because the reflection appears to be very
continuous between these two points. How does one know
that the time difference is not due to a change in the
average velocity between the two points? He doesn't, so
he must consider this a possibility as well as normal
dip.

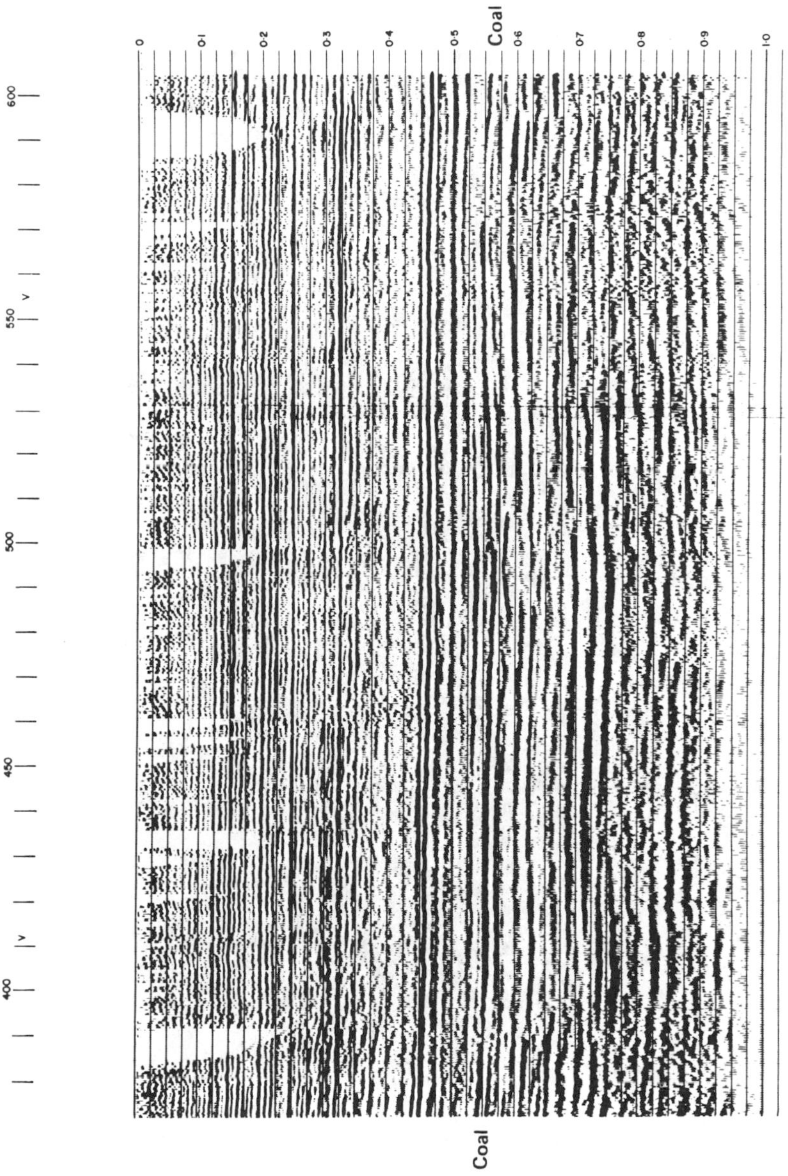

FIGURE 5

AN EXAMPLE OF GOOD DATA

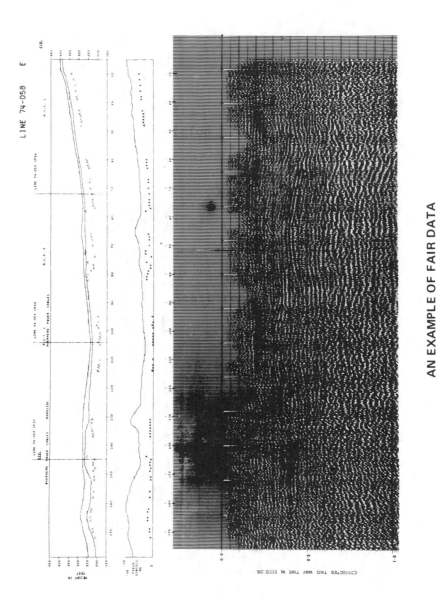

LINE 74-058

AN EXAMPLE OF FAIR DATA

FIGURE 6

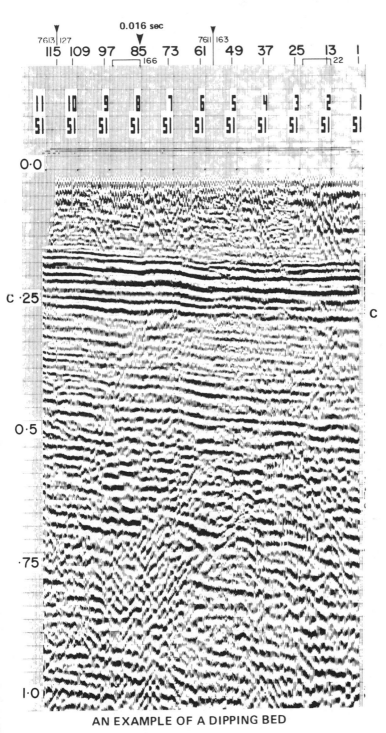

AN EXAMPLE OF A DIPPING BED

FIGURE 7

SEISMIC METHODS FOR DEPOSIT DELINEATION 205

If this time difference was due to faulting, one would expect that the reflection would show a break in continuity. Figure 8 shows a typical and very well defined faul It will be noted that Horizon "C" is at a time of .450 secon on the upthrown side of the fault and .493 second on the down thrown side of the fault. Therefore, it is said that the fau has .043 second of throw. How much is this in feet? That depends on the average velocity of course, but if one assumes a velocity of 10,000 feet per second, we have a throw of 215 feet. In this particular instance, the discontinuity of the reflections is due to faulting rather than an abrupt change in the LVL. However, it may be more difficult to differentiate between discontinuities caused by faults with a small amount of throw and those by changes in the LVL. Figure 9 shows a typical example of poor static (Elevation and LVL) corrections between points A and D. This strongly suggests a static correction error since there are very few vertical faults in nature.

If there is good continuity between two points and on wants to be sure that the change in time between two points i a reflection of dip rather than a change in velocity, how could this be determined? Basically, the only truly accurate method of determining what the average velocities are doing in a certain area is to acoustically log bore holes and to shoot velocity calibration surveys. This is a relatively simple procedure, but it does require drilling bore holes. One simply lowers, at the same time as other wire line logging operations, an acoustic logging sonde and measures repetitively the transit time in each formation for continuou uniform increments of depth. These incremental intervals of time and thickness are integrated and plotted as a function of total time versus depth. As there are instrumental variants that offset the acoustic log; hole diameter, temperature, etcetera, the sonic log is calibrated by lowering a receiver down the bore hole and initiating an acoustic impulse at the surface. Measurements of travel time at a number of depths are recorded.

If several velocity surveys are available in an area, one may compute the average velocities to a common horizon and contour these velocities. This will show the direction and rate of change (gradient) and one can then determine what effect the changes in velocities have on the apparent dips shown in time. It is quite possible in areas of relatively flat dip that the apparent direction

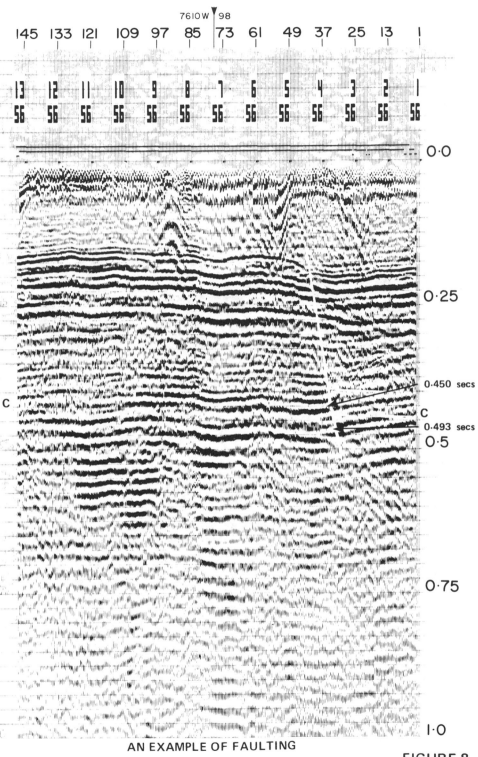

AN EXAMPLE OF FAULTING

FIGURE 8

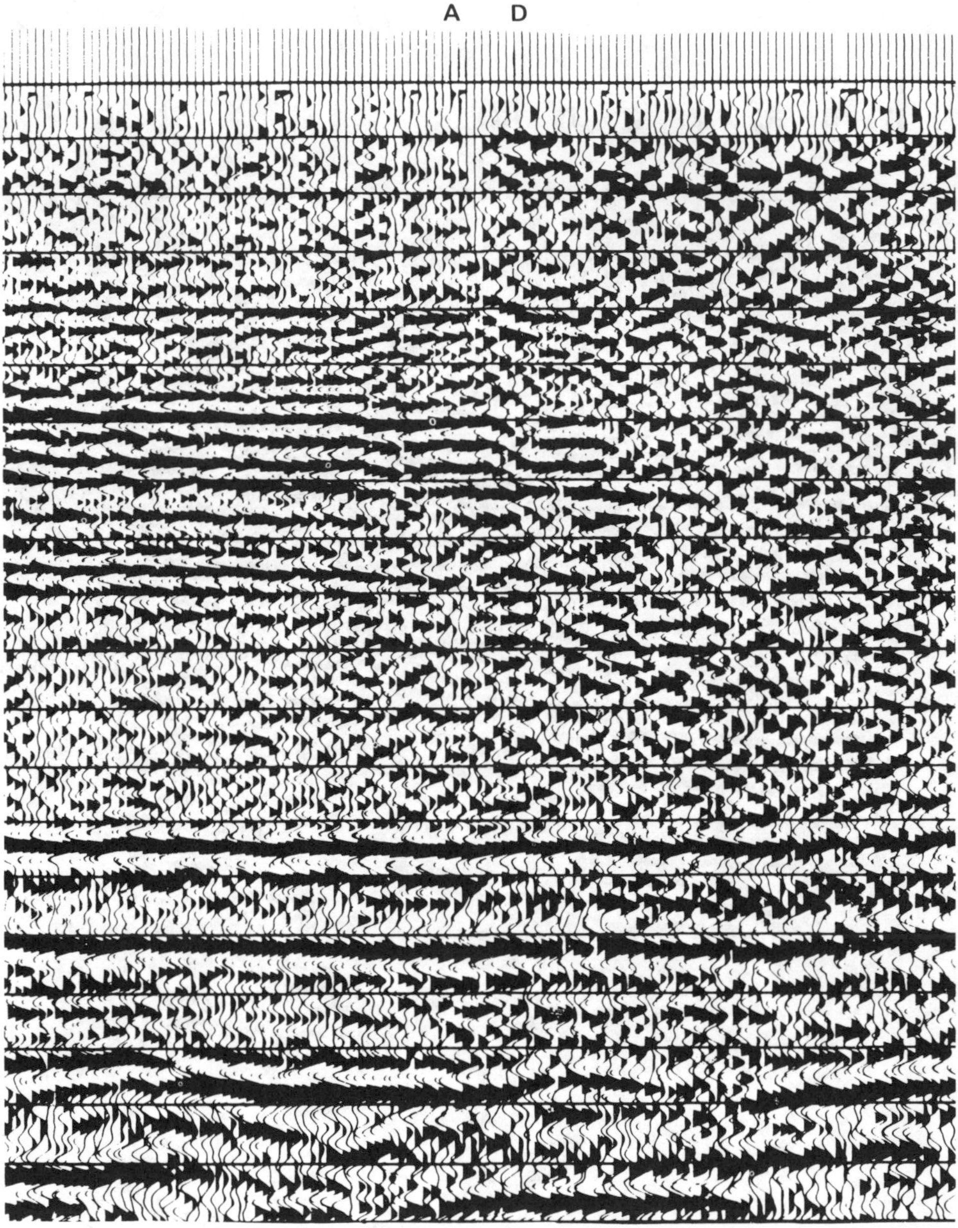

AN EXAMPLE OF POOR STATIC CORRECTIONS

FIGURE 9

of dip shown by a time map will be reversed when appropriate corrections have been made for velocity gradient.

How does one identify faults with the reflection seismograph? One fault in Figure 8 has already been seen, but let us examine a seismic record section across a known fault in another area. This particular fault was encountered in a bore hole, so one knows it is present. The seismic record section (Figure 10), which presents a two dimensional representation of the geology shows the fault is quite easily detected on this section. One notes that the good reflection labeled "C" shows an abrupt interruption at point 40 and that the good reflection appears at the lower position to the left of this point. This section suggests that it is a very simple matter to locate faults with the seismograph. However, let us now examine the throw of this fault. The time of reflection "C" on the upthrown side of this fault is 0.425 second, while the time on the downthrown side is 0.455 second. If the average velocity to Horizon "C" is 9840 feet per second, the throw of this fault is computed as follows: Depth of reflection on each side of the fault is computed using the following equation: $\frac{TR}{2}$ x Vav. The depth of the reflection on the downthrown side of the fault is - 2239 feet, while the depth on the upthrown side of the fault is - 2091 feet. The fault therefore has 148 feet of throw. On the other hand a fault with 15 feet of throw would only show 0.003 second difference between the times on the upthrown side and downthrown side, and it is obvious that faults of this magnitude would be much more difficult to recognise. Indeed, they may not be recognised at all. What then are the present limits of the resolving power of the reflection seismic method? It is believed that one is able to recognise faults with as little as .004 second of throw (approximately 20 feet), and in some cases perhaps as little as .003 second of throw or approximately 15 feet.

It is obvious from these examples that if geophysicists are to approach the degree of resolution that is needed by the coal industry, it is essential that they try to obtain data of the highest quality. Those factors that affect data quality must be examined. There are many factors that affect data quality, but the one factor that has the most profound effect on the quality of the data is the geology of the area. This includes not only the subsurface geology, but the surface geology as well. Changes in surface conditions can cause significant changes

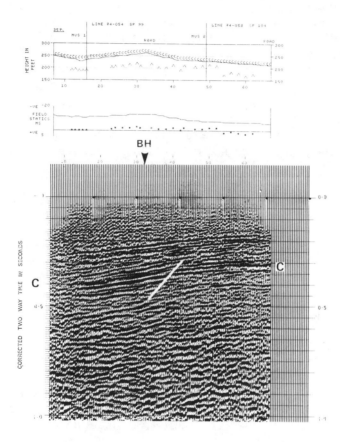

KNOWN FAULT IN BOREHOLE (BH) AT 0.430 - 0.450 SECONDS
(Compare with Figure 15)

FIGURE 10

in data quality, because surface conditions have a very pronounced effect on the signal to noise ratio. For example, in those areas where limestones, gypsum, anhydrite, gravels or peat are on the surface, poor quality records are to be anticipated.

There are a number of ways of improving the signal to noise ratio, but before discussing these techniques let us examine what noise looks like on a record section. When moving into a new area, it is usual to do at least a limited amount of experimental work to determine what problems will be encountered in the area and how these problems may be solved. One method of defining some of the problems in an area is to shoot a "noise spread." Figure 11 shows the layout of a typical noise spread. Basically, the purpose of this spread is to record and analyse the various signals that are generated by the introduction of energy into the earth. It will be noted that the geophones or receivers are usually closely spaced, and that a series of energy input stations are used to introduce energy into the earth. This energy is recorded by the receivers and the results are analysed.

Figure 12 shows a series of noise spreads that show very significant changes from one to the other. From these noise spreads, it is possible to determine amplitudes, frequencies, wave lengths, velocities, etc. of various signals and devise field techniques that tend to minimise the unwanted signal. This is much easier to do when conducting a seismic survey for the petroleum industry than it is if the survey is for the coal industry. Very briefly, it is possible to design source and receiver arrays to cancel unwanted signals. This method is almost universally used in the petroleum industry. However, the natural wave length of most seismic frequencies is such that relatively long spreads and large patterns of geophone or energy source patterns are required in order to cancel unwanted signal. Since the mining industry is interested only with the uppermost portion of the sedimentary section, a long spread is usually not practical. Remember that the distance between the energy source and the receiver determines whether or not a reflected wave or a refracted wave is recorded. If the spread becomes too long, the recording of a reflected wave from a shallow zone of interest will become technically impossible.

There are other methods of enhancing the signal to noise ratio, and one of the most commonly used is what is

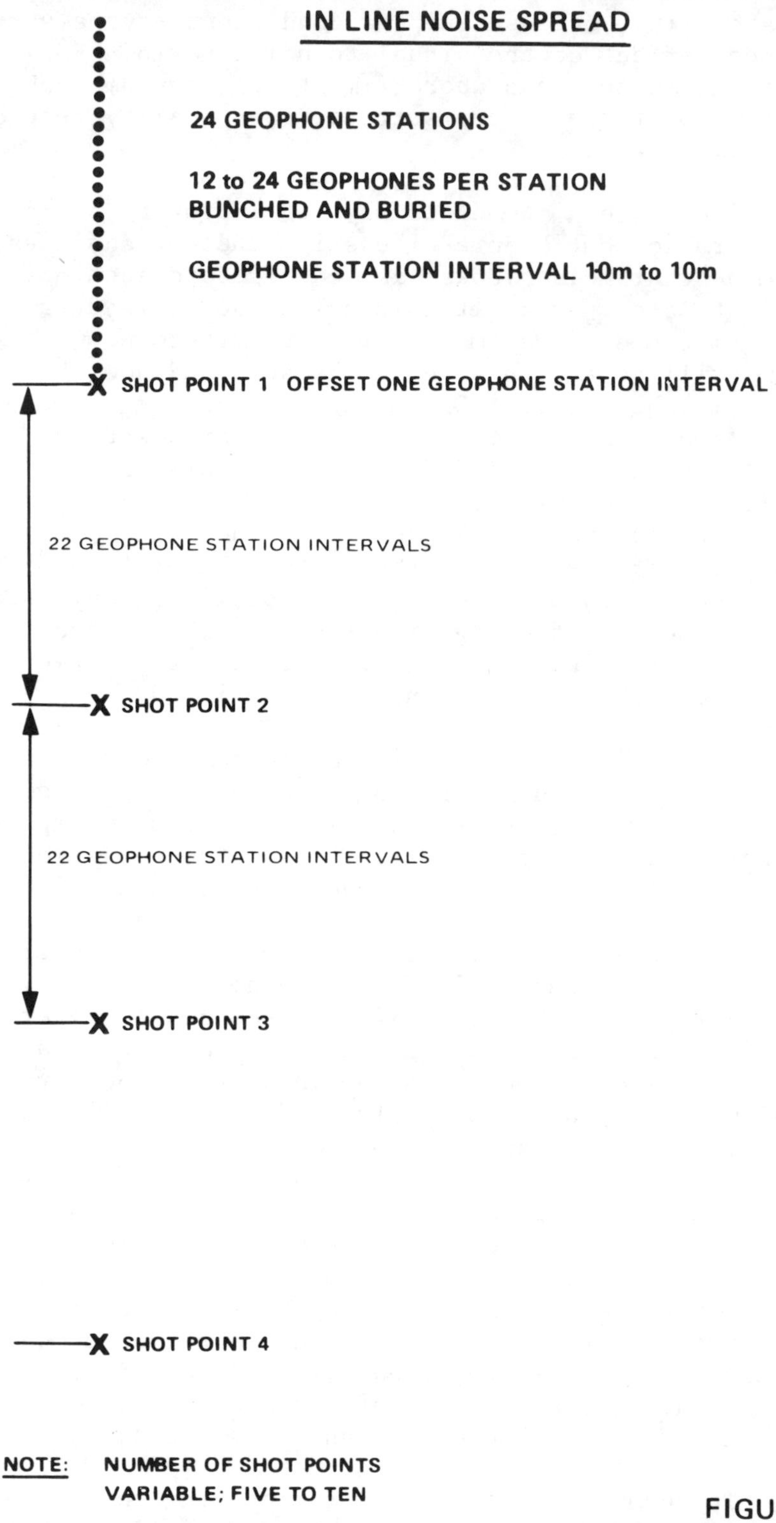

IN LINE NOISE SPREAD

24 GEOPHONE STATIONS

**12 to 24 GEOPHONES PER STATION
BUNCHED AND BURIED**

GEOPHONE STATION INTERVAL 10m to 10m

X SHOT POINT 1   OFFSET ONE GEOPHONE STATION INTERVAL

22 GEOPHONE STATION INTERVALS

X SHOT POINT 2

22 GEOPHONE STATION INTERVALS

X SHOT POINT 3

X SHOT POINT 4

NOTE: **NUMBER OF SHOT POINTS
VARIABLE; FIVE TO TEN**

FIGURE 11

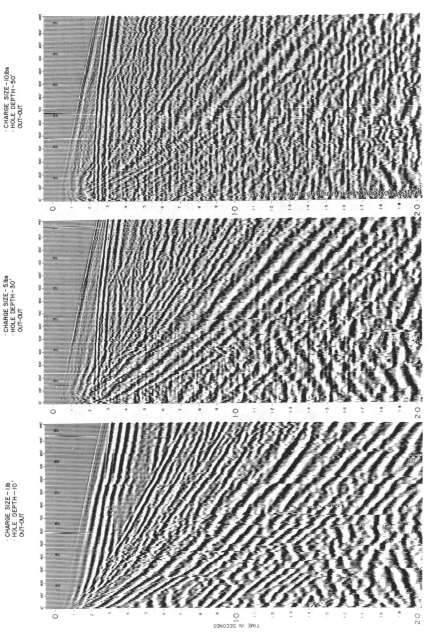

· CHARGE SIZE – IIb.
· HOLE DEPTH–10'
OUT–OUT

· CHARGE SIZE – 5 lbs
· HOLE DEPTH – 30'
OUT–OUT

· CHARGE SIZE –10lbs
· HOLE DEPTH–50'
OUT–OUT

TIME IN SECONDS

TYPICAL 'NOISE SPREADS'

FIGURE 12

SEISMIC METHODS FOR DEPOSIT DELINEATION 213

usually referred to as the Common Depth Point method of recording. The theory behind this method is relatively simple. There is a law which states that if a series of electrical signals that are in phase (peaks coincides with peaks) are combined with each other, the signals will tend to reinforce each other. On the other hand, combination of signals that are "out of phase" tend to cancel each other This then forms the basis of the common depth point method of enhancing the signal to noise ratio. If two or more reflections that are in phase are added, they will tend to reinforce each other. On the other hand, noise, which tends to be random, will probably be out of phase and tend to cancel. This method of signal to noise enhancement is in almost universal use today.

A third method of improving the signal to noise ratio is to introduce acoustic waves at various depths below the surface of the earth. The amount and type of noise generated is very strongly affected by the media in which the energy source is located. Figure 13 shows the different types of noise that were generated by simply changing the depth of the charge of dynamite (our energy source in this case). It is very obvious in this particular case that the best signal to noise ratio occurs when the charge was at a depth of 50 feet. It will be noted that when the charge was at a depth of 10 feet, the reflections which are shown at a time of 0.490 second and 0.640 second are almost completely obliterated by noise.

Electronic filtering is another method of improving the signal to noise ratio. The great majority of the signals of interest will occur in the range from 20 to 140 cycles per second. Therefore, if the noise being generated has a frequency of less than 20 cycles, or more than 140 cycles, it is possible to design an electronic filter that will effectively screen out these higher and lower frequencies.

There are a number of other methods of dealing with noise problems, but an exhaustive treatment of the noise problem is beyond the scope of this paper.

As can be seen, the acquisition of the seismic data is of critical importance. Up until now, this paper has discussed how improvement in the signal to noise ratio is based on analysis of the signal and noise generated at one particular point in a seismic survey. As mentioned

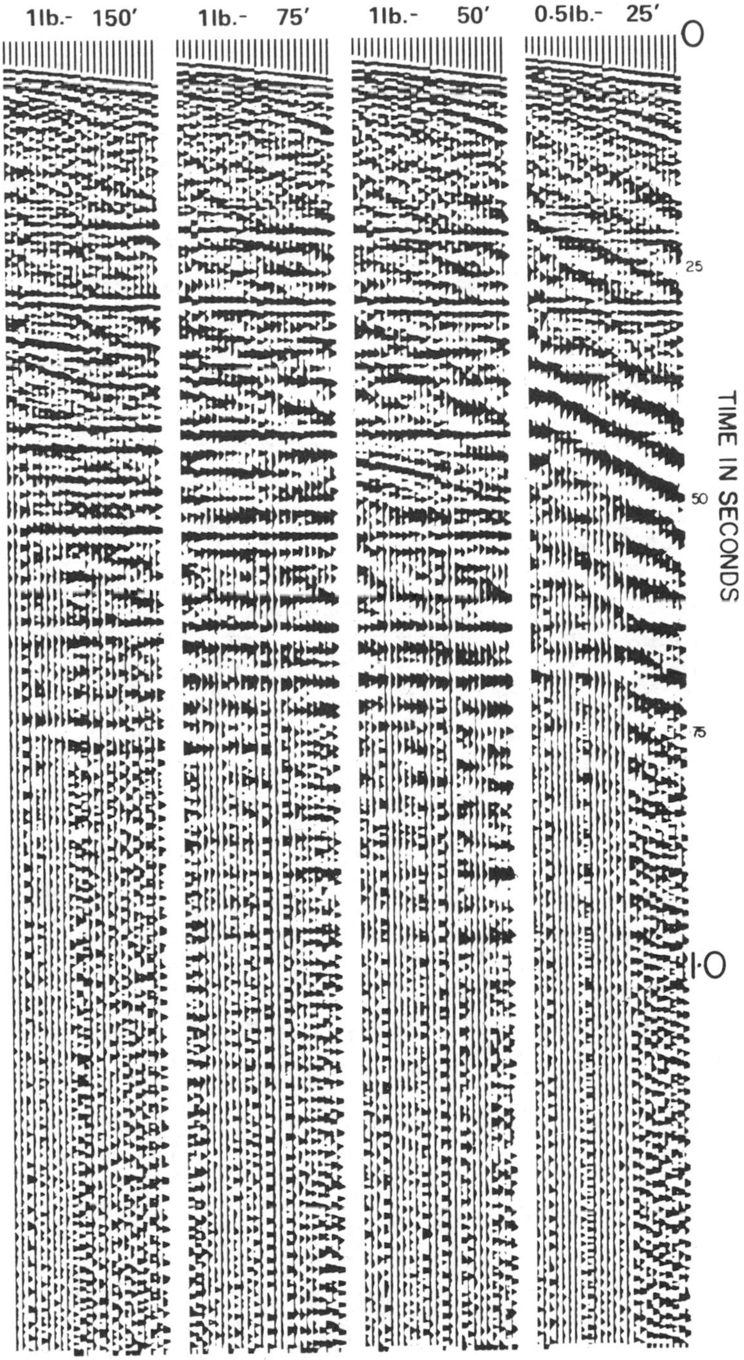

1lb.- 150'   1lb.- 75'   1lb.- 50'   0.5lb.- 25'

TIME IN SECONDS

EFFECT OF CHARGE DEPTH ON 'NOISE'

FIGURE 13

previously, the signal and noise will change as the shooting
and recording media change, so if there are any changes in
the surface geology, the technique which gives the optimum
signal to noise at one point may be completely ineffective
at another point in the survey area. As a result, very
close attention must be paid to the apparent quality of the
data as the survey progresses. If the quality deteriorates,
it may become necessary to conduct additional experimental
work in order to obtain usable data.

        Equally important, zones of high velocity contrast
between the surface and the zone of interest can have a
severe effect on the quality of the data at the zone of
interest. Large velocity contrasts produce strong reflection
because a large amount of the energy is reflected back to
the surface. Since a larger proportion of the energy is
reflected back to the surface, it follows that there is
less energy to be refracted into the lower layers. Several
such high contrast layers can effectively "use up" much of
the original energy and allow very little energy to reach
the zone of interest. At the same time, these high
contrast zones tend to produce what are called multiple
reflections.   "Multiple" reflections exist in many forms
and configurations. Every interface that results in a
reflection from the downward radiating ray will be an
interface to rays returning from deeper reflectors. So
just as in a "hall of mirrors" at an amusement park;
repeated images of the reflector may appear on the seismic
record. The seismic ray is reflected and re-reflected
by every change in the sedimentary section until the energy
level becomes imperceptible. The seismologist must realise
that multiples occur, whether the multiples are significant
or not. It is sufficient to say that multiples are not
always a problem but they may be and must be considered in
data acquisition, processing, and interpretation.

        Is it possible to do anything about the multiple
reflection problem? Perhaps. There is a law of physics
that states if two signals, exactly 180 degrees
out of phase are added together, the resulting signal will
be zero. Therefore, if a signal is generated that is
exactly the opposite of the signal from the multiple
reflection and introduced into the system, it should cancel
the multiple reflection. This is an overly simple way
of describing what happens in a process called deconvolution. Deconvolution is commonly used in processing the
seismic data, and if it is properly applied, it is an
effective way of eliminating multiple reflections. A point

of interest however, if the deconvolution signal happens to be exactly out of phase with the primary reflection of interest at any particular time, it will eliminate that reflection just as effectively as it will eliminate a multiple reflection. Therefore, deconvolution must be used with care.

What other things can be done to improve the quality of the data in the processing centre? A very commonly used method of improving the quality of seismic data in the petroleum industry is the use of what is called a residual statics programme. In actual practice, it is often very difficult to properly determine and correct for changes in thickness or changes in velocity of the low velocity layer. Improper corrections give the seismic record section a very uneven appearance. Figure 14A is a typical example of a record section that probably has problems with the static corrections. If a residual statics programme is applied to this section as a part of the processing, the programme will average out the differences between traces and the result appears to be a better quality record section as shown in Figure 14B. However, keep in mind that one of the primary objectives of a seismic survey for the coal mining industry is to locate faults that may show up as minor time changes between adjacent recording points. Therefore, if any programme that averages the difference between recording points is utilised, it is obvious that such a programme will severely limit the interpreter's ability to identify small faults. Therefore, it is recommended that any processing programme that averages between recording points or a series of points not be utilised in processing data for the coal industry. They make the already difficult job of interpretation more complicated.

Let's examine some of the problems of interpretation. As already discussed, the degree of resolution the coal industry wishes to achieve is beyond the present limits of the "state of the art" seismic methods. Moreover, a considerable adjustment is required on the part of those interpreters who have only been associated with the oil industry. This is not only a problem of adjusting to the vastly different scales between the two industries, but to the problem of the very different objectives as well. Nevertheless, the basic techniques utilised are the same in both applications.

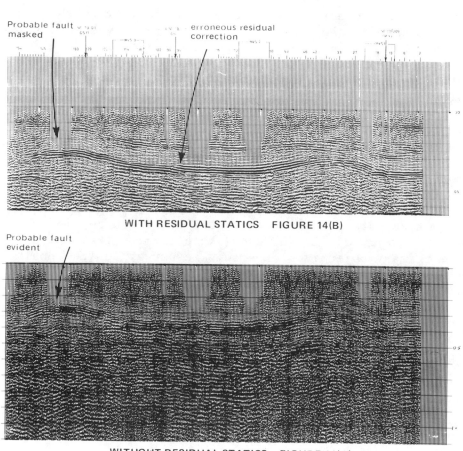

Probable fault masked

erroneous residual correction

WITH RESIDUAL STATICS   FIGURE 14(B)

Probable fault evident

WITHOUT RESIDUAL STATICS   FIGURE 14(A)

AN EXAMPLE OF STATIC CORRECTION PROBLEMS

FIGURE 14

The first thing the seismic interpreter needs to do is study every available bit of geological information in and around the area being surveyed. This information may come from surface outcrops, old workings, water wells, bore holes, gravity surveys, magnetometer surveys, etc. This information should be evaluated, and pertinent information should be noted on the shot point location map of the seismic survey.

The second thing the interpreter needs to do is identify, if possible, the origin of the various reflections shown on the seismic record sections. If a bore hole is present in the area to be surveyed, at least one seismic line should be shot across this bore hole. If a velocity survey has been made in conjunction with the acoustic log in this bore hole, the identification of the reflecting beds is usually a simple matter. The travel time has been measured to various depths in the bore hole, so a comparison of the velocity survey travel times and the reflected travel times will allow one to identify the reflection.

Once the interpreter has examined all available geological information and identified the beds causing the reflections on the seismic record sections, he is ready to start preparing a map at the level of interest. If there is a reflection from the level of interest, the problem is simpler, of course. If not, he must prepare a map based on the nearest conformable mappable reflection and then adjust this map upward or downward to the level of interest. If there is no reflection conformable to the level of interest, there is little point in trying to utilise the seismic method in the area. Fortunately, coal seams will generally produce reflections providing they are thick enough.

Since the physical measurement made in a seismic survey is time, it stands to reason that the basic interpretation will be a "Time" map. The times coming from a given reflection are noted on a map, properly corrected of course, as they were measured at various points and these times are contoured. However, as already seen, it is very difficult to determine if time differences, or lack of time differences have any geological significance, so the record sections must be examined in detail as these times are posted on the map, and notations of any apparent faulting should be made at the appropriate location on the map, together with the time of the reflection on the upthrown side and the time of the same reflection on the downthrown side. This assumes, of course, that one can identify the same reflection on each side of the fault, and this is not always a simple matter.

Unfortunately for the interpreter, the earth will pa lower frequencies more easily than high frequencies. That means that the acoustic energy, which may have started with a very wide spectrum of frequencies, tends to lose its higher frequency components with depth. As a result, the reflections from deeper horizons tend to have a lower frequency than reflections from shallower beds. It will be recalled that in an effort to increase the apparent signa to noise ratio, filters were used to eliminate unwanted noise. It stands to reason therefore that those filters which are most useful in this respect would be filters that change the frequencies they pass as the depth of the reflector increases. These are called time varying filters. Such filters can be very effective at improving the apparent quality of the seismic data, so they are used extensively.

Every reflection has what is described as "character by the geophysicist. This is another of the many virtually undefinable terms. It means that the reflection has a certain look about it that the interpreter learns to identif within a particular area. In general, "character" is a composite of frequency, apparent amplitude, the number of peaks or troughs present, and phasing. This "character" is a function of the entire sedimentary section traversed by the reflected energy. Now, suppose that a large fault is crossed. A minimum of two things have happened that will affect the "character 'of the reflection on different sides of the fault. The most obvious, of course, is that the sedimentary section through which the reflected energy passes may be different on the two sides of the fault. The second factor is the application of the time varying filters Obviously more lower frequencies will be passed on the down thrown side of the fault, and since frequency content is a very important part of "character" this also changes the appearance of the reflection. It is usually fairly easy to identify the same reflector on each side of a fault if the throw is small, but the difficulty of identification increas as the throw increases. The identification of reflections across a fault can be extremely important in the oil industr but in many cases it may be sufficient for the mining engineer to know that a large fault is present at a particul point. The fact that one cannot really say if the fault has 400 feet of throw or 800 feet of throw may be very unimporta to the engineer. On the other hand it may be very important for mine planning purposes to know if a fault has 15 feet of throw or 50 feet of throw. Reflection identification across faults is not usually a problem with throws of small magnitude.

As mentioned earlier, if beds are dipping, the reflection point is no longer located midway between the Shot Point and the receiver. That means that if the interpreter is to make an accurate map he must "migrate" the reflection point into its proper subsurface position. At first glance this would seem to be a relatively simple matter since the degree of dip can be determined from the seismic record sections and a line normal to the angle of dip will locate the true reflecting point (see Figure 3). However, keep in mind that the seismic record section gives a two dimensional look at a three dimensional world, so things seen on the record section are not always what they seem. For example, Figure 15 shows what appears to be a very low angle fault across this record section. Figure 10 shows the same fault on a line at right angles to Figure 15. The apparent low angle of the fault plane in Figure 15 is due to the fact that the strike of the fault is nearly parallel to this line of profile. These two figures point out the importance of using the map and the sections together in order to see the three dimensional picture.

The small segment of an interpretation shown in Figure 16 shows some interesting results of the use of the maps and section together. Several small east-west faults cutting the seismic profile in the northern part of the map will be noted. Since these faults only cut one line, it would seem that the strike of these faults would certainly be subject to question. However, the apparent dip of the fault planes as shown on the seismic profile approaches 70 degrees in every case. It is known, from previous mine workings in this area that most of the fault planes dip at an angle of 60 to 70 degrees, so it can be assumed, with a fair degree of certainty, that the strike of the faults will be very nearly at right angles to the line of profile and must therefore be approximately east-west. On the other hand, if the apparent dip of the fault plane across the record section had been 30 to 40 degrees, it would have been difficult to determine whether the strike of the faults was northwest-southeast or northeast-southwest, or for that matter the faults could still strike east-west and the angle of the fault planes was simply less than "normal" for the area.

This paper has touched very briefly on field techniques, processing techniques and interpretation

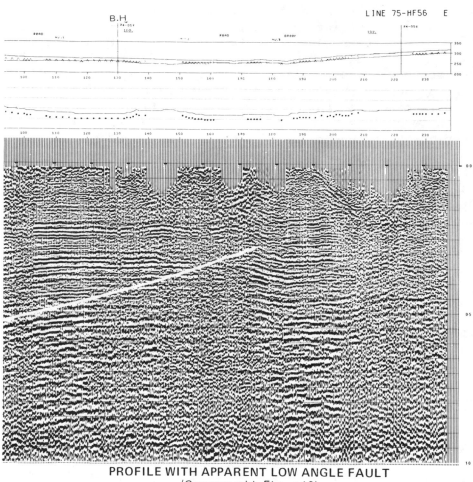

PROFILE WITH APPARENT LOW ANGLE FAULT
(Compare with Figure 10)

FIGURE 15

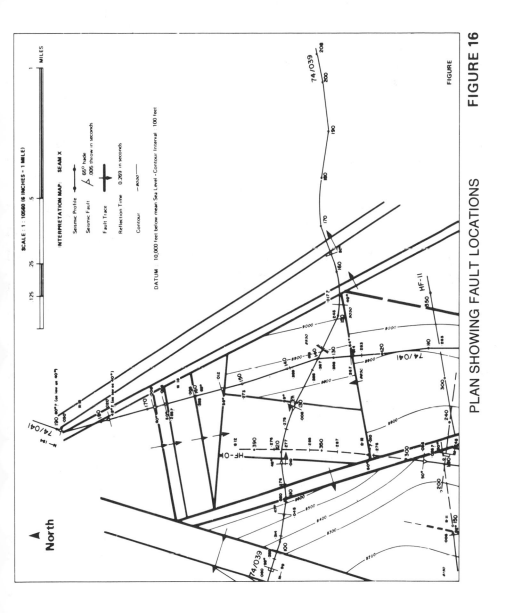

PLAN SHOWING FAULT LOCATIONS

FIGURE 16

SEISMIC METHODS FOR DEPOSIT DELINEATION 223

techniques. One very important point that has not been discussed is how does one go about planning a seismic survey once it has been decided that such a survey is warranted? Figure No. 17 shows a typical area here in England and some of the problems that must be considered when planning a survey. One would usually start by trying to plan a survey to solve as many geological problems as possible with the amount of money budgeted for the survey. Therefore, one starts planning the survey by drawing up a more or less idealised pattern of seismic lines to solve the geological problems. The lines should be straight in order that one can use the CDP technique.

The second step is to superimpose this idealised pattern on a good topographic map showing the culture and land ownership if possible. In a highly developed country such as England, a large amount of adjustment and compromise will be required in order to locate the lines so that they do not pass through a town, do not follow electric transmission lines, avoid rivers, etc. etc. Once these problems have been avoided, and the compromise lines located the next step is to obtain permission from landowners to shoot these seismic lines. Needless to say, there may be someone in the centre of the most important line who will not let work be carried out on his land under any circumstances, so a compromise in the position of the line will be required again. As you can see, the planning of a seismic survey can be somewhat complicated.

Some of the limitations of the present state of the art have already been discussed, but in general they may be summarised as follows:

1. The present accuracy of the seismic method is not adequate for coal exploration needs.

2. Modified data acquisition techniques are needed in order to enhance the quality of the basic seismic data.

3. Data processing techniques must be re-evaluated in light of the objectives of the coal industry.

4. Interpretative techniques need further refinement and interpreters trained in the requirements of the coal industry.

5. Mining engineers and management must be trained in the use of the seismic method as a geological

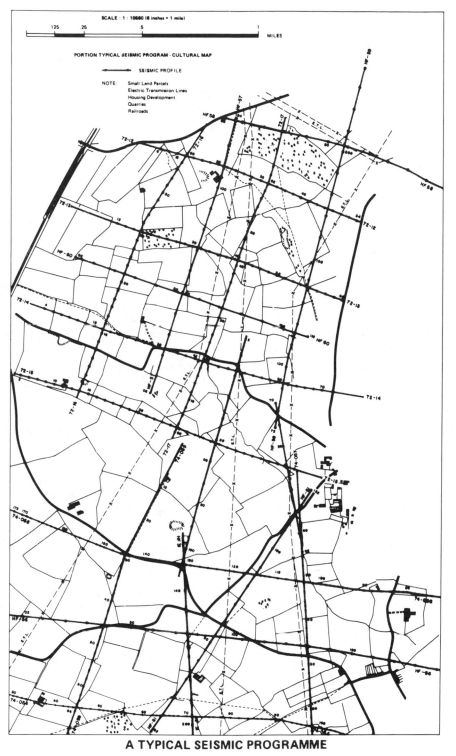

A TYPICAL SEISMIC PROGRAMME

FIGURE 17

SEISMIC METHODS FOR DEPOSIT DELINEATION 225

tool. Above all, they must be taught that
the seismic method cannot resolve all their
problems. Even under the best conditions,
it will still be an indirect method of
looking at the subsurface, and as any mining
geologist can tell you, it is not always
easy to interpret the geology even when one
can see it directly in the mines.

These five items probably summarise the major curren
limitations of the seismic method relative to the coal
mining industry. On the other hand, attention is called to
the fact that the demand for seismic exploration by the
coal industry is very recent and has been pioneered by the
National Coal Board here in England. It is believed that
the worth of the method has been proven in spite of the
existing limitations, and now that a demand for improvements exists, one can be confident that many of the present
limitations will be overcome. For example, the sampling
rate and consequently the ability to time the arrival of a
reflection have already been greatly improved. A considerable amount of research is being done in the higher frequenc
ranges which, in theory, should greatly improve the
resolution of the seismic method. Much has been learned
about noise generation and cancellation. Different types
of energy sources, better adapted to use by the coal
industry, will be perfected. New processing methods will
be developed that will reflect the subsurface geology more
accurately. As interpreters gain experience in the interpretation of seismic surveys for coal exploration and
exploitation, their ability to provide the mining engineer
with an accurate picture of the geology ahead of the coal
face will undoubtedly improve.

In closing, we wish to thank the National Coal Board
for allowing us to use many examples of their work as
figures for this paper. In addition, our company is honoure
and grateful that we have had the opportunity to work with
the National Coal Board in what we believe has been a
pioneering effort to use the seismic method as a standard
approach to the solution of geological problems. We are
firmly convinced that, although the method will not solve al
geological problems, it will probe to be extremely useful
to the coal industry in future years.

8

IN-SEAM SEISMIC EXPLORATION TECHNIQUES

Theodor C. Krey

retired Chief Supervisor of PRAKLA-SEISMOS GmbH, Hannover,

Professor at University of Hamburg

Introduction

It was about 1960 when I was told about the difficulties arising in the working of coal in deep mines if the seams are interrupted by small faults or other minor irregularities. I was asked whether reflection seismics, carried out at the surface of the earth or within the mines using normal seismic body waves, could detect faults with a throw of say 3 to 6 feet (1 to 2 meters) only. Of course, I had to negate this question. In reflection seismics from the earth's surface the smallest possible wave lengths would still be far too long (Figure No. 1) . In fact, they are about 120 to 160 feet (40 to 50 meters), and body waves produced and observed underground within the mines at a face or in a gallery would have bad chances of being efficiently reflected from a fault plane, because reflected energy would only arise from the narrow strips where coal is adjacent to country rock, and this would only be a small percentage of the fault plane (Figure No. 2). Only unhealed fractured zones in the vicinity of the fault plane would improve the reflectivity as a whole.

After having studied the physical properties of coal and rock a little, I realised that coal has a low velocity and low density, as compared to the country rock, at least in the Ruhr area, and it

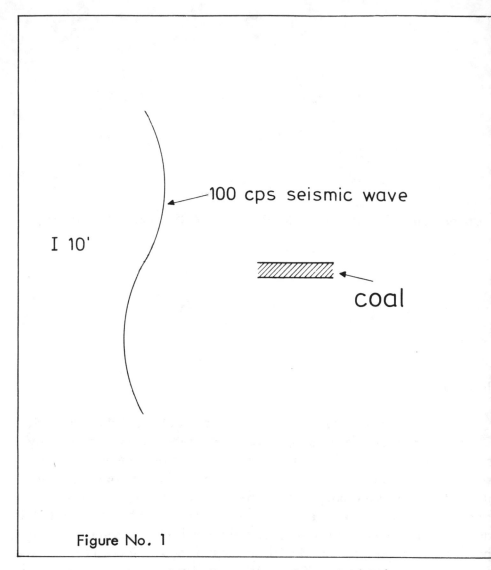

I 10'

100 cps seismic wave

coal

Figure No. 1

Comparison of the dimensions of a typical Ruhr area
seam and a 100 cycles per second seismic wave
originating from a shot at the surface.

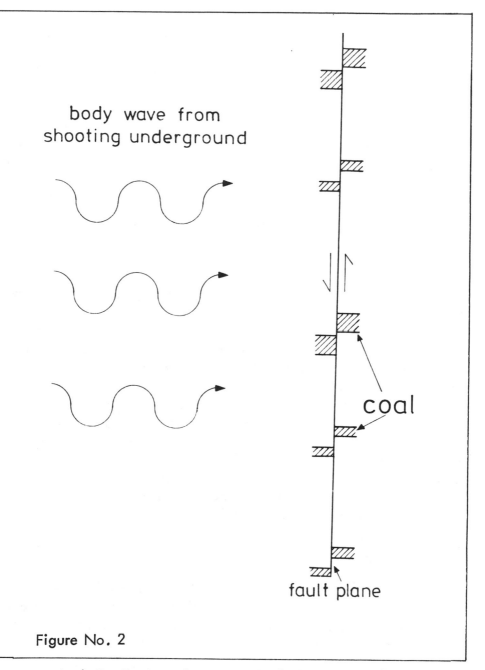

body wave from
shooting underground

coal

fault plane

Figure No. 2

Typical distribution of seams on both sides of a fault plane
of minor throw. Total thickness of seams is about 10 percent
of total rock thickness.

occurred to me that a coal seam should be a wave guide, at least for waves of Love type, where rock and coal particles move parallel to the seam and perpendicular to the direction of the rays, i.e. the movement occurs within the wavefronts, which are vertical to the seam and to the rays (Figure No. 3).

Such guided waves have their largest amplitudes in the seam. In the adjacent rock the amplitudes decay exponentially with the distance from the seam. This decay depends on the wave lengths, i.e. on the frequencies as can be seen in Figure No. 4, which refers to the so-called fundamental symmetrical mode. This mode is the most important one in the following. Obviously, the higher the frequency, the more rapid is the decay. This involves another important fact, i.e. the percentage of energy within the seam increases with the frequency. Another important feature of guided waves is dispersion (Figure No. 5 which results in typical long wave trains as a response to short impulse sources (1) .

There is still another type of waves guided in seams, i.e. the Rayleigh type (Figure No. 4). In this case rock and coal particles move in a plane perpendicular to the seam and to the wave fronts. These waves have similar qualities as those of Love type as far as dispersion and exponential decay of amplitudes off the seam are concerned. But genuine guided waves of Rayleigh type, i.e. so-called non leaking modes which do not permanently lose energy into the country rock, can only exist when the shear wave velocity of the country rock exceeds the dilatational velocity of the coal, and this condition is not always fulfilled (2).

Period of Tests

My feeling at that time was : Waves guided by a seam, or simply seam waves, should be reflected by faults more effectively than body waves because the percentage of energy in the seam is much higher in the case of seam waves.

A series of underground test surveys in mines was agreed upon, first with Tasch, Rheinstahl, then with Kneuper, Saarbergwerke, and finally with Brentrup, Steinkohlenbergbauverein, Essen. The second test ordered by Saarbergwerke and carried out with governmental financial assistance first proved the existence of seam waves, and reflected such waves could be observed too (Figures Nos. 6, 7)

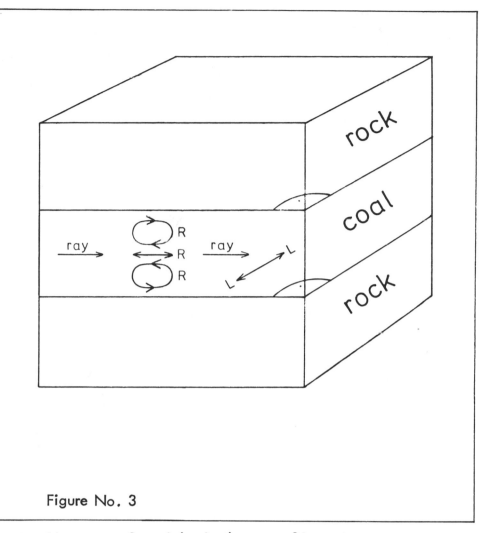

Figure No. 3

Movement of particles in the case of Love-type
seam waves (L) and Rayleigh-type seam waves (R) .

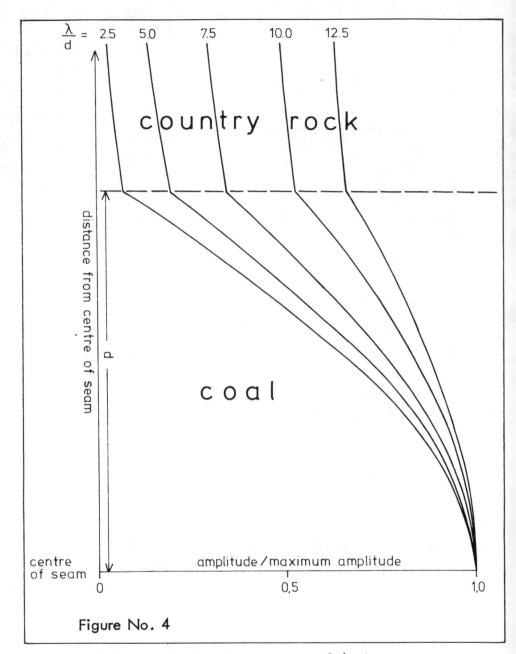

Figure No. 4

Relative amplitudes of seam waves of the Love-type
fundamental symmetrical mode for various wave lengths λ .
d = half seam thickness.
Ratio of velocities 1 : 2.
Ratio of densities 1 : 2.

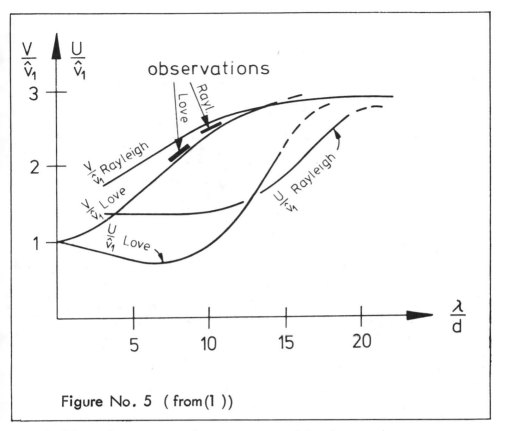

Figure No. 5 (from (1))

Dispersion curves of seam waves of fundamental
symmetrical mode.
λ = wave length
d = half seam thickness
V = phase velocity
U = group velocity
$\hat{v}_1$ = shear velocity in coal

ratio of coal to rock density   1.1 : 2.6
ratio of coal to rock velocity   1 : 3
Poisson's ratio = 0.25.

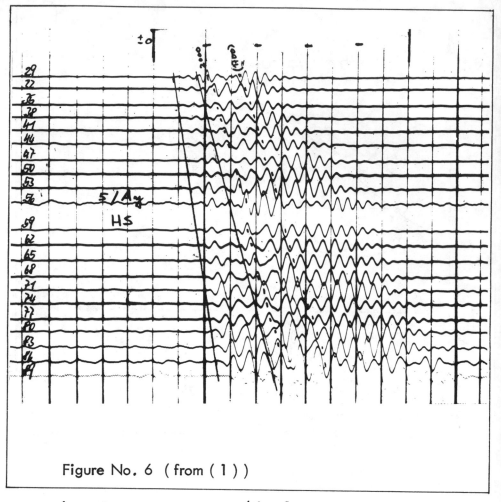

Figure No. 6 (from (1))

Love-type seam waves resulting from a
hammer blow.

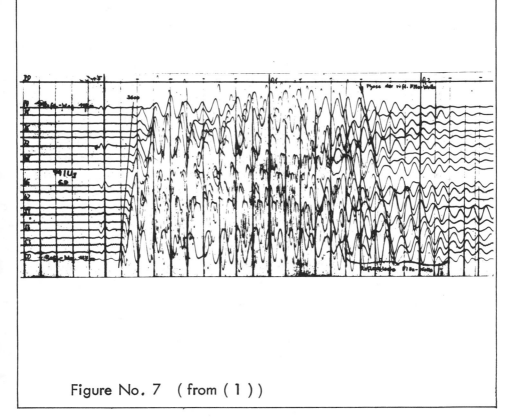

Figure No. 7 (from (1))

Dynamite shot with first observed reflected
Love-type seam wave .

(3), (1), (4). Therefore it was decided to built a recording unit which could be testified to be fire-damp proof. Much to our regret this took several years. But then, in 1966, we could start the 3rd series of tests in the Ruhr area by order of Steinkohlenbergbauverein, also financially assisted by the government. Seam waves and reflected seam waves could be observed in most cases. Agreement between predicted and observed faults was 66 percent according to Brentrup's investigations (5). This means, interpreted reflections were counted positive when they coincide with a fault which lies within an errorzone of + one wave length. All others and missing reflections at a place of fault were counted negative provided the throw of the fault was at least about equal to the thickness of the seam. Distances from the site of measurement which exceeded 100 times the thickness of the seam were excluded from this comparison. The result of 66% was quite encouraging, but not perfectly satisfying.

Very soon it was found out that seam waves could be used in yet another manner, i.e. as transmitted waves (Figure No. 8) ;(5), (6). In case an area of seam, planned for being worked is partly surrounded by either two parallel galleries or by a working face and a gallery more or less at right angles, the continuity of seam waves transmitted through this area can be checked. Seam waves are not observed or become very weak when a fault displaces the seam between source and receiver by an amount greater than or comparable to the seam thickness. The percentage of positive predictions of faults by transmitted seam waves was 83 as stated by Brentrup (5) . This was still more encouraging than the reflection results.

Routine Surveys

As a consequence of these tests some 250 routine surveys have currently been carried out underground in many mines in France, Great Britain and Germany since 1967 by SEISMOS GmbH and PRAKLA-SEISMOS GmbH. Both, reflected and transmitted seam waves were used for these investigations ahead of face (5) , (6) , (7).

Before presenting to you some case histories of this activity the main technical details of this seam-wave method should be outlined in short

The seismic source is normally a shot fired in a hole drilled to a length of 6 to 10 feet (2 to 3 meters) in the centre of the seam and parallel to the seam. In preference explosive of low detonation

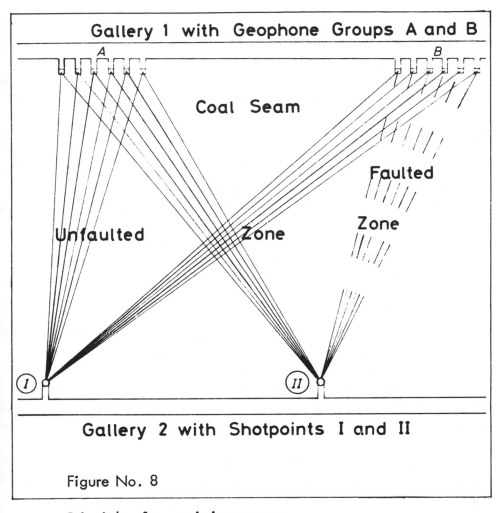

Gallery 1 with Geophone Groups A and B

Coal Seam

Faulted

Unfaulted Zone Zone

I

II

Gallery 2 with Shotpoints I and II

Figure No. 8

Principle of transmission surveys.
No seam waves pass through the faulted zone.

velocity is used. But also hammer blows upon a bar put into the "shothole" are effective in many cases. Pairs of moving coil geophones of a natural frequency of about 27 cycles per second are also planted in holes in much the same way as the shots. They are sensitive for two directions parallel to the seam but perpendicular to each other. Thereby it is possible to discriminate between the various kinds of waves observed, e.g. whether they are of Rayleigh or Love type.

The recording unit (Figure No. 9) used hitherto has 24 amplifiers, and wiggle trace paper seismograms are produced by an oscillograph of 24 traces. Frequency range is up to 500 cycles per second, and 5 high and 5 low-cut filter settings are possible with 2 slopes. No magnetic recording, either analogue or digital, was used due to the difficulties in making such equipment fire-damp proof and due to the lack of commercial availability of smaller sampling rates than 1 millisecond.

In Figure No. 10 you see the interpretation of a seam-wave survey using reflected waves, mainly of Love type. The faults encountered when working the coal agreed pretty well with the locations of the reflecting planes. This includes the splitting of the extended fault running nearly parallel to the gallery of measurement. Reflection from this fault zone are partly masked by other reflecting faults.

In certain mines of northern France working of coal can be seriously impaired by a special kind of washouts called "puits naturels". Here predictions by transmitted seam waves were quite successful. According to Figure No. 11 an area of seam planned for being exploite was nearly completely surrounded by roadways. Thus, the transmission method could well be applied. In this figure the space between shotpoint and geophones is bounded in different manners according to the quality of the transmitted seam waves observed. Straight lines stand for good, dashed lines for fair, dotted lines for questionable to not observe Areas through which good to fair transmitted waves never passed point to serious discontinuities of the seam, either to faults with a throw comparable to the seam thickness or higher or to "puits naturels" as in the present case. Such areas are indicated in Figure No. 11 by heavy and minor dotting. Figure No. 11 shows too that the exploitation of the are really encountered 2 "puits naturels", as indicated by hatching and N.Sch., at about the predicted locations. Here, too, the seam waves used for interpretation were mainly of Love type (8).

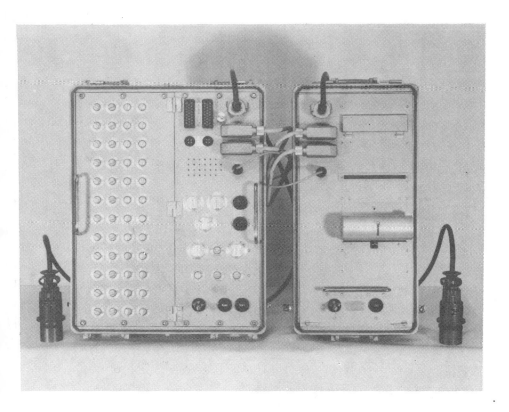

Figure No. 9

Fire-damp proof seismic recording unit GSU,
constructed by SEISMOS GmbH.
Left side module of 24 amplifiers,
right side oscillograph.

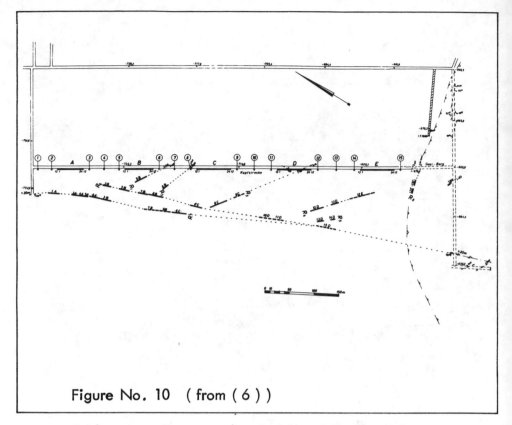

Figure No. 10 (from (6))

Exploration of a zone ahead of face by reflected
seam waves. As to details see text.

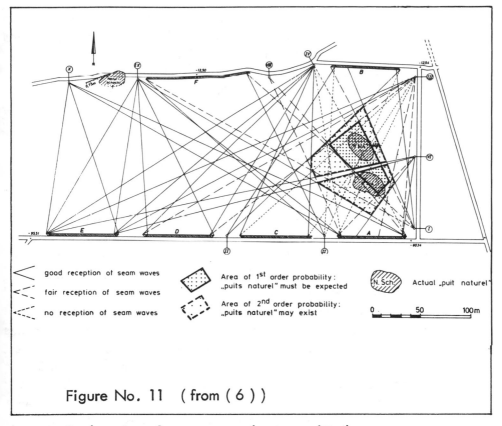

Figure No. 11 (from (6))

Exploration of a seam **area** by transmitted seam waves
using three road ways.

Figure No. 12 presents an example of a combined reflection and transmission survey. In the left part of the figure several reflections detected a fault which proved to be an overthrust with a throw decreasing towards the right. Obviously the transmission of seam waves to spread C was impaired to a certain degree. Here, the throw of the fault mentioned still comprises about half the thickness of the seam

Transmission of guided waves cannot only be used from gallery to gallery or from gallery to face, but also from a hole to a face or from hole to hole. Several years ago we carried out a transmission survey in the Saar region from a hole to a gallery (9) . The hole had been drilled from the surface to a depth of about 3700 feet (1100 meters). The gallery was at a distance of ~ 4000 feet (1200 meters) from the hole and at a depth of about 2500 feet (750 meters). A seam encountered at the bottom of the hole was believed to be identical with that which was worked at the face. Shooting was carried out in the hole and geophones had been planted in the gallery in such a manner that the reception of Love type guided waves was favoured. Some geophones however, were enhancing Rayleigh-type waves for comparison purposes. The result of the survey was that a pronounced transmission of guided waves could be observed. The upper part of Figure No. 13 shows a transmission seismogram shot in the rock of the borehole about 100 feet (30 meters) above the seam. No seam waves of appreciable amplitude can be recognized. In the second seismogram the shot was fired in the central part of the seam. Seam waves of Love type appear pretty well between 525 milliseconds and 650 milliseconds in spite of a considerable amount of noise. For comparison purposes you see in the lowest part of Figure No. 13 how seam waves look at corresponding distances when transmitted from gallery to gallery through a fault-free part of the same seam.

Thus it could be concluded with a high degree of probability that the seam in the hole and the seam in the gallery were identical and that no faults would interrupt the continuity of the seam between the two locations. In the meantime this conclusion has been proved by the mining activity, but by the way, when proceeding beyond the hole without applying any additional geophysical measurements a bad fault was struck which gave the miners some serious problems due to water invasion. The fact that guided waves did not appear when shooting in the country rock at a certain vertical distance from the seam (Figure No. 13 upper part) agrees well with result of the doctor thesis of Guu (10).

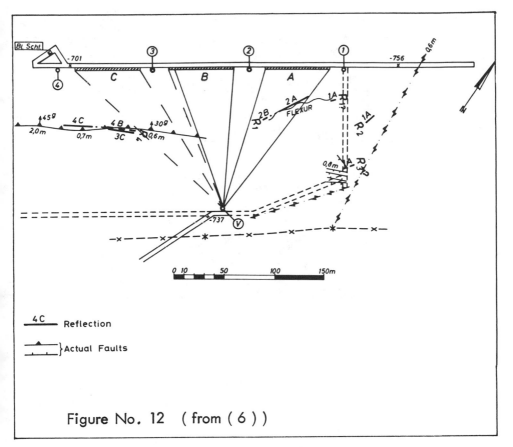

Figure No. 12   (from (6))

Combined reflection and transmission survey.
As to details see text.

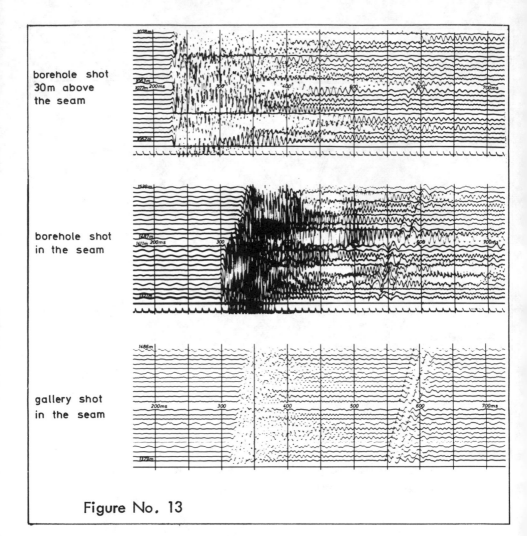

borehole shot
30m above
the seam

borehole shot
in the seam

gallery shot
in the seam

Figure No. 13

Various transmission records.
Geophones in the seam of a gallery.
Uppermost record shot in a borehole from
about 100 feet (30 meters) above the seam.
Central record shot in the same borehole
from the central part of the seam.
Lowest record shot in a gallery from the
central part of the seam.

Some Problems Connected with the Method

After having realized the merits of seam waves in the planning of the exploitation of coal, we now draw our attention to deficiencies and limits of the method, and then we shall discuss what is going on and what can still possibly be done to improve the method.

When assessing the precision of any seismic method we have first of all to consider the effects of diffraction, which depend mainly on the wave length. According to our observations these wave lengths are of the order of 20 to 80 feet (6 to 25 meters). This implies uncertainties in the positioning of discontinuities of the seam of about a similar order of magnitude, though there is always an important dependance on the geometry of the special case at hand too.

Another problem arises from the dispersion of guided waves, i.e. the velocity of seam waves is a function of frequency. Therefore it is difficult to determine a precise arrival time for transmitted and reflected seam waves, and the question is : which velocity has to be attributed to it. One approach to the problem is deconvolution as favoured by our British colleagues (personal communication of Clarke, National Coal Board (NCB)). Arnetzl and myself (11) tried to crosscorrelate transmitted and reflected seam waves with each other provided that both events have approximately the same traveltime. The small difference of the traveltimes should then be equal to the time delay τ_{max} at which the maximum crosscorrelation value occurs. The length of the reflected raypath S_r (Figure No. 14) would then be equal to the length of the transmitted raypath S_t augmented by the product of the time delay τ_{max} just mentioned and the velocity V, but now this velocity must be known approximately only because it is multiplied with a rather small time delay. Therefore any average phase velocity V_{phase} can be used. Of course, such processing presupposes digital recording for routine application. An additional adventage of such a procedure is that it should also improve the recognisability of reflected seam waves because it is nothing else but matched filtering. A drawback is that transmitted seam waves must perhapt additionally be observed at the desired distances, as indicated in Figure No. 14.

We could test this crosscorrelation method at some sites of the Ruhr area, partly after digitizing paper records, which is rather troublesome, partly after digital recording at a place free of CH_4-gas, but this was with a sampling rate of 1 millisecond only. The results were

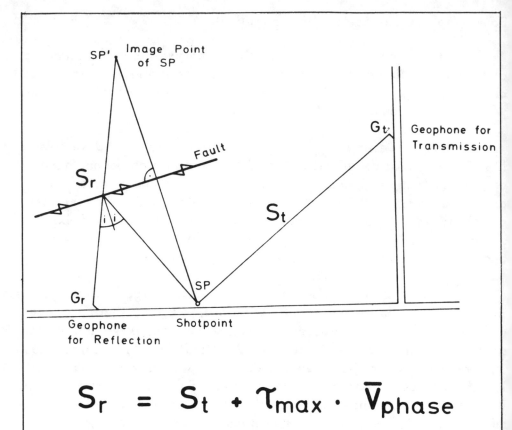

$$S_r = S_t + \tau_{max} \cdot \overline{V}_{phase}$$

S_r length of reflection ray path

S_t length of transmission ray path

τ_{max} time of maximum crosscorrelation amplitude

$\overline{V}_{phase}$ average phase velocity

Figure No. 14   (from (11))

Determination of the lengths of reflection ray paths
by means of cross-correlation.
For details see text.

partly encouraging (Figure No. 15) and partly not too promising. The reason is probably a considerable frequency dependence of the reflected energy of seam waves as proved explicitly with the help of physical models by Dresen (12). Thus, crosscorrelation can probably become more effective when combined with additional filtering by which the reflected energy is enhanced.

We shall now give some more consideration to the reflected energy as a function of frequency. Let us suppose that a seam ends at a major fault and that discontinuous physical properties are only occurring where the seam ends and nowhere else at the fault plane. Then it is quite suggestive that the percentage of reflected energy is mainly dependent on the percentage of energy travelling within the seam, and this percentage increases with frequency as mentioned earlier (Figure No. 4). On the other hand, the attenuation of seismic waves is frequency dependent too, higher frequencies show stronger attenuation than lower ones. Thus, both effects act in opposite directions, resulting in an optimal frequency range for reflected seam waves. This optimal frequency range is, of course, dependent on the length of the reflection ray path, i.e. there is an unfavourable shift towards lower frequencies with increasing length of ray path. These facts are probably the main reasons why reflected guided waves could never be discerned when the reflecting discontinuity, say fault, was farther away than 600 to 900 feet (200 to 300 meters) from the shot-geophone array, even with 3 geophones per trace and 3 hole shots. On the other hand transmitted seam waves can be observed up to a considerable distance from the source, say 6000 feet (2 kilometers). These observations refer to seam thickness of 6 to 8 feet (2 to 2.5 kilometers).

There is another point which calls for being improved, i.e. the source of seismic energy (11). The main useful seam waves are the fundamental mode of the symmetrical Love-type wave. But, with normal shooting in a single hole the percentage of this desired energy is small very often at the beginning. But we found out that subsequent shooting in holes at a distance of 2 to 3 feet (50 to 100 centimeters) from the first shot hole resulted in higher amplitudes of the Love-type seam waves, probably because the first shot destroyed the cylindrical symmetry of physical properties around the second shot hole. This technique can on principal be extended to the enhancement of higher modes too, if desired. E.g. (Figure No. 16), when the first higher symmetrical Love-type mode is to be enhanced, 3 holes for the production shot are arranged in a row perpendicular to the seam at those places where the am-

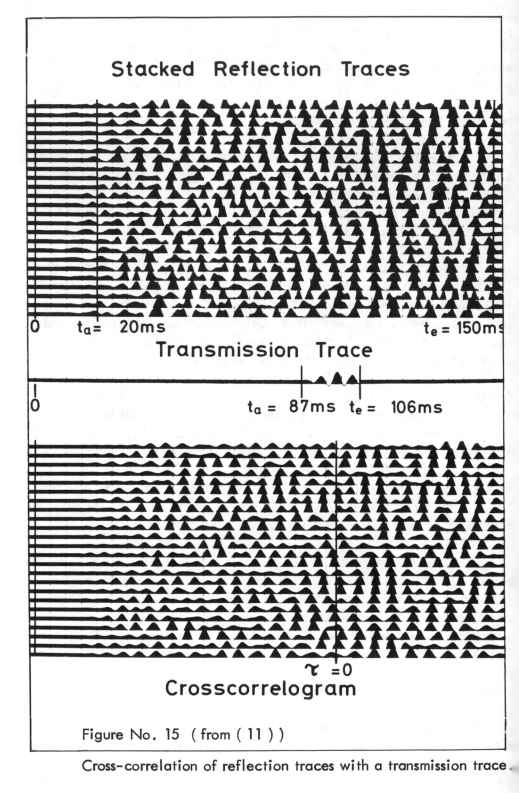

Figure No. 15 (from (11))

Cross-correlation of reflection traces with a transmission trace.

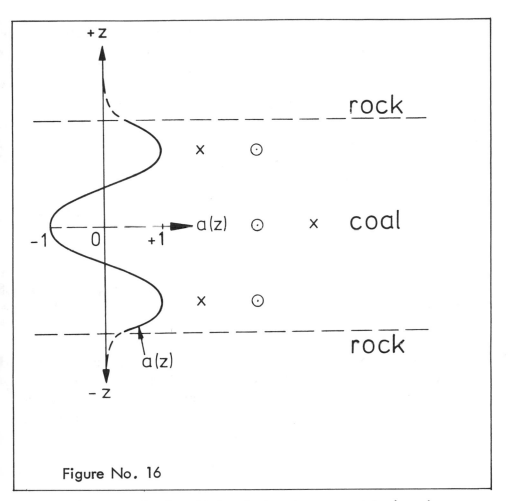

Figure No. 16

Enhancing the first higher Love-type symmetrical mode.

X   = hole for preparatory shot
⊙   = hole for genuine production shot
a(z)   = amplitude of desired mode as a function of the
           distance z from the centre of the seam.

plitudes of the desired first higher mode exhibit extreme values. Then 3 preparatory holes are drilled at the same distances from the centre of the seam as the production shot holes but with a lateral offset of say 2 feet (0.6 meters) applied alternatively to the left and right side of the production shot holes. After having shot the 3 preparatory holes with the purpose to produce the desired inhomogeneities the production shot should preferably produce the first higher mode.

Trends for Improving the Method

Now, what can be done and what has been done to improve the efficiency of the seam wave method ?

As to the detection of reflected waves it is obvious from the preceding considerations that digital recording with a sampling rate of 0.5 milliseconds or less is necessary for any progress. This unit must be fire-damp proof.

The National Coal Board in Great Britain was the first to build such equipment. The main features of this unit are preamplification, multiplexing and frequency modulation underground, transfer of the data by wireline to the surface of the earth with different carrier frequencies then analogue to digital conversion and digital recording in real time. It is possible to have a sampling rate of 0.8 milliseconds. (Personal communication of Clarke, NCB.)

In Germany the Steinkohlenbergbauverein at Essen started another development. A DFS V instrument from Texas Instruments, which enables the recording of 24 traces with a sampling rate of 0.5 milliseconds, is being made fire-damp proof by PRAKLA-SEISMOS GMBH, Hannover, and will probably be used underground in the autumn of this year (1976). This development is sponsored by the government of Nordrhein-Westfalen and the European High Commission.

The additional possibilities arising from the introduction of this digital equipment do not only comprise filtering and crosscorrelation, as mentioned earlier, but also horizontal and vertical stacking, two-dimensional, i.e. inseam, migration, migration-stack, deconvolution and many others. It is hoped that the signal-to-noise ratio can be effectively improved for reflected seam waves and that the range of the detectibility of such waves can be increased from 600 to 900 feet (200 to 300 meters) now to 1000 to 2000 feet (300 to 600 meters) in future. Much to our regret the demands of our mining clients have increased too in the meantime, i.e. to 3000 to 4000 feet (900 to 1200 meters). Experience will have to show whether this range can be reached by the reflection method in future. Anyhow such distances are within the range of observable transmitted waves.

Little has been done to improve the efficiency of the source. Hasbrouck from the US Geological Survey in Denver, and Guu, Colorado School of Mines, developed a mechanical device to enhance Rayleigh-type seam waves in vertical holes drilled from the earth's surface into a seam (13).

It would, of course, be very thrilling to use the Vibroseis method underground at the faces with vibrations parallel to the face and to the seam. The sweep could then be adapted to the range of maximum amplitudes of the reflected Love-type seam waves. The main difficulty would be to build such a vibrator and to make it fire-damp proof.

As to geophones a considerable progress has been achieved quite recently. Rüter of the Westfälische Berggewerkschaftskasse in Bochum, Ruhr area, developed a two-component system measuring the total movement parallel to the seam. This system is planted in holes which are 2 meters long and parallel to the seam. The diameter of the hole has to be about 2 inches (50 millimeters). The geophone is clamped to the wall of the hole by compressed air. The improvement effected by the new system can be seen in Figure No. 17. We hope that this new geophone and digital recording will render a considerable progress in the utility of the seam-wave method.

Rüter also carried out experimental investigations underground attributing much to understanding the nature of seam waves (14), (15).

I am very glad that much laboratory and mathematical research too has been carried out as to the topic of seam waves. This activity is still going on and will hopefully continue. I can only mention some of it.

Tests with physical models were carried out by Klusmann (16), Freystätter (17) and Dresen (12) in Western Germany and are going on in Great Britain (Clarke (NCB) , personal communication). They proved the existence of waves guided by a seam as well as the possibility

IN-SEAM SEISMIC EXPLORATION TECHNIQUES 251

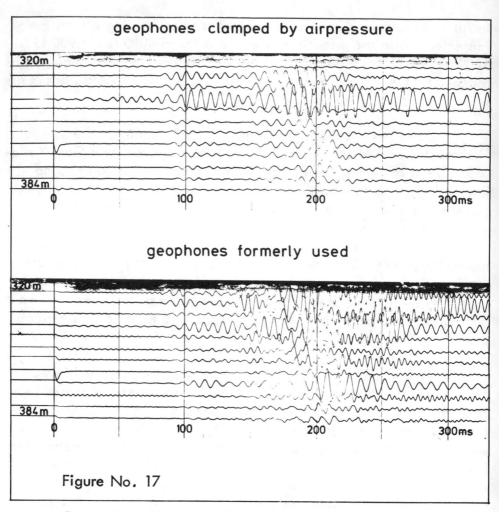

Figure No. 17

Comparison of new geophones (upper 12 traces),
clamped by airpressure, with geophones formerly
used (lower 12 traces).

of getting reflected such waves. The vanishing of seam waves behind a fault could also be shown. But these tests also point to the fact that the reflection amplitude is depending on frequency as mentioned already (12).

Model computations using the finite difference method were carried out by Guu (10). They proved that transmitted seam waves can well be used to test the continuity of a seam.

Mathematical investigations were carried out by Masson (18) in France and by Halliday at the NCB Great Britain, and are going on in London at the Imperial College (Clarke (NCB), personal communication).

Summary

In summarizing, seismic waves guided by a seam, especially waves of Love type, have proved to be useful in the planning of working coal. They have currently been applied in underground seismic surveys for a decade. Both, reflected and transmitted waves have been interpreted. Scientific research and technical progress will certainly improve the method in the future, just as normal exploration seismics carried out at the earth's surface has undergone a tremendous development, and yet it has been useful right from the beginning.

Acknowledgment

I have to thank PRAKLA-SEISMOS GmbH for the permission to present this paper, and I am indebted to various mining companies for the possibility to include results of some surveys. Finally, I thank my colleague H. Arnetzl for his assistance in compiling this paper.

REFERENCES

1.   Krey, Th., "Channel Waves as a Tool of Applied Geophysics in Coal Mining", Geophysics, Vo.XXVIII, 1963, pp.701-714.

2.   Schwaetzer, T. et R. Desbrandes, "Divergences constatees dans la mesure de la vitesse des ondes acoustique longitudinales du charbon", Rev.Inst.Française Petrol 20, 1965, pp. 3-26.

3.   Schmidt, G. and G. Kneuper, "Zur Frage nach der reflexions-seismischen Ortung von tektonischen Störungen in Steinkohlenbergwerken", Glückauf 98, 1962, p.43.

4.   Kneuper, G. and Th. Krey, "Neue Untersuchungen zur reflexionsseismischen Ortung von tektonischen Störungen in Steinkohlenbergwerken", Bergbau-Wiss. 14, 1967, pp.428-430.

5.   Brentrup, F.-K., "Die Reflexionsseismik als Hilfsmittel zur Ortung tektonischer Störungen in Steinkohlenflözen", Glückauf 106, 1970, pp.933-938.

6.   Arnetzl, H., "Seismische Messungen unter Tage", Tagungsbericht "Mensch und Maschine im Bergbau" der Gesellschaft Deutsche Metallhütten- und Bergleute, 1971, S. 133-141.

7.   Baule, H., "Vorfelderkundung mit geophysikalischen Mitteln" Mitteilungen aus dem Markscheidewesen 74, 1967, S.205-228.

8.   Decherf, J., Oral communication at the International Scientific Symposium on Mine Surveying, Mining Geology and the Geometry of Mineral Deposits, Prague, August 1969.

9.   Brentrup, F.-K., "Flözdurchschallung aus Tiefbohrlöchern", Glückauf 107, 1971, S.685-690.

10.   Guu, J.-Y., "Studies of Seismic Guided Waves: The Continuity of Coal Seams", 1975, Doctor Thesis, Colorado School of Mines.

11.   Arnetzl, H. and Th. Krey, "Progress and Problems in Using Channel Waves for Coal Mining Prospecting", paper presented at 33rd EAEG-Meeting, Hannover, 1971.

12.   Dresen, L. and St. Freystätter, "Rayleigh Channel Waves for the In-Seam Seismic Detection of Discontinuities", Journal of Geo-physics-Zeitschrift für Geophysik , 42, 1976, part 2.

13.   Hasbrouck, W. P. and J-Y. Guu, "Certification of Coal-Bed Continuity Using Hole-to-Hole Seismic Seam Waves", paper presented at 45th SEG-Meeting in Denver, 1975.

14.   Rüter, H., "Messungen zur Bestimmung des Frequenzinhaltes von Flözwellen im Flöz O der Schachtanlage Auguste-Victoria im Auf-trage des Steinkohlenbergbauvereins",Institut für Geophysik, Schwin-gungs- und Schalltechnik der Westfälischen Berggewerkschaftskasse, Bochum, 1973.

15.   Rüter, H., "Messungen zur Bestimmung des Frequenzinhaltes von Flözwellen im Flöz Gudrun, Schachtanlage Prosper IV", Institut für Geophysik, Schwingungs- und Schalltechnik der Westfälischen Berg-gewerkschaftskasse, Bochum, 1973.

16.   Klusmann, J., "Untersuchung über die Ausbreitung elastischer Wellen im flözartigen Schichtverband", Dissertation, Clausthal-Zeller-feld, 1964.

17.   Freystätter, St., "Modellseismische Untersuchungen zur Anwendung von Flözwellen für die untertägige Vorfelderkundung im Steinkohlenbergbau", Bericht Nr. 3 des Instituts für Geophysik der Ruhr-Universität Bochum, 1974.

18.   Masson, J.-L., "Les Ondes de Couche - Théorie et Appli-cations", Publication Cerchar No. 2207, 1972.

9

GEOPHYSICAL EXPLORATION TECHNIQUES

APPLIED TO WESTERN UNITED STATES COAL DEPOSITS

Wilfred P. Hasbrouck

Coordinator of Coal Geophysics Research Project

U.S. Geological Survey, Denver, Colorado USA

Frank A. Hadsell

Professor Geophysics, Colorado School of Mines

Golden, Colorado, USA

ABSTRACT

Coals at strippable depths in the Western United States
generally are thick, low grade, often severely parted and
split, and variable in thickness within short distances.
Also, outcrops over vast stretches of the Western Plains are
either concealed or burned. For deposits such as these, dri
ing and stratigraphic correlation have been the mainstays
of most pre-mining programs. Now, however, geophysical techniques present an economically attractive means of augmentin
geologic and borehole data: by reducing the number of drill
holes required to delineate and evaluate a deposit, geophysical techniques can decrease overall costs of coal exploration. Results from the joint coal geophysics research program of the U.S. Geological Survey and the Colorado School o
Mines indicate that for strippable coals of the Western Unit
ed States: (1) high-precision gravity surveys can be used t
locate cutouts of thick coal seams; (2) burn facies can be
mapped effectively and quickly with magnetic methods; (3)
seismic seam waves can be observed when seam boundaries are
well defined; and (4) combination of borehole logging, seism
seam-wave certification, and shallow seismic reflection tech
niques is the preferred geophysical exploration method when
precise mapping is required.

INTRODUCTION

Low-sulfur coals of the American West are becoming increasingly important in electric power generation, a trend
expected to continue at least to the year 2000. Many look
to them as major sources of synthetic gas, liquid fuels,
lubricants, and a host of hydrocarbon compounds for centuries
to come (1). The demonstrated-resource tonnage of the
strippable-depth deposits for the tier of eight western
states southward from North Dakota is 9×10^9 metric tonnes
(Paul Averitt, U.S. Geological Survey, oral communication).
Thus a present and potential market exists, and a resource
ample to meet the demands of the market lies waiting - a
condition not unnoticed by many energy companies.

Although coals of the Western States are abundant, their
mode of occurrence makes accurate determination of the reserve base challenging and costly. Generally these coals
are distributed discontinuously in isolated structural basins, Cretaceous and younger beds often are parted and split,
seam thickness may vary markedly within short distances,
and outcrops are scarce across vast featureless stretches
of the high plains. Where lower-rank coals have been exposed by rapid erosion, they often have burned spontaneously
and only clinker beds remain as evidence of their former
outcrops. Surface geological examination is thus limited,
causing much reliance to be put upon drill data obtained at
fairly short spacings, an expensive and time consuming
operation.

This paper describes the results of our research effort
over the last few years to develop geophysical techniques
to reduce the number of drill holes required for evaluation
of both coal resources and reserves within the Western United
States.

LOCATION OF CUTOUT EDGE WITH A HIGH-PRECISION GRAVITY SURVEY

Because the density of coals is low relative to the density
of the rocks which surround them, gravity surveys appear to
offer simple, straightforward means of locating the edges
of cutouts at strippable depth. Only a few drill holes
would be needed to verify the geophysical interpretation.
Density data obtained from logging these holes with a
compensated-density tool could be used to refine the interpretation and build confidence toward extending the surveys
into areas of meager drill-hole control.

An elliptically shaped cutout in Section 17, T. 51 N., R. 72 W., Campbell County, Wyoming (2) was selected for a feasibility study of cutout-edge detection by gravity method Coals in this area are approximately 30 metres thick and 60 metres deep. Density logs from comparable sections indicate a density contrast of -1.0 grams per cubic centimetre is reasonable. The anticipated maximum gravity anomaly under these conditions is 1.3 milligals with quarter-anomaly point located about 75 metres to either side of the inflection point on the gravity profile.

Results of the gravity survey are shown in figure 1. The station spacing was 30.48 metres along all traverse lines, and readings were taken with a high-sensitivity gravity mete whose scale constant was 0.0755 milligals per scale division The meter was read repeatedly at each station until the sample standard deviation within a set of observations did not exceed 0.1 divisions (less than 0.01 milligal). The maximum traverse closure error in station elevation was 0.12 metres, equivalent to an elevation error of 0.027 milligals. Difference in terrain effect along each profile was insignifica relative to the size of the anomalies measured. Since the combined maximum instrumentation and reduction error would shift the contours by only one half their 0.1 milligal contour interval, the map would be little affected by these errors.

The maximum gravity anomaly observed was within 0.1 milli gal of what was predicted, suggesting that the assumptions of density contrasts, depth to the top of the seam, and seam thickness were realistic.

From visual inspection of the gravity map it appears that the boundary between the coal and no-coal regions roughly parallels the 8 gravity unit (0.8 milligal) contour. Decrease in slope of gravity values along the west border of the area could be produced either from thickening of overburden or loss of abruptness of the coal's leading edge. Regardless of the cause of loss of slope, the alignment of inflection points nevertheless has a definite trend which on the west and southeast edge of the area is more suggestiv of a channel rather than an elliptical cutout.

From the gravity study in this area we can draw the following conclusions: (1) gravity mapping appears capable of delimiting the edge of a thick-seam cutout, (2) the edge of a seam one-third the thickness of the existing seam would

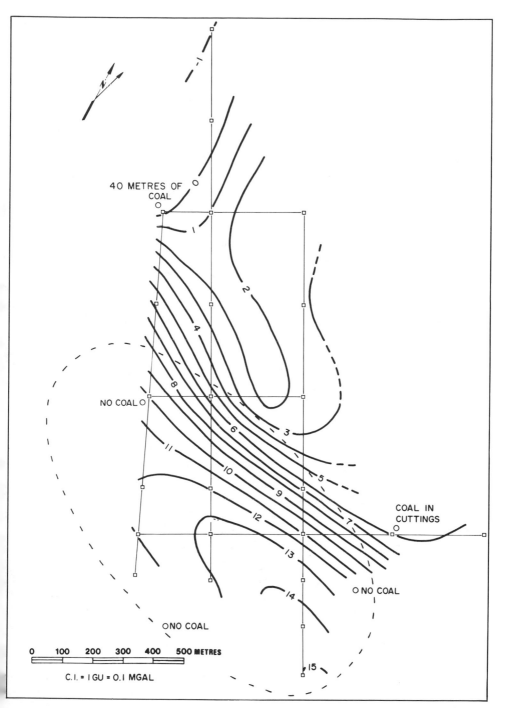

Figure 1. Gravity map spanning the north edge of a suspected elliptical cutout in Campbell County, Wyoming. Cutout edge mapped from drill data is outlined by the dashed line. Station spacing was 30.48 metres, and stations shown by squares were those used to affect drift corrections.

be marginally detectable, (3) the gravity map can be used as
a guide for optimizing subsequent drilling, and (4) a station
spacing of 60 metres would be sufficient to reveal the anoma

LOCATION OF BURN FACIES WITH A MAGNETIC SURVEY

Spontaneous burning of coal beds along and back from thei
outcrops has produced large masses of clinker throughout the
Powder River Basin of Montana and Wyoming. These baked rock
may occur in patches several kilometres wide or they may be
localized to ridges and buttes. Their characteristic red-
brick color and their resistance to erosion make them stand
out in what would be otherwise featureless terrain. To the
coal geologist, the clinker serves as an indicator that coal
is (or at least, has been) in the area, as a marker for mapping, and as an approximator of the thickness of the burned-
out seam; the thickness of the baked zone generally ranges
from three to five times the original thickness of the burne
coal bed (W. J. Mapel, U.S. Geological Survey, oral communication).

Presence of clinker does not necessarily condemn a coal
prospect. Many of the Western State coals occur in multiple
seams with only the upper seam having been burned. Also,
with the exception of thin overburden deposits for which the
entire seam might be taken by burning, the invasion of the
fires inward from the outcrop usually is limited, stopping
when the supply of oxygen is exhausted. As the coal burns,
its volume is reduced, and the roof above it weakens and
finally collapses. New passageways are thus opened through
which air is supplied to maintain the burning. Near the bur
front an oxygen-starved or reducing environment may exist;
but close to the surface, oxidation takes place freely and
produces the characteristic reddening of the clinker beds.
If the collapse fracture system is extended almost verticall
upward from the advancing burn facies, then soil reddening
can be used to map the distance of burning back from the
outcrop. However, if the roof rock is highly competent,
burning may proceed much further than the soil-color-change
line would indicate. Also, erosional debris may subsequentl
cover the reddened area and mask surface evidence of the bur
limit. Near Kemmerer, Wyoming a reddish sandstone overlies
some of the coal measures thus reducing the reliability of
colored aerial photographs as aids to mapping burn facies.

Baking not only changes the color of the rocks but also
alters their magnetic properties. Laboratory tests by

Donald E. Watson (U.S. Geological Survey, oral communication) indicate that when common sediments in the area are baked and then cooled a 6000-fold increase in magnetization may occur. Rocks nearest to the burn front and thus subjected to a reducing atmosphere appear to be most magnetically changed, some become more magnetic than basaltic lavas. Watson's experiments also show that it is not necessary to heat all the rocks above the Currie temperature of magnetite (580°C) before significant changes in their remnant magnetization are produced upon cooling. Rather, some of the samples acquired high magnetizations when cooled from a temperature of only 200°C. Watson's work therefore indicates that a large volume of highly magnetized material can be created by baking. Burning may continue for a century or more and temperatures of over 1100°C may be attained, as evidenced by flow structures seen in those rocks that were originally part of the roof. Therefore, with sufficient coal and time for a 200°C isotherm to have moved a considerable distance outward, a large volume of overlying material can be altered magnetically. That a process somewhat like the one outlined above has occurred is indicated by the large magnitude of the anomalies (some as much as 2000 gammas) that we have observed in the field.

Figure 2 shows a total-field magnetic profile obtained near a clinker quarry. There is little doubt that the left side of the magnetic profile differs significantly both in magnitude and spatial wave number from its right side. At a traverse distance of 125 metres soil reddening disappeared. The burned-area edge would have been mapped at this position if only surface observations had been used. Magnetic interpretation places the burn facies at a traverse distance not in excess of 250 metres. Clearly the burned area extends further than the position of the color-change line would indicate. The distance between the burn facies location using surface instead of magnetic evidence is about 125 metres. This translates into a difference in reserve estimate for a 20-metre thick coal seam along an outcrop distance of 300 metres of about one million metric tonnes. Topography in this area is gently rolling, and the base of a hill rising to the north is located at a traverse distance of about 100 metres. Material from this hill could have washed down and covered the baked area out to a traverse distance of 125 metres, but we have no drill data to support this speculation.

Figure 3 shows a total field magnetic profile obtained at an active mine near Kemmerer, Wyoming. The objective of

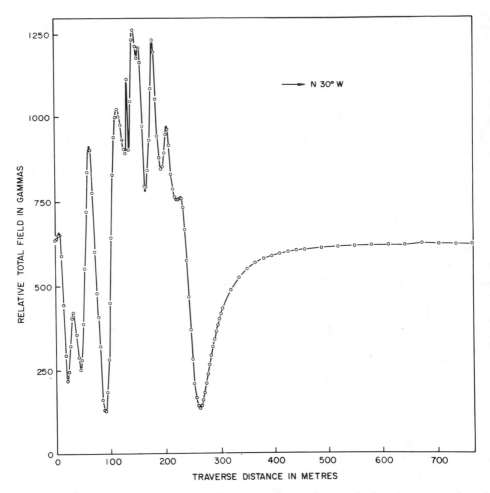

Figure 2. Magnetic profile across a burned facies north of a clinker quarry. Station locations are shown by open circles.

this study was to see if the down-dip edge of the burned zone could be located magnetically. Surface evidence for the 3-metre thick burned bed in this heavily brush-covered area consists of a break in topographic slope, a scattering of clinker detritus at the base of the hill, and a slight reddening of a sandstone overlying the coal. Thus, with careful geologic field work, the burned outcrop can be located.

The standard procedure at this mine for determining the extent of burning is to locate the burned-zone outcrop with surface mapping and then drill until the down-dip edge of the burned section is found. Our results indicate that a

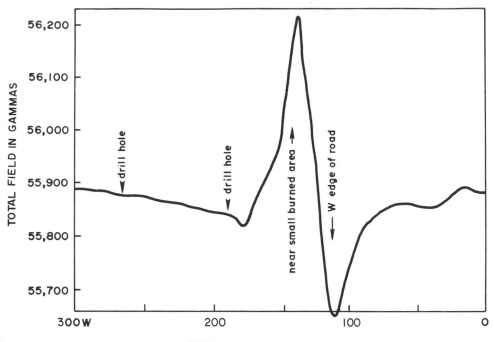

Figure 3. Magnetic profiles spanning the outcrop and down-
    dip extension of a burned zone within a sedimentary
    sequence which dips westerly at 18 to 20 degrees. Station
    spacing was 7.6 metres and traverse orientation was east-
    west.

better way would be to use magnetic profiles to guide the
selection of the optimum drill sites.

The position of the down-dip extension of the burned zone
along the traverse of figure 3 was estimated upon quick inspection of the magnetic profile to be at a distance of no
more than 180 metres west of the traverse origin. Upon return to the mine office, it was determined that the drill
hole at a traverse distance of 190 metres did not encounter
any burned material, only coal. For preliminary planning
of the mine, confirmation of the magnetic interpretation
within 17 metres was acceptable.

Data for the results displayed on figure 3 were obtained
with a proton magnetometer with digital readout in gammas.
To lay out the 41-station traverse, take and reduce the
readings, plot the data in a field notebook, and make an
initial interpretation took less than 2 hours. With experience in the application of the magnetic method to a
specific target, the number and length of traverses can be
optimized (for example, on the profile of figure 3, about

half of the traverse could have been omitted without affecting the interpretation) and thus productivity can be increase
greatly. Magnetic surveys are made on foot; therefore, they
can be conducted over terrain and under weather conditions
which would severely limit other geophysical operations.
The cost of a magnetic-survey is minimal, particularly at a
developing mine where surveying control exists. One-month's
rental of a magnetometer roughly equals the cost of drilling
one 60-metre hole. Assuming traverse lines have been previously laid out and surveyed, a production of 128 kilometre
of profiles per month could be anticipated, sufficient to
examine 16 kilometres of outcrop with 244-metre long profile
separated by 30 metres.

Currently a set of borehole assaying experiments is being
developed to study the variation in coal quality as a function of distance from the burn front. From these tests we
hope to obtain empirical relations which will allow prediction of how far the drill hole is situated from the burn
facies. If this scheme works, then after logging the drill
hole whose site was selected from the magnetic profile, only
this one hole would be needed to establish the burn front
position. Test holes also are to be logged magnetically.

To obtain an areal tracing of a burn-front boundary,
magnetic data were taken along a set of seven parallel profiles whose profile-separation distance was approximately
equal to the station spacing. Figure 4 shows the magnetic
profiles along each traverse and the positions of these
traverses relative to the surficial expression of the burned
zone. With the exception of traverses E and F, comparison
of the edge of the burned zone as mapped geologically (on
the basis of change in soil color) to that interpreted con-
servatively from the magnetic data is in agreement with 15
metres. The cause of the discrepancy between the magnetic
and geologic locations of the burned-unburned interface
crossed by the E and F traverses cannot be ascertained without drill data; however, the suggestion from the magnetic
profiles is that a multiple-burn section may exist northeasterly from the dotted line position. The dot-dash line
connecting the inflection points at the north edge of the
anomaly represents an interpretation of the least distance
of burning. Along each profile the change in soil color
occurred down the hill (northeasterly) from the position
shown by the dot-dash line suggesting that some of the reddish ground in this semi-arid environment had been transported down slope by rain-splash erosion or by sheet flooding. The line joining the last minima outward from the

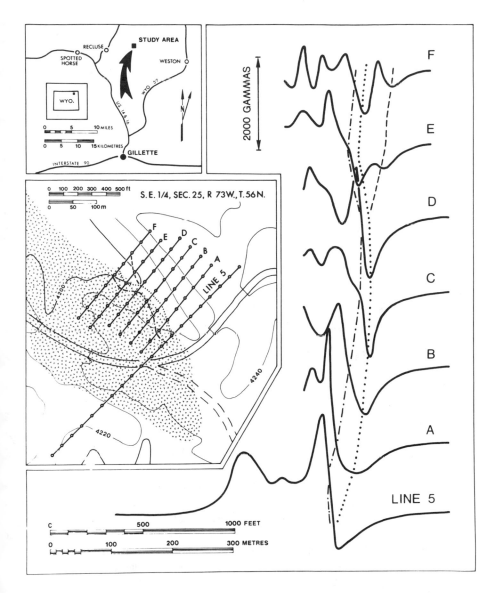

Figure 4. Series of magnetic profiles in a clinker region within the Powder River Basin of Wyoming. Surface expression of the burned zone is shown by the stippled area. Dotted line represents the furthermost indicated extent of burning from the magnetic profiles; dot-dash line is the interpreted edge of the burn zone. The dashed line through profiles E and F indicates the burn-facies position in a possible partial- or multiple-burn region.

anomaly (the dashed line on profiles E and F and the dotted line on the other profiles) represents the interpreted amount of the maximum possible extent of burning. It is characteristic of magnetic anomalies in the northern middle latitudes that their north sides show a sharp minimum and their south sides exhibit a gentle rise. This behavior is illustrated on profile 5 (fig. 4) which overlaps the entire burned section. Interpretation on the south side of the anomaly assumes that the position of the inflection point is an acceptable approximator for marking the southerly boundary of the burned zone.

Although many more magnetic profiles than those shown in this paper have been run, the three sets displayed are sufficiently representative to illustrate the following conclusions: (1) a large magnetic effect is produced by the baked rocks, anomalies in the hundreds of gammas (nanotesslas) are common; (2) the spatial wave number (the number of peaks or troughs within a given distance) is considerably higher over the clinker than off to the side, thus the usual assumption of uniform magnetization with the anomaly-producing body does not hold; and (3) the magnetic method can be used to locate burn facies.

THE SEARCH FOR HOLE-TO-HOLE SEISMIC SEAM WAVES

Seismic seam waves, also known as trapped or channel waves are seismic waves trapped within a layer with a low velocity and or density, such as a coal seam. Beyond a near-field region, which we call "the spawning ground", the seismic waves are totally reflected back into the seam from both the top and bottom of the seam, and particular sets of the multiply reflected waves constructively interfere. Therefore in the far field, waves of specific dispersive characteristic are produced. Trapping is most efficient when the seismic source is within the seam; detection is accomplished best when the seismometers are also within the seam.

When continuity of the seam is broken (say by a mine working, or by a cutout, or by total faulting of the entire seam), modeling shows that the seam wave will be disrupted (3). Intuition indicates that if seam boundaries are poorly defined, then some of the seam-wave energy will escape. Other expectations are: minor blockage of the seam will produce partial reflection of the seam wave, thus less of its energy will be transmitted down the seam; and, if a seam thins or thickens, then the dominant frequency of the amplitude spectrum of an observed seam wave will shift upward or downward respectively and an attendant loss in amplitude may occur. Therefore, the seam waves may contain diagnostic properties which when detected and interpreted might lead to reasonable predictions of the behavior of coal seams which have suffered minor faulting and have undergone changes in thickness.

There is a great need for a geophysical technique with as much promise as that held forth by the hole-to-hole seismic seam-wave concept. The expectations that channel-wave methods will be developed and put into routine practice appear to be well founded because: (a) elastic wave theory indicates channeled, or trapped, waves can exist (4); (b) model studies, using both physical devices (3) and the digital computer (5) and (6), support the theory and extend the investigations into those areas for which closed-form solutions become intractable but for which trapped waves nevertheless may exist; (c) channel waves have been reported, first in the well known SOFAR channel (7) and then later underground in coal mines by Krey's classic work (4); and (d) digital seismic recording equipment of sufficient dynamic and frequency range to capture seam waves now is commercially available. Seam-wave technology has been applied underground by Prakla-Seismos at more than 120 mines with outstanding success, 90 percent of seam-continuity predictions having later been verified (H. Arnetzl, written communication). However, we know of no instances where hole-to-hole seismic seam-wave methods have been used to assist in the solution of coal problems in the western United States.

The prime objective of the U.S. Geological Survey and Colorado School of Mines cooperative program is to develop hole-to-hole seismic seam-wave methods to the point that they will be accepted as standard techniques. Review of the history of geophysics indicates that no method, regardless of how well it performs at one location, will necessarily do as well at all locations. Thus, the secondary objective of the program is to identify those areas where seam-wave methods are not effective, and then find the reasons why.

The joint program is an integrated mixture of model, theoretical, and field studies. True to the traditions of research, it has taken more time than we had originally planned to develop the necessary tools, mathematical as well as instrumental. After two years, we now believe we have a good grasp of how to proceed. In the sections of this paper which follow, the results of this preliminary work are summarized.

Digital Computer Model Studies

Model studies made on the digital computer by Guu (5) employed a finite difference computer program whose basic structure was developed by Leitinger (8) and Darken (9), using algorithms of Alterman and Karal (10). Guu's modeling technique is applicable to a limited number of layers of infinite extent, and the characteristics of the source function are selectable. Two source types are used, one simulating an explosion and the other simulating a vertically directed hammer blow.

Figure 5 shows seismic arrivals as if they had been detected by a vertical array of both vertical and radial component seismometers 20 metres from the hole containing a seam-centered explosive source. The seam waves are the highe amplitude events contained within arrival times from 20 to 35 milliseconds. Note the considerable difference in amplitude detected by seismometers located inward from the edge of the seam as contrasted to the signal levels seen by seismometers located at and beyond the seam boundaries. This effect is even more strikingly demonstrated with radial component geophones. The "first arrivals" of exploration seismology though barely visible have an arrival time of about 8 milliseconds. These body wave arrivals are apparent when the printer output of the program is viewed; however, if the amplifier gain of the model, so to speak, were increased enough to show these early arrivals, the seam-wave amplitudes would be so great as to extend off the paper.

Use of a simulated vertically directed hammer-blow source produced the results shown in figure 6. Maximum peak-to-peak amplitude detected by the in-seam radial component seismometers exceeded that from a seismometer located 0.3 metres outside the seam by a factor of 26. In contrast to the almost complete cancellation of arrivals detected by a vertical component seismometer at the center of the seam excited by an explosive source (fig. 5), with the vertical source, there is a noticeable amount of seismic activity recorded from the seam-centered position.

These results from the model reinforce the need in future work to construct a down-hole source whose nature of force application and whose frequency content can be controlled. In the limited space of the borehole, 10 centimetres wide for example, and at the depths to which some of our experiments have been made, as deep as 150 metres, the development of these special sources will be a challenge. Note on

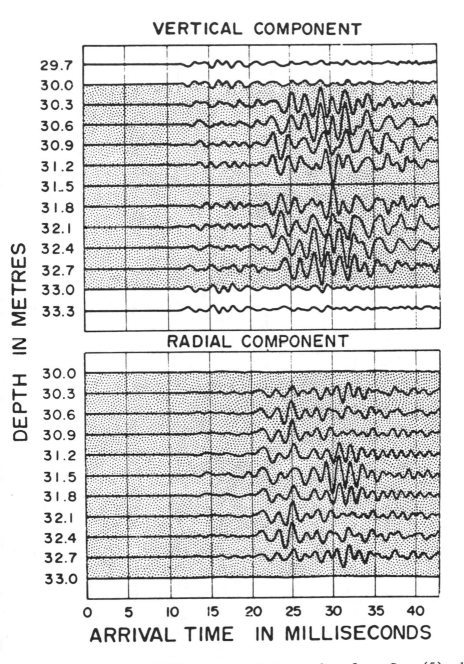

VERTICAL COMPONENT

RADIAL COMPONENT

DEPTH IN METRES

ARRIVAL TIME IN MILLISECONDS

Figure 5.   Finite-difference model results from Guu (5) showing seam-wave events detected at vertical arrays of seismometers 20 metres from a seam-centered explosive source.
   Density, longitudinal velocity, and shear velocity of
   coal and bounding layers are respectively: 1.2 and 2.6
   grams per cubic centimetre, 1200 and 3600 metres per
   second, and 693 and 2080 metres per second.

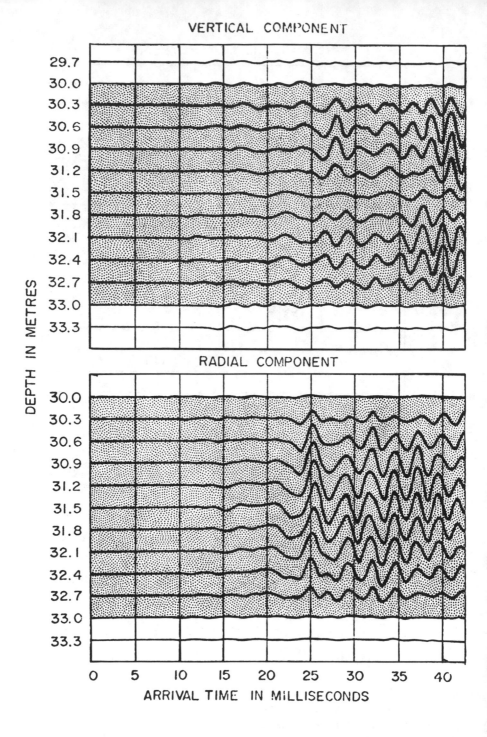

Figure 6.   Finite-difference model results from Guu (5) as
            generated by a vertically directed hammer-blow source.
            Other model parameters are identical to those used to
            produce the model results shown in figure 5.

figures 5 and 6 the source-generated differences in response, implying that either the frequency characteristics of the source must be monitored or a seismic source of highly reproducible output must be developed. Our experience indicates that hanging a sturdy detector in the same hole and above the source is not a satisfactory means of obtaining needed data on the nature of the output source. Not only do tube waves enter to distort the source spectrum, but also in the region immediately surrounding the source, the material may very likely behave non-elastically; that is, it would act as a non-linear element. Though it would be scientifically interesting to monitor source characteristics in a neighboring hole, development of a reproducible source is more economical.

That seam thickness, keeping all other model parameters the same, has a linear effect on observed frequency of the seam-wave arrivals is illustrated in figure 7. In this model, seam thickness was 9 metres; in the previous models, seam thickness was 3 metres. A simulated hammer-blow source, placed at the center of the seam, was used. Energy was again trapped within the seam, the amplitude of the arrivals dropping from inside to outside the seam by a factor of 19 and 10 for radial and vertical displacements respectively. The dominant frequency, about 100 Hz, corresponds to the second mode of the channel wave and it is approximately one-third that observed for the 3-metre thick seam. Higher frequencies, those near 400 Hz, are unreliable and are symptomatic of convergence problems due to the low shear velocity of coal, a problem which has been solved by using larger computers and faster data handling subroutines.

Figure 8 shows seismic arrivals for a system composed of two 3-metre thick coal seams separated by a 3-metre parting. A simulated explosive source was placed in the middle of the upper seam, and both radial and vertical seismometers were 20 metres from the hole containing the source. This is an encouraging display for it shows the seismic seam wave is to a large measure trapped within the seam containing the seismic source. No arrivals with either the characteristic frequency or the anticipated phase velocity are visually evident in the lower seam.

Although it is certainly true that Guu's finite difference models do not exactly depict real-earth responses (no one thought that they would), they nevertheless are an extremely significant step toward our understanding of seam-wave

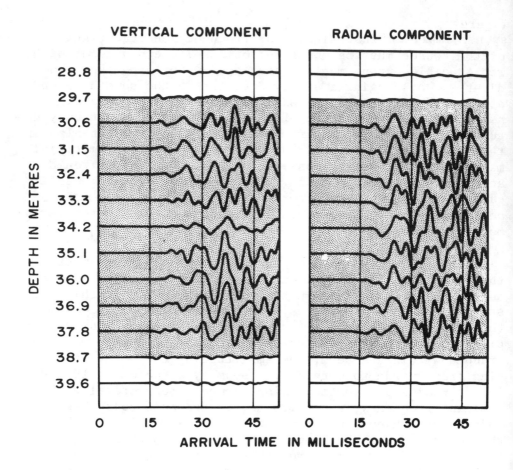

VERTICAL COMPONENT   **RADIAL COMPONENT**

DEPTH IN METRES

28.8
29.7
30.6
31.5
32.4
33.3
34.2
35.1
36.0
36.9
37.8
38.7
39.6

0   15   30   45   0   15   30   45

ARRIVAL TIME IN MILLISECONDS

Figure 7. Finite-difference model results from Guu (5) for
a 9-metre thick coal seam. Density and velocity parameter
are the same as those used in the models whose results
were shown in figures 5 and 6. A vertically directed
hammer-blow source was positioned at a depth of 34.2 metre
and at a hole-to-hole distance of 19 metres from the
vertical array of seismometers.

behavior. It is also equally true that if the model results
did not show the seam waves, then the incentive to go to the
field and look for them would be dulled.

Three doctoral candidates (Messrs. Su, Yang, and Peterson)
at the Colorado School of Mines are preparing theses on com-
puter-assisted elastic wave modeling of those waves thought
to have potential usage in coal geophysics.

Fu-Chen Su has incorporated the finite-element approach
into the previously mentioned finite-difference program to

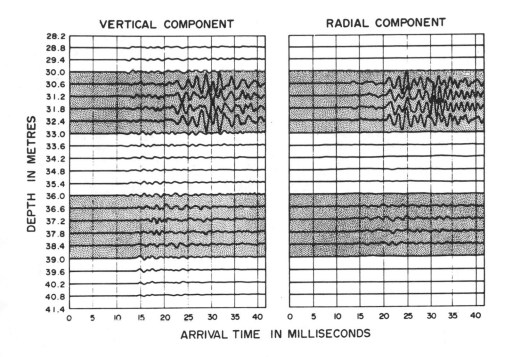

VERTICAL COMPONENT

RADIAL COMPONENT

DEPTH IN METRES

ARRIVAL TIME IN MILLISECONDS

Figure 8. Finite-difference model results from Guu (5) showing seismic waves within two 3-metre coal seams split by
a 3-metre parting. Seismic energy was generated explosive-
ly from a source 31.5 metres deep, the center of the upper
seam; shothole and receiver-hole separation was 20 metres.
The source and its position, the densities, and the velocities are the same for this model as for the one of figure
5.

obtain a hybrid modeling program which has most of the advantages of both methods with little increase in cost. The
upper part of figure 9 shows the preliminary model used by
Su to test his modeling technique and to begin his study on
seam-wave behavior within a partially faulted bed. Velocity
and density parameters are the same as those used by Krey (4)
and later by Guu (5). Seismic records obtained from the two
vertical arrays of radial component seismometers at offset
distances of 8 and 14 metres are shown in the lower part of
figure 9. A simulated explosive source was used. Note the
higher amplitude arrivals within the seam boundaries. From
visual inspection of the seam-contained arrivals, there appears to be little difference in frequency content between
the data obtained ahead of and behind the fault. The seismic
traces displayed on figure 9 are normalized within each data
set against the trace with the maximum range of excursion;

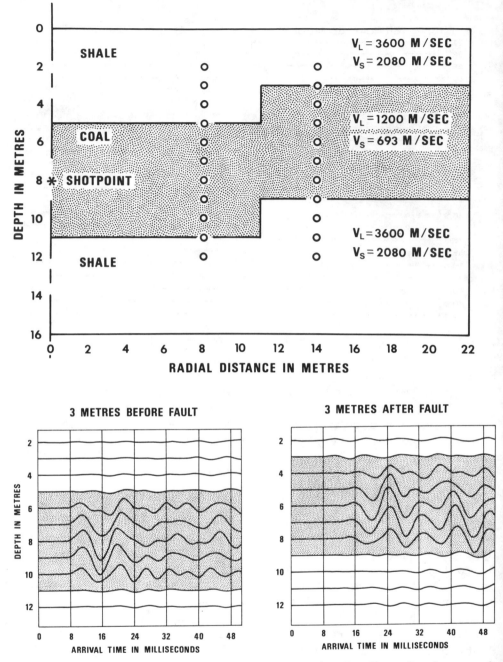

Figure 9. Model and results obtained by Fu-Chen Su in testing his hybrid modeling program. Densities of coal and bounding shale are 1.2 and 2.6 grams per cubic centimetre respectively. Horizontal-component seismometers were positioned along the vertical arrays (shown by the open circles) at 8 and 14 metres from a seam-centered explosive source.

therefore, though relative amplitudes are maintained within each 11-trace record, amplitude comparison cannot be drawn between the two separate records. From maximum-range data listed on the computer print-out, the seam-centered seismometer beyond the fault was shown to have detected a decrease in maximum amplitude range of 57 percent between it and the seam-centered detector ahead of the fault. Without the fault, the loss in amplitude detected between the same radial distances was 31 percent. Thus, the presence of the fault decreased the amplitude by a factor of almost two. These computer-model results suggest that amplitude could be very important in the study of seam waves; thus, seismic instrumentation designed to look almost exclusively at shape and arrival time of events would be inadequate, particularly if automatic-gain-control circuits were employed.

Using his hybrid modeling programs, Su also has begun investigations of the effects on seam waves produced by changing model parameters. In his first test series, he fixed the coal and shale densities at 1200 and 2600 kilograms per cubic metre, respectively; maintained the longitudinal velocity of the shale at 3600 metres per second; and kept the shear velocity of coal at 693 metres per second, the same densities and velocities as used by Krey (4) and Guu (5). Only the shear velocity in the shale and the longitudinal velocity in the coal were changed from their Krey values of 2080 and 1200 metres per second (test A) to 1700 and 1800 metres per second (test B) and then to 2080 and 3600 metres per second (test C), see figure 10. Although seam waves are not as well developed as when the Krey parameters are used (test A), the simulated seismic records of test B and C nevertheless contain arrivals with the appearance of seam waves. With the Krey parameters, the ratio of maximum amplitude range inside the seam to that just outside the seam is 19. This ratio is 5 when the shear velocity of the shale is less than the longitudinal velocity of the coal, and it is 8 when the longitudinal velocities of the coal and shale are equal. Thus, even though model parameters were varied widely, Su's results indicated that seam-wave phenomena are robust; that is, their existence is not dependent on a highly specialized and restrictive set of conditions which may or may not be found in nature.

Chieh-Hou Yang is applying the multilayer version of the generalized ray theory of Cagniard (11), Pekeris (12), and Sobolev (13) to the problems of coal seismology. His modifications of, and approximations for, this time-domain theory

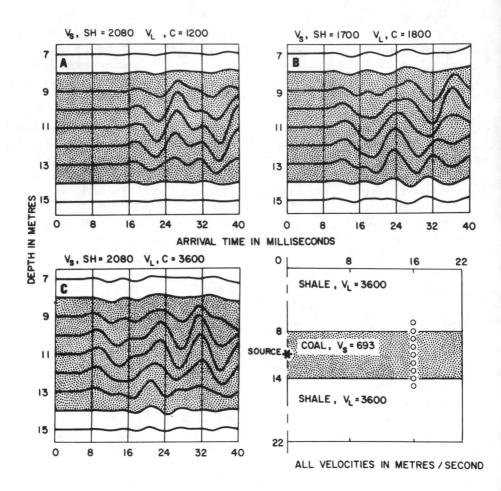

Figure 10. Seismic arrivals as functions of different velocity conditions as detected by a vertical array of
horizontal-component seismometers 16 metres from a simulated explosive source. Densities of the coals and the
surrounding shales, shear velocity of the coal seam, and
longitudinal velocity of the shale layers were the same
in all three cases.

yield a viable approach currently capable of computing arriva
times, amplitudes, and waveforms of first arrivals. Yang
has demonstrated that it is highly unlikely that Stoneley
waves contribute to the seam wave.

Steven Peterson is using the spectral or mode approach
of Haskell (14) as generalized by Harkrider (15) to develop
an interpretation scheme which is applicable when the generalized rays become too numerous. The main program that
he is using was written at Pennsylvania State University by

P. Glover and D. McCowan. He will investigate the effects of partings and clay floors on seam waves. Preliminary studies indicate that for the fundamental Rayleigh-like wave of Krey (4), the dilatation will be more confined to a seam than either component of the particle velocity. This result suggests that piezoelectric transducers, immersed in a fluid-filled borehole, may be the optimum detectors of seam waves.

Because the hybrid modeling approach is prohibitively expensive for day-to-day interpretation, current studies are directed toward finding an adaptation of either the generalized-ray or the mode approaches which could be used with minimal computer facilities.

Seismic Seam-Wave Sources

Design of sources for generation of seismic seam waves must meet the following requirements: (1) they must produce sufficient seismic energy so that upon data processing the seam-wave arrivals can be extracted, (2) they must have a highly reproducible output, (3) they must generate an output signal rich in those frequencies required to excite the seam wave, and (4) they must fit easily within 7.62-centimetre diameter holes. The first and second requirements are linked, for if the source output is not reproducible, usual signal-enhancement methods are ineffectual. Reproducibility is also needed when the number of detectors and recording channels are limited and analytical comparisons of results are to be made between sets of drill holes within an array. Attainment of reproducibility is dependent on the source itself and on meeting the condition that the neighborhood surrounding the source not be excessively disturbed by it. Therefore, a compromise must be reached between building a source strong enough to generate an enhanceable signal, but weak enough not to significantly alter the materials around it.

Two sources of seismic energy have been constructed, one mechanical and the other explosive. Field tests confirm that both meet the design criteria. Sufficient energy was generated to produce clearly recognizable body-wave first arrivals at hole-to-hole separations of 125 metres with the mechanical source and at 500 metres with the explosive source. If the sources are powerful enough to produce body wave first arrivals, then model studies indicate they should be capable of generating seam waves. High reproducibility apparently was obtained, for when recordings from a series of shots were laid on top of one another, they were congruent within the thickness of their traces. Frequency of body waves recorded from the output of three-component seismometers ranged to 250 Hz, and sharpness of the rise time with piezoelectric transducers indicated frequencies of well over 300 Hz had been generated.

Figure 11 shows the mechanical source, in essence a down-hole hammer, in both its armed and fired configurations. The tool weighs 50 kilograms, its diameter is 7.3 centimetres and it is 2.5 metres long. Eight sturdy leaf springs compressed by a hydraulic piston tightly hold the tool at the required depth. Inside the watertight cylinder, a 10-kilogram hammer is driven by a 160 kilogram spring. The device has several interesting features: (1) because it is basically a sealed hammer, it has no permissibility problems, that we can see, to prohibit its use in underground mine workings; (2) it can be cocked and fired repeatedly without bringing it to the surface; (3) it can be used in the horizontal position, in contrast to gravity-operated devices; and (4) because it is rigidly attached to the borehole wall, fluid in the hole is not required to effectively couple seismic energy, as is the case with an explosive source.

The explosive-source tool is shown in figure 12. By changing barrels, either 12-gauge or 410-gauge shotgun shells can be used. A plastic wafer near the end of the barrel keeps borehole fluid out of the chamber. The firing sequence is initiated by a switch closure keyed to the magnetic recording system, and closing of an inertia switch on the tool is used to provide the zero-time pulse on the recording. Presently we are operating this device on a 150-metre long cable. In an attempt to develop increased horizontal directivity of the source, a barrel-termination unit, figure 12, was constructed.

Seismic Seam-Wave Detectors and Recorders

Piezoelectric transducers and oriented three-component seismometers were used as detectors in the search for seam waves. The pressure sensitive detectors function only when immersed in borehole fluid; the three-component seismometers provide useful data only when rigidly held against the borehole wall.

The pressure units were concentrically mounted barium titanate piezoelectric cylinders molded at a separation of 3.05 metres into a special cable whose velocity was held

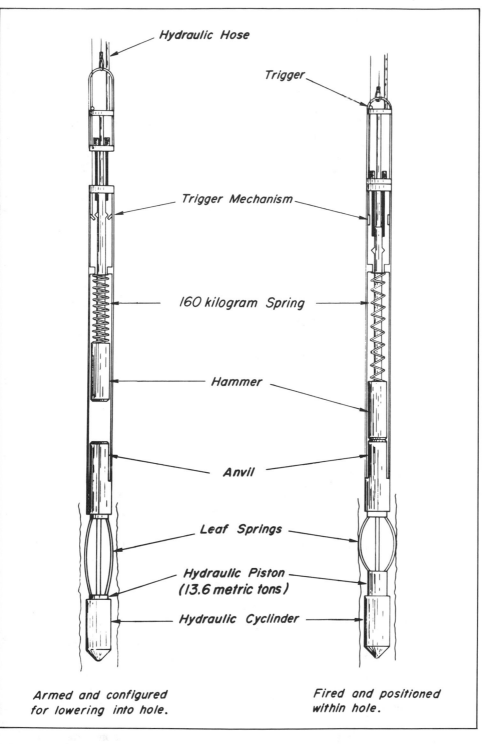

Figure 11. Down-hole-hammer source shown in its armed and
fired configurations. Diameter of the tool is 7.3
centimetres, and its length is 2.5 metres.

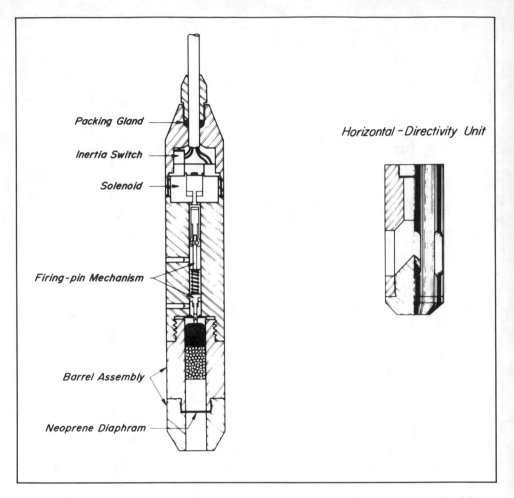

Packing Gland

Inertia Switch

Solenoid

Firing-pin Mechanism

Barrel Assembly

Neoprene Diaphram

Horizontal - Directivity Unit

Figure 12. Down-hole explosive-source tool shown with 12-
   gauge shotgun shell in the chamber. Diameter of the tool
   is 6.4 centimetres, it is 35.6 centimetres long, and it
   weighs about 4.5 kilograms. The horizontal-directivity
   unit is shown to the right.

below 1200 metres per second. These assemblies, known as
velocity cables, are commercially available.

   The prototype magnetic orientation device (shown with its
protective rubber sheath removed) and one of the seismometer
holders are depicted in figure 13. Operations with these
units begin by setting the azimuth preset unit to the angle
between the magnetic declination and the direction of the
line joining the source and receiver holes. The orientation
rods are then pushed through holes in four 3-component seismometer holders separated by 1 metre. Upon reaching the
desired depth, this assembly is twisted until a maximum
reading is obtained on the flux-gate magnetometer. Pneumatic

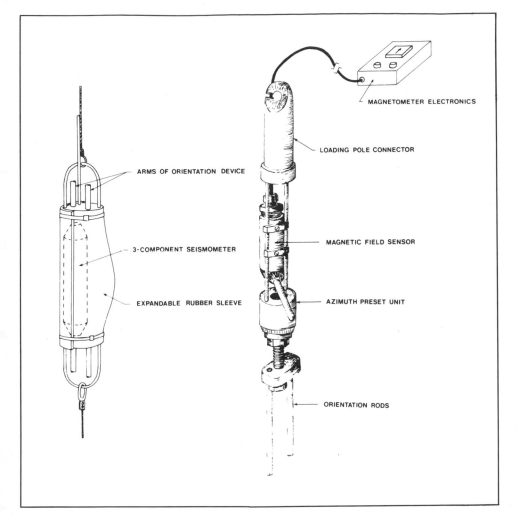

ARMS OF ORIENTATION DEVICE

3-COMPONENT SEISMOMETER

EXPANDABLE RUBBER SLEEVE

MAGNETOMETER ELECTRONICS

LOADING POLE CONNECTOR

MAGNETIC FIELD SENSOR

AZIMUTH PRESET UNIT

ORIENTATION RODS

Figure 13. Seismometer holder, one of four strung together at 1-metre separations, and the magnetic orientation device. Each holder is about 35 centimetres long and 7 centimetres across. When the rubber sleeve is expanded with a differential pressure of 7031 kilograms per square metre, it is capable of holding the 3-component seismometers rigidly within a 15-centimetre diameter hole.

pressure is then applied until the expandable rubber sleeve is sufficiently inflated to push the holders firmly against the wall of the borehole. Next, in order to prevent seismic coupling between seismometers, the orientation device is raised until its rods are only within the upper holder. Although this is a more complex operation than obtaining data with a velocity cable, it has the advantage that it can be performed in a dry hole and it provides three-component information, data of great interest during the early stages of research.

Seismic recording systems for seam-wave studies must have broad frequency response. The bandwidth of seam-wave frequencies observed by Krey (4) for a 2-metre thick seam was from 150 to more than 200 Hz. Since bed thickness and seam-wave frequency are inversely related, Krey's results imply that seam-wave frequencies within 30- to 0.5-metres thick coal seams could range from 10 to 800 Hz and beyond if the higher modes are excited. The Colorado School of Mines analog seismic recording system has a frequency range from 10 to 600 Hz, thus it should be sufficiently broadband for most initial studies.

Field Studies

Early in the program, experiments were conducted to determine if a set of seismometers emplaced on the outcrop of a coal seam could detect whether a seam wave had been generated by an explosive charge detonated in a down-dip hole which encountered that seam. The field site contains multiple-bed seams. We sought to demonstrate that through the use of seam waves we could search and find that outcrop of that particular seam which contained the shot. In this way, hole-to-surface results could be used to establish bed correlations. If this idea worked, we then planned to extend it to develop a procedure to locate the extent of abandoned underground workings which were poorly mapped and which are now probably flooded and partially collapsed.

Vertical-component seismometers were deployed along the highwall of an exposed 3-metre thick seam. One group of detectors was planted within a 10-centimetre clay parting approximately 1.3 metres down from the top of the seam; the other was set into a narrow ledge in the coal seam at 0.9 metres from its top. Shotholes were located approximately 100 metres downdip from the outcrop.

Seam waves were obtained. However, their frequency was substantially lower than theory had predicted, 30 Hz instead of 120 Hz. Using modal analysis and velocity data obtained from other tests in the area, Professor Maurice W. Major (Colorado School of Mines, personal communication) explained the anomalously low frequency in terms of a thicker wave guide composed of underclay and coal.

Other experiments at this strip mine showed that within a day after the seam face was exposed a low-velocity zone was produced at the face. Delay time through this zone could be as much as 13 milliseconds, and we suspect these delays are not uniform along the face. This result implies that seismic methods dependent on arrival-time analysis to reveal cutouts within a coal panel may produce unreliable results unless the seismometers are set well into the face.

Field studies to produce and detect seam waves have begun at a lignite area east of Denver, Colorado. The lignite zone is approximately 10 metres thick and has overburden of 10 to 50 metres. Our initial hole-to-hole experiments produced no recognizable seam waves, but did serve to evaluate the equipment. First shots were positioned on the basis of driller's logs. This was a mistake, for upon subsequent borehole logging it was discovered that the source had been placed in a thin seam (30 centimetres) separated from the major lignite seam by a 0.5-m thick high-velocity layer. The condition was therefore somewhat like the two-seam case of figure 8. A seam wave generated in the thin layer would have a frequency perhaps as high as 1200 Hz, well beyond the limit of our recording system. Also, the detectors would have had to have been set with great precision to guarantee they were within the shot-contained thin seam. From this experience, we are convinced of the necessity of having high quality borehole well logs prior to searching for seam waves in a highly parted zone.

Detection of seam waves within thin beds (defined by the U.S. Geological Survey, Averitt (1), to be from 35 to 71 centimetres for higher rank coals) will thus require equipment capable of detecting frequencies as high as 1000 Hz. Using the practical rule of sampling (50 percent more samples than are indicated by Shannon's sampling theorem), digital seismic equipment with a quarter-millisecond sample interval would be needed.

Research on the use of shallow seismic reflection methods for exploration of Western States coal deposits is proceeding concurrently with seam-wave studies. This is a natural course because the equipment needed for both investigations has almost the same set of design requirements and because the information gained in these twin sets of experiments is mutually useful. For example, by shooting from the surface and recording at a vertical array of seismometers spanning a seam, not only are velocity data on the coal seam and its neighboring sediments obtained but also the time from the surface to the seam, half the reflection time, is determined. Recordings made from a surface array of seismometers deployed between

the hole containing the seam-wave detectors and the shothole
also provide information on the variation of seismic conditions between the paired holes. When a sufficient number of
recording channels are available, these surface-array data
are obtained contemporaneously with seam-wave data.

CONCLUSIONS

The work reported in this paper shows the following:
(1) high-precision gravity surveys appear capable of mapping
cutout edges of thick coal seams; (2) close estimates of the
locations of burn facies can be obtained with magnetic methods; and (3) hole-to-hole seam-wave methods have a tremendous potential which additional research has every likelihood of bringing to fulfillment. Use of these methods
should reduce the number of drill holes required in coal exploration. For example, from the gravity map one could select the optimum number of drill sites to outline the edge
of a cutout; and from the magnetic map, one could choose
drill locations so as to approach the burned zone from a
starting position closer to the facies.

Within the limits of our knowledge, the transition of
the hole-to-hole seismic seam-wave concept to a working
field procedure has as yet not been accomplished. However,
we are confident that this step will be made. Encouraging
theoretical and model studies, successful application of
seam-wave methods in underground mines, and detection of a
seam wave in a hole-to-outcrop experiment all constitute
favorable signs. The economic incentive for the eventual
development of hole-to-hole seam-wave methods clearly exists.
Consider, for example, that in evaluation of Western States
coal reserves, drill-hole spacings as small as 40 metres
have been used. Now, if we accept the assumption that seam-
wave propagation distances of 400 metres can be obtained
(Arnetzl (16) shows seismograms exhibiting seam waves to
1486 metres), then only a 9-hole pattern of seam-wave holes
(a 3 by 3 array with spacings of 400 metres) within an 800-
metre square would be needed to certify bed continuity in
contrast to 121 holes 80 metres apart using only drill data.
Also consider the problem of precisely locating the extent
of old mine workings, areas of potential hazard to subsequent mining and areas of lost reserve. Model studies indicate major breaks in coal-seam continuity caused either
by open rooms or by collapsed rooms will effectively

terminate the seam waves. To guarantee drill penetration of all rooms, spacing between holes must be less than the anticipated dimensions of the rooms. With seam-wave methods to guide the drill program, drilling at close intervals would be required only within those areas bounded by holes in which seam waves were not observed.

The three geophysical methods described in this paper are by no means the only exploration techniques available. Shallow seismic reflection methods and in-situ assaying procedures (the determination of coal quality from well logs), are now being used. Research on application of the large diverse families of electrical prospecting and remote sensing methods to coal exploration is now in progress. We see no reason why all of these methods will not be used routinely within the current decade.

ACKNOWLEDGMENTS

The down-hole shotgun device was designed and constructed by Orville L. McKim, and the magnetic study of figure 4 was conducted by Manton L. Botsford, both of the U.S. Geological Survey. We are especially appreciative of the useful suggestions freely given by graduate students and members of the Department of Geophysics at the Colorado School of Mines, in particular Professor Maurice W. Major. We also want to thank the Mintech Corporation and the Kemmerer Coal Company for their cooperation and their kind permission to work on their properties.

REFERENCES

1. Averitt, Paul, 1975, Coal resources of the United States, January 1, 1974: U.S. Geol. Survey Bull. 1412, 131 p. 11 Figs., 10 Tables.

2. Denson, N. M., and Keefer, W. R., 1975, Map of the Wyodak-Anderson coal bed in the Gillette area, Campbell County, Wyoming: U.S. Geol. Survey Misc. Investigation Series Map I-848-D.

3. Freystätter, Stefan, 1974, Modellseismische Untersuchungen zur Anwendung von Flozwellen for die utertagige Vorfelderkundung in Steinkohlenbergbau: Berichte des Institutes fur Geophysik der Ruhr-Universitat Bochum, no. 3, June.

4. Krey, T. C., 1963, Channel waves as a tool of applied geophysics in coal mining: Geophysics, v. 28, no. 5, part 1, p. 701-714.

5. Guu, J. Y., 1975, Studies of seismic guided waves: the continuity of coal seams: Colorado School of Mines Ph.D. Thesis no. T-1770.

6. Lagasse, P. E., and Mason, I. M., 1975, Guided modes in coal seams and their application to underground seismic surveying: 1975 Ultrasonics Symposium Proceedings, Inst. Elec. and Electronic Engineers, no. 75 CHO 994-4SU.

7. Ewing M., and Worzel, J. L., 1948, Long range sound transmission in Propagation of sound in the ocean: Geol. Soc. America, Mem. 27.

8. Leitinger, H., 1969, Investigation of displacement steps in a layered half-space by the finite difference method: Colorado School of Mines Ph.D. Thesis no. T-1243.

9. Darken, W. H., 1975, A finite difference model of channel waves in coal seams: Colorado School of Mines Masters Thesis no. T-1729.

10. Alterman, Z., and Karal, F. D., 1968, Propagation of elastic waves in layered media by finite difference methods: Seismol. Soc. America Bull., v. 58, no. 1, p. 367-398.

11. Cagniard, L., 1939, Reflection et Refraction des Ondes seismiques progressives: Gauthier-Villars, Paris, also, 1962 Reflection and refraction of progressive seismic waves: translated and revised by E. A. Flinn and C. H. Dix, New York, McGraw-Hill Book Company.

12. Pekeris, C. L., 1940, A pathological case in the numerical solution of integral equations: Proc. Nat. Acad. Science of the U.S., v. 26, p. 433-437.

13. Sobolev, S., 1932, Application dela theoris des ondes planes a la solution du probleme de H. Lamb: Trudy Inst. Seism. Akad. Nauk, no. 18.

14. Haskell, N. A., 1953, The dispersion of surface waves in multilayered media: Seismol. Soc. America Bull., v. 43, no. 1, p. 17-34.

15. Harkrider, D. G., 1964, Surface waves in multilayered elastic media, I. Rayleigh and Love waves from buried sources in a multilayered elastic half space: Seismol. Soc. America Bull., v. 54, no. 2, p. 627–680.

16. Arnetzl, H., 1971, Seismische Messungen unter Tage: Tagungsbericht der Gesellschaft Deutscher Metallhuttern – und Bergleute, May 8, 1971, p. 133–141.

Section Four Discussion

H.S. SCOTT: (Argonaut Mineral Exploration Inc. Philippines).
This morning Dr. Hasbrouck showed us magnetic profiles
across coal seams in the Powder River basin, wherein
in the coal had been partly burnt. The resulting anomalies
are very strong and irregular and you mention in your paper
that they are caused by the development of clinker. I want
to just point out that my interest in asking this question
is, in the Philippines, one of our problems in getting coal
mining on its feet, is the old Komoti mining system as it is
called. The Komoti is a local sweet potato and so this is
just an extension beyond farming, but they sometimes go into
the seam as much as three or four hundred feet. These seams
are actually filled with water and cause a great deal of
hazard, and for other reasons we want to know where they
are, so, my question is - is this phenomenon sufficiently
widespread to be of general use? Does it occur only when
the coal bed is overlain by shale or might we expect to find
it in sandstone and possibly limestone, and also, as part of
the same question, and part of the same answer, would you
indicate how strongly magnetic is the resulting alteration?
For example, could we expect to find it by running a magnetometer 30 metres above the seam, or must we be closer to the
seam. I know this is a very general question, it depends on
the thickness of the seam of course, but could you give us
some measure of the strength of magnetism of this clinker?

W.P. HASBROUCK: You can take either sandstones or shales,
(these are tertiary sediments from the Fort Union Formation),
and look at them before heating and baking. They are just
about as non-magnetic as this wooden table. When they are
heated, and particularly in a reducing atmosphere, you can
increase the magnetisation by a factor of as much as 6,000.
So, it is very highly magnetic. I would think that, yes,
you could see the effect at 30 metres above in the dipping
beds, so I wouldn't think there would be any problem there.
I find that in the use of the magnetic technique, the best
thing to do is to work as closely as you can with the geologist. We have several cases where the magnetics and geology
agree just beautifully and that is the case in an area
liable to geophysics. If you find a place somewhere that
geophysics is needed, then go ahead and use it.

P.K. GHOSH: I heard with interest about the possibilities
of mapping effectively and quickly the burnt faces of the
288 COAL EXPLORATION

coal seams with magnetic methods. Our experience in this respect in India may be worth noting. Many a coal seam, especially of the superior quality ones, are occasionally typified by intermittent outcrops of burnt rocks along the strike of the seam or seams. Though this burning has depleted some quantity of coal, it invariably furnishes a reliable clue for tracing the outcrop position of the seam or seams, especially in the area which is under a cover of alluvium. We thus welcome such "outcrops" when in search of coal seams!

India is an ancient country. There is no folklore immortalising the spectre caused by such fires on coal outcrops as in respect of many other phenomena. It is thus logical to conclude that all these happened before the advent of human habitation in the areas concerned. The cause for initiation of such burning had probably been spontaneous combustion or lightning or forest fire. Any way, it has been proved that such burnt outcrops of coal seams, though they may be extensive on the surface, do not extend much at depth. At the water table or even earlier, fresh coal is invariably encountered.

W.P. HASBROUCK: That is very interesting. In parts of Wyoming, the only time you see an outcrop is when it is burned. As for the depth of burning, I have seen it from the magnetic records a couple of hundred metres at least, and I know of one place confirmed by drilling, where it is well over a mile back from the outcrop. One thing I probably did not bring out this morning was how irregular the edge of that burned surface can be. You may have an outcrop even along a river front and the outcrop itself is almost straight and you think it has got to burn back the same distance from the outcrop, but it doesn't work out that way. It could be highly variable.

R.E. CARLILE: (Petty-Ray Geophysical Inc. U.S.A.) I have four very quick questions Mr. Clarke. In your applications of seismics to the delineation of coal beds, first, what have you found has been the shallow limit of bed delineation? Secondly, what has been the minimum seam thickness that you have been able to repeatedly define? Thirdly, have you examined station spacings less than 25 ft. and if so, with what result? And, lastly, in your applications deriving figure 7, I am interested to know how many phones per recording channel that you used there.

A.M. CLARKE: Four quick answers to those questions. A shallow limit we haven't investigated. The shallowest target horizons we have shot have been about 1500 feet. That is areas of interest - perhaps 1,000 feet in some cases. We cannot of course, as you know, go shallower than that. I think there are some members of this audience here who have been down recently to about 100 feet. With regard to minimum seam thickness, don't forget if you read Van Reel's paper or any of the others cited in the texts, that we are not looking at coal seams. We very rarely do, except as Tom Daly said, we do get events in some combinatio We are actually looking at internal reverberations in groups of beds that get into phase. Most of the coal measure beds are too thin to give a reflection, but collectively they behave like a reflector. Thank goodness, because being deltaic they change rather rapidly. Your third question, had we shot less than 25ft. station spacing? Not to my knowledge. Six meters is about the smallest. And the number of phones per channel on figure 7, is 12. There is one general point I would like to make to remind you of what you have been listening to this afternoon. Don't forget the enormous gearing of the coal industry and that, in other words, one makes one's capital, one makes one's profit, out of margins. They are very much controlled by marginal output. What you have been listening to this afternoon, is a system. In my paper, I point out that at the moment we are losing about 10% production due to unexpected geological disturbances, (about half of those might be detectable by the methods you have been hearing about this afternoon), that is, 10% of just under £15,000 million per year, (that is our revenue) and we are concentrating now in the less disturbed areas. Once what you have been listening to in geophysics gets going, it is going to revolutionise producti from the more disturbed coal fields. It will take a decade to get round the world coal mining industry. When it does, watch out because we make our profits on our margins and it is those profit margins we are boosting with this system.

DR. T. KREY: I think Mr. Clarke and Mr. Daly have very wel outlined the conditions for useful application of geophysics in coal mining. These conditions may be very serious in one province, not in another one. For instance where deep drilling is necessary to explore a new area, reflection seismics may more easily be advantageous than in shallow areas. In this context, it may be worthwhile to mention that Ruhrkohle and Prakla Seismos tried out a new approach recently, a rigorous aerial survey was carried out, covering about 16 square kilometers (about 7 square miles I think) resulting in a 50 meter square grid of reflection points in the underground. On account of the many sections available, one section every 50 meters in both directions the fault pattern can very easily be outlined without ambiguities, but of course, not very small faults, medium faults of about 50 feet (10 to 15 metres). But those faults could really be outlined and it was sometimes fairly astonishing how these faults ran.

A.M. CLARKE: May I comment on that quickly. I think this emphasises the complimentality of the in-seam wave work and the surface wave work and with regard to not being able to detect small faults from the surface, don't forget that what we are after is the pattern of faulting if you use this gaming or geosimplan method. If you know the pattern in the larger faults, you can say a great deal about how the minor faults are going to affect mining and what kind of insurances you put in and thus make coal mining cheaper.

DR. A. ZIOLKOWSKI: (National Coal Board) I was very interested in the remarks you were making about automatic static corrections. I believe you said that automatic static corrections which use an averaging technique from trace to trace, are dangerous because they can eliminate apparent faults. I don't quite understand how this can be because static correction programmes simply apply time shifts to various traces. Each trace can be shifted up or down relative to the traces either side of it and the shift, as calculated by the programmes, is based on cross-correlations from trace to trace, and what you are trying to correlate are horizons. Unless the fault is actually a vertical fault, which I believe you said is rather a rare occurrence, I don't understand how a vertical shift in time can eliminate an apparent fault.

T.E. DALY: Primarily, I think this would occur if you are working in an area where you have essentially one good reflector. This is the reflector that we are going to concentrate on when we make our time shifts to account for static errors and, unless you have a second reflector to see this verticality, you run the very big risk that you are shifting out faults as you make these changes in your corrections for what we hope, are changes in the low velocity layer or related changes. It is primarily in this context that you run the risk of averaging time differences between traces that may actually reflect a fault rather than a static correction problem.

A. ORHEIM: Mr. Daly, you and the rest have been suggesting that the seismic reflection technique cannot be used in any way for finding the coal seam and deciding how it runs. Secondly, I have a question from a geologist's point of view. Can you say anything about the relative vertical distance between shot point and the recorder point on the surface when you do the shallow reflective measurements for coal. In other words, this is the vertical difference that is mostly a question of mathematical and computer correction Is there a limit to this correction you can do when you do coal measurements?

T.E. DALY: Yes, there certainly is a limit on how far you can move. The basic problem is tied up in several factors. If you recall, the reaction of the seismic wave is dependent on three things: it is dependent upon the distance from the source to the receiver, it is dependent upon the distance from the surface to the reflecting point, the change in velocity at that point, and it is dependent upon the velocities. at that point. A change in any of these three things will change the critical distance, if you will, that can be maintained between the shot point and the receivers. Basically, it really resolves into a problem of defining the critical angle. To answer your question, I think about the shallowest reflections that we have recorded to date are in the neighbourhood of 150 milliseconds and the horizontal distances I can't remember. However that distance is a function of these three factors and how these three factors tie together. We can figure it out mathematically but I can't give you any set distance for any particular case at the moment. I would be happy to go into it with you later.

R.P. HOLLIS: (National Coal Board) As a colleague of Mr. Clarke, I would like to congratulate him on an excellent paper, and to make one observation and ask two questions.

In the early part of his paper, Mr. Clarke commented that frequently the mine planner was unable to specify his needs. I think it should be realised that the use of seismic survey in the British mining industry is a comparatively new tool and I think that perhaps the shortfall there is that the technique has not been well sold. I note also Tom Daly's reservations with regard to its application. The first question revolves around the particular experience we have in North Yorkshire where we are working a single seam in a coal field covered by Permian and have no prior information. The loss of a longwall tunnel in such circumstances unexpectedly through the interruption of a fault, would cost us something in the order of half a million pounds. The question is really, and perhaps Mr. Clarke can give me some guidance about this, whether or not there is any international experience of the success of seismic surveys under such circum-- stances. The last question: in some cases, would Mr. Clarke agree that the resources would be better employed in underground drivages rather than on seismic surveys insofar as the underground drivages would in part be providing additional pit room.

A.M. CLARKE: Perhaps one point I didn't make clear, but it is made clear in the text, that is that the mining engineer or mine planner more specifically, was unable to specify his needs to the geophysical contractor in a form that the geophysical contractor understood. In other words, they may say yes, we want better resolution or something. We haven't been able to say "and what you have got to do is so and so". It was lack of communication which caused these problems. With regard to your second question, is there any international experience on the success of seismic surveys, the NCB have pioneered this so there isn't going to be very much experience elsewhere anyway. I am not too sure how pioneering we have been when one listens to Prof. Krey by the way. I think we must include Germany to a very great extent, so I am sorry the answer to your second question is - we can't answer it because we haven't got the data and even if we could, it wouldn't mean much. The third question is: couldn't the resources be better employed on productive exploration which was one of the things I was hammering. Whenever we can, we should do so. The whole point that I was trying to make was that our object is not to find more coal but to make the same stuff cheaper, and it is a matter of balancing how much you spend in forward development spare capacity and so on, compared with what you save by making the coal cheaper to mine. In certain circumstances, it would be better to spend on underground drivage and really it is a matter of getting the right mix. I would like to give a public vote of thanks to Mr. Hollis. The feed back from him when the Board was working on one of his prospects saved the Board an awful lot of money.

Cores obtained from offshore drilling in the North Sea are packed for laboratory examination. Photo courtesy World Coal

SECTION 5

10.
Evaluation of Geologic, Hydrologic, and Geomechanic Properties Controlling Future Lignite Open Pit Mining
by Rudolf Voigt

11.
Geotechnology: An Integral Part of Mine Planning
by Richard D. Ellison and Allen G. Thurman

Section Discussion

10

EVALUATION OF GEOLOGIC, HYDROLOGIC, AND GEOMECHANIC
PROPERTIES
CONTROLLING FUTURE LIGNITE OPEN PIT MINING

Rudolf Voigt

Hydrogeologist, Rheinbraun Consulting GmbH

Cologne, W. Germany

1. INTRODUCTION

Brown coal or lignite is poorly carbonized coal
with calorific values ranging widely from about
16oo-45oo btu/lb (9oo to 25oo kcal/kg). Most of
the known deposits in Europe and Asia Minor are
of Tertiary or Lower Quarternary ages. They are,
therefore, embedded in sequences of unconsolidated
to weakly consolidated sediments. Both the charact
of such strata and the low calorific values of the
lignite allow profitable large scale mining only i
huge open pit mines. The most advanced mining syst
employing the bucketwheel excavator-conveyor belt-
stacker techniques, has been developed in the open
pits of Rheinbraun Company in the Rhenish Mining
District near Cologne, Germany (1). Fig. 1 shows
the principal parts of an operating mine.

It is the lithologic, tectonic, and hydrologic
features of the strata top and bottom of the coal
seams that largely control the layout of a mine
and the design of the machinery. A thorough evaluation of the geologic, hydrologic, and geomechanic properties of these rocks is a must for
these reasons. The methods to compile and evaluate

Fig. No. 1, Main units of a modern open pit lignite mine: Bucketwheel excavator-conveyor belt-stacker.

the relevant data do not resemble those of traditional hard coal mining. Rather, they are identical with procedures that have been developed
through the last few decades to explore ground
water resources or to enable the construction of
large earthen dams and road cuts in soft rock areas.

The aim of this paper is to give a general
review of the various ways to gather and evaluate
pertinent data. However, due to the immense volume
of information on these subjects, this paper will
present a general overwiew rather than an indepth
study.

2. GEOLOGIC PROPERTIES

2.1 Preliminary Remarks

There are two major types of lignite deposits
that reflect the environmental conditions in which
they have been formed:

Paralic deposits: Coal fields of this type
have developed in wet lowlands between a rising
mainland and a shelf sea. Detrital sediments,
often even of marine origin, have been deposited
in phases of rapid subsidence. The coal seams
have been formed during times of extremely slow
sinking. Example: The large Rhenish lignite fields,
having developed between the Rhenish Mass and the
Miocene North sea (fig. 2).

Limnic or lacustrine deposits: These coal
fields owe their existence to fossil lakes within
intramontane basins that are connected with alpidic mountain ranges. Examples: The isolated coal
fields of the Pyrenees, the Balkan Peninsula and
Anatolia along the huge alpidic mountain ranges
north of the Mediterranean (fig. 3).

Paralic deposits are distinguished by well developed sequences that consist of beds of sand and
gravel, silt and clay, and coal, that are separated
from each other by marked partings. It is an easy
task to tell aquifers from confining beds in such
sequences. Likewise, an overall geomechanic characterization is possible at relatively low expenditures.

298 COAL EXPLORATION

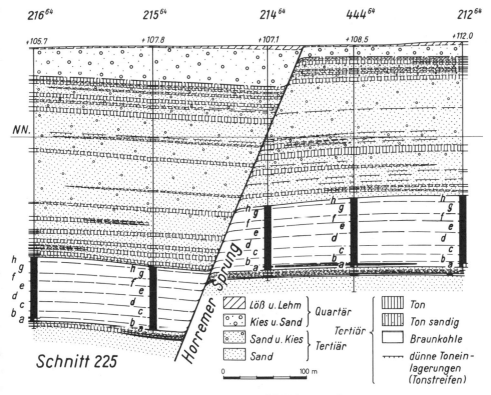

Fig. No. 2, Paralic lignite deposit:
Planned Bergheim mine, near Cologne, W. Germany

On the contrary, limnic deposits usually do
not allow hydrologic and geomechanic evaluations
that are beyound all doubt. Topwall and footwall
strata of the lignite seams are made up of sandy
and silty-clayey sediments with varying contents
of organic matter and carbonates, thus forming
lenticular bodies of fine sands, marls, clays,
and organic muds. Hydrologic and geomechanic properties of sediments like these can vary unfore-
seeably in both horizontal and vertical directions.

As shown by LÜTTIG (2), virgin lignite fields
can be discovered not only by using the indica-
tive method of prospecting, e.g. by prospecting
at random, but also by applying the deductive
method. By carefully analyzing all available
data of a region with known Tertiary strata, the

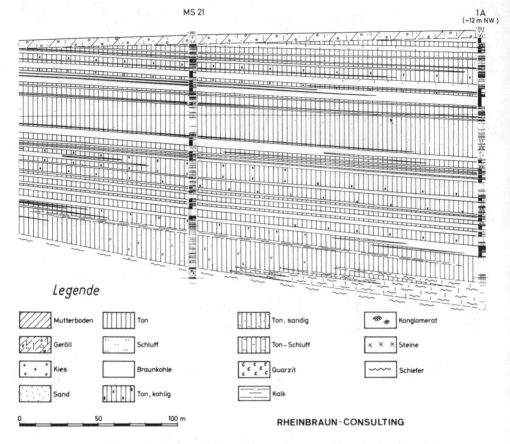

Legende

Mutterboden	Ton	Ton, sandig	Konglomerat
Geröll	Schluff	Ton – Schluff	Steine
Kies	Braunkohle	Quarzit	Schiefer
Sand	Ton, kohlig	Kalk	

0   50   100 m

RHEINBRAUN-CONSULTING

Fig. No. 3, Limnic lignite deposit:
Puentes de Garcia Rodriguez, Spanish Pyrenees

skilled geologist is able to exclude from further
investigation those areas that are expected to be
free of lignite seams.

Of course, only bore holes will eventually
corroborate the theoretical perceptions. Before
deciding on drilling the exploratory holes, the
geologist must have become sure that three
prerequisites to the development of any coal deposit have been met in that particular region:

a. the possibility of formation of biogenic
   sediments in a humid environment;

b. the possibility of accumulation of large
   volumes of these sediments in subsiding
   forelands, troughs, and basins,

c. the possibility of preservation of the accumulated bioliths in stratigraphic traps, and,
thus coalification.

A systematic application of this deductive
method resulted in the discovery of the Elbistan
lignite deposit with $3 \cdot 10^9$ t now the largest coal
field in Turkey (3).

2.2 Evaluation of geologic data

Core samples from all over the virgin lignite
field are analyzed in coal laboratories to obtain
values of:

the ash content of the raw coal,

the water content,

the percentage of volatiles,

the calorific value, and

the chemical composition of the ash.

In addition, the bore logs provide information
on the coal-overburden ratio, which is expressed
by two different values
the ratio of coal thickness   (in meters or
feet) to overburden thickness (in meters or feet),
and
the ratio of coal volume   (in m^3 or cft) to
overburden volume to be removed (in m^3 or cft),
a value which takes the design of the future slopes
into consideration.

2.3 Processing of Data

The stratigraphic and tectonical data of the
driller's logs are converted into geological cross
sections and maps. It is advisable to have the
stratigraphic data of the early stage of any
exploration processed by a computer and plotted
as three-D representations of selected bedding
planes. Due to their lucidity, these three-D
models facilitate the geologist's task of recognition of fault zones or, simply, errors in
geologic logging.

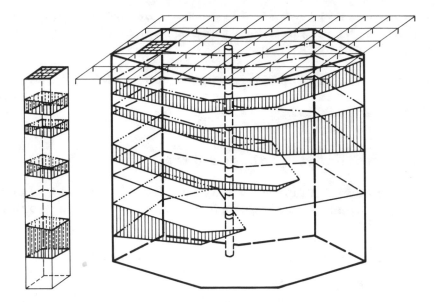

Fig. No. 4, Schematic diagram showing part of a
lignite deposit model with four seams. The central
borehole represents the averaged properties of the
prismatic body. Grid lines discretisize the continuous deposit.

Provided the interpretation of the logs has
proven correct, cross sections and maps are drawn
which any further planning of the layout of the
future mine is based on. For example, slopes and
high walls should be cut perpendicularly to the
strike of tectonical fault zones to minimize the
danger of slope failures.

The next and most important step is the calculation of the volumes of overburden and coal that
will be excavated in the course of mining operations. As depicted in fig. 4, each of the driller's
logs of the lignite field in investigation is
weighed. That means, a prismatic body of the deposit is assigned to the log the interesting
properties of which are assumed to be truly represented by the bore hole. Now, the continuous
deposit system is replaced by an equivalent set
of discrete elements, i.e., by a system of grid
lines. The intersections of the grid lines are
called nodes and are referenced with column and
row coordinates that coincide with the x and y

directions. Assigned to each node are the values
of the mining parameters: top and foot wall ele-
vations of the lignite seams, coal:overburden
ratios, quality of coal etc. By feeding the computer with data on number and elevation of working and dumping benches, the planner can determine the coal and overburden production per unit
time for each bench (4).

As the result of such calculations, the planner
is able to prescribe the optimum operational
design of the future mine: slewing or parallel
operation or a combination of both.

3. HYDROLOGIC PROPERTIES

3.1 Preliminary Remarks

Hydrologic as well as geomechanic investigations
ought to be conducted right from the beginning of
the exploration of the coal field. The additional
costs involved during this stage are low compared
to those that would later become neccessary in the
course of this kind of field and laboratory work.
As self-evident such a statement may appear, as
seldom is a combined optimum investigation
actually carried out on the international scene.

Paralic deposits are predominantly made up of
well developed sequences of alternating water
bearing and confining strata, and the hydrologic
properties of both can generally be evaluated
at little cost. This is mainly due to the
generally large lateral extent of these deposits
with only minor variations in hydrologic properties. On the other hand, it is just these
strata that can require enormous dewatering
measures prior to and during excavation (fig.2).

Limnic deposits, on the contrary, are predominantly poor aquifers or even aquicludes, that
require very intensive exploratory field and
laboratory work. Quite commonly, these lake deposits are overlain by alluvial fan and talus
cone deposits that have been transported into
the basins from the adjoining mountains. Sediments
of this origin display sudden changes and gradual

transitions from coarse and permeable strata to finegrained impervious loamy beds, generally towards the centers of the basins. In exploring these hydrogeologic conditions, the hydrologist often has to apply a trial and error method; a tedious and costly business. The overall expenditures to dewater strata of this type, however, are generally low.

3.2 Evaluation of Hydrologic Properties

Any hydrologic investigation aims at obtaining the information, which is required to properly design measures which are neccessary to completely dewater the top wall aquifer(s) of the lignite seams and/or to lower the potentiometric surfaces of the footwall aquifers sufficiently for mining. By designing these measures, e.g. number, drilling sites and operational schedules of wells, the hydrologist must also calculate the investments involved and assess the operating costs.

The following properties are to be evaluated:

a. aquifer parameters: Hydraulic conductivity, and transmissivity respectively describe the ease with which ground water flows through a given aquifer. The storativity, indicated by the storage coefficient, tells the volume of ground water that is held in storage by a unit volume of this aquifer (see annex I for definitions).

b. boundary conditions describe the geometric patterns of the aquifer. Barrier boundaries, such as impervious mountain fronts, prevent the inflow of ground water into the pumped aquifer. Recharge boundaries, such as normally gaining streams, bogs etc., make up for the loss of the withdrawn water by feeding into the pumped aquifer.

c. water balance: The portion of rain-and snowfall that eventually enters the ground water body, must also be withdrawn to prevent it from entering the mine. To determine this infiltration rate, the hydrologist must set up a

water budget of the whole area that will finally
be affected by the dewatering process.

Values of the intrinsic permeabilities, that is,
the solid matter constants of conductivity coeffi-
cients, are evaluated by determining the grain size
distributions of clastic sediments from screen or
wet mechanical analyses. Compiling the values,
the hydrologist obtains the horizontal and vertical
variations of the hydraulic conductivities (and
transmissivities) in the aquifers. It is advisable
to carry out such work in the course of the early
exploratory phase, when cutting samples can be
taken directly at the drill sites.

Estimates of the storativity of aquifers, that
is the storage coefficient in the case of confined
aquifers and the specific yield in the case of
water table aquifers, can be made by taking the
saturated thickness and the grain size distribution into consideration (5).

To check the degree of accuracy of such determinations, in situ pumping tests are conducted,
which yield the most reliable data on transmissivity, hydraulic conductivity, and storativity. Because these tests require a considerable
amount of money and time, their number will always be limited.

Standard bore hole geophysics, such as self-
potential, resistivity, and γ-ray logging, allows
the hydrologist to distinguish between water
bearing and confining beds. Geoelectrical field
surveys enable him to trace zones of resistivity
anomalies, for example, buried sandy-gravely stream
beds near the surface in intramontane basins.

Winding buried stream beds, having higher
conductivities than the neighboring strata, are
kept track of also by recording the temperatures
in piezometers or shallow test pits. Ground water
in shallow aquifers tends to keep a nearly constant temperature which is identical with the
mean annual air temperature. Therefore, negative
temperature anomalies, measured in summer, indicate zones where the water flows, that means

zones with higher hydraulic conductivities.

Differences in the hydrochemical composition of
various water samples can be used to map the travel
paths of ground water and therefore different
recharge areas. In addition, the analyses may hint
at corrosional or incrustational problems that
will be encountered some day in pumps and pipes.

An indispensable requirement is the installation
of piezometers or observation wells. Best suited
are common exploratory holes, which can be properly designed to allow for water table measurements. These readings are used to draw water table
contour line maps. Maps, that are to be prepared
at intervals, record the successive stages of
drawdowns from the start of the ground water withdrawal through the lifetime of the mine, and, eventually the recovery of the water tables.

Climatic data, needed to compute the water
budget, can be obtained from the local metereolo-
gical observatories, provided there are any. Of
even greater importance is to know the recurrence
intervals of major flood events in streams in
the vicinity of the mining area. Depending on the
scheduled period of mining operations, the surface drainage facilities must be designed to handle
2o or even 5o year floods.

3.3 Processing of Data

The withdrawal of ground water in and around
a mine is intended to
dewater the aquifers on top of the seams, that will
be exposed by the high walls of the mine, to prevent its flooding, or
to depressurize the confined aquifers, beneath the
seams, i.e.,
to lower their potentiometric surfaces sufficiently
to prevent blowthroughs.

A similar problem is to withdraw water from
behind a fault near a high wall that is likely to
fail.

The consumers in the wake of the future mine,

such as thermal power plants or fertilizer plants, need water at a sufficient rate as long as they operate. The water has to be delivered from the dewatering installations. Finally, the violated ground water rights of other users and a possible adverse environmental effect of a large scale ground water withdrawal must be reckoned with. A sound hydrologic planning of the dewatering measures is expected to satisfy the divergent demands of the mine as well as of the water users.

Traditionally, brown coal mines in Germany have been dewatered by means of drain adits, driven into the seams. Drain holes, that had been spot-drilled into the tunnels from the ground surface and screened in the aquifer portions of the topwall, discharged the ground water into the tunnels, from where it was pumped or could run off by gravity flow. Well points, driven from the adits into the confined aquifers beneath the seams acted simply as freely flowing de-pressurizing wells (6, 7).

In spite of its technical simplicity, this dewatering method is not used in modern mines, because its labor-intensive design interferes with the highly mechanized mining operations.

Dewatering by means of tube wells is the now optimum method to withdraw ground water from the aquifers and it is less costly and fitting better the operational demands of an open pit. The wells are equipped with submersible turbine pumps of widely varying capacities (fig. 5).

Having determined the hydrologic properties as described above, the hydrologist has to com-pute number, design and drilling sites of the wells as well as their mode of operation, that are required to achieve the necessary drawdowns at the open pit mine in the available time.

One-dimensional approximative methods, that have been developed by Rheinbraun (8), are applied to calculate the rate of discharge, Q, (the number of wells) and its variation in time, the draw-down, s, at the mine perimeter and the radius, r,

Fig. No. 5, Installation of a submersible turbine pump into a dewatering well (Type KSB, DPG 554/111, Q: 265o-5ooo gpm (1o-19 m^3/min), H: 125o-72o ft (38o-22o m).

of the cone of depression. Despite inherent errors these methods have proven well applicable to the required job. To carry out the computations, a programmable desk computer is sufficient.

A more versatile tool is the digital modelling of aquifers, a method that has been developed through the last decade (9). Similar to the way of computing the coal and overburden volumes (see chapter 2.3), here the continuous aquifer is replaced by a system of discrete volumes. Then, having written the differential equations governing the flow of ground water in finite difference form, a digital computer program is used to solve the resulting set of finite difference equations.

In the phase of calibration, the previously eval ated parameters are assigned to the respective node Then the accuracy of the hydrologic parameters is checked against the output of the first few runs which are expected to reflect the water table configuration of the undisturbed real aquifer. As soon as there is a satisfactory conformity, the model is assumed to be a true image of the aquifer. Well and operational data will then be fed into the computer to evalute the effects of pumping in space and time.

The use of such models enables the planning staff to observe the optimum development of the drawdown in an aquifer by simulating the effects that varying sites of well lines or galleries and variable pumping rates would exert. To give an example, it is possible to achieve the same drawdown along the perimeter of an open pit mine either by running a few wells for long a time or by pumping many wells for a short time. This allows the planner to choose between low investments and high working costs, or high investment and lower operating costs.

It is obvious that numerical models can yield far better results than the one-dimensional approximative computations, provided that there is sufficient and accurate data to be processed. The more sophisticated the method used, the higher quantity of data required. This means in turn higher costs during exploration.

Fig. 6 shows one of the results of the computations, the sequential change of the submersible pumps in function of the drawdowns in the wells.

Beyond the actual technical design of the dewatering process, a thorough knowledge of the ground water flow regime, that is revealed by the numerical model, enables the hydrologist to predict the recovery of the water table after termination of pumping. This natural recovery is significant to the future use of the remaining hole and to the reclamation of the devastated land.

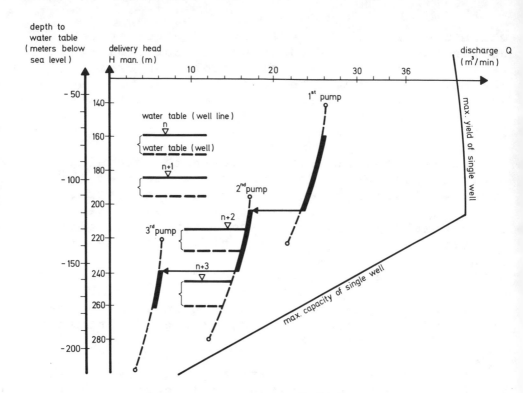

Fig. No. 6, Schematic diagram of sequential changes
of submersible pumps in a well.
To achieve the required drawdown, several pumps
with different characteristics must be installed
one after another.

4. GEOMECHANIC PROPERTIES

4.1 Preliminary Remarks

The rhythm of work in an open pit lignite mine,
- excavating overburden (and coal) along the
working face,
- transporting the bulk material by train or belt
across the mine,
- dumping it along the rear side of the mine,
results in a thorough breakdown of the structure
of the unconsolidated rock. A sapropelic mud such
as gyttja, being thixotropic, assumes a secondary
structure during excavation and conveyance, that,
after dumping, will change into the former one
(transition from gel to sol and to gel again).
In case of cohesionless soft rocks such as sand
and gravel, the new stable structure will develop

after dumping, a process which can last for years. In the course of mining, the strata below the lignite seams are released from pressure temporarily and reloaded again by the inside dump. This process is accompanied by a reversible change of various geomechanic properties (fig. 7).

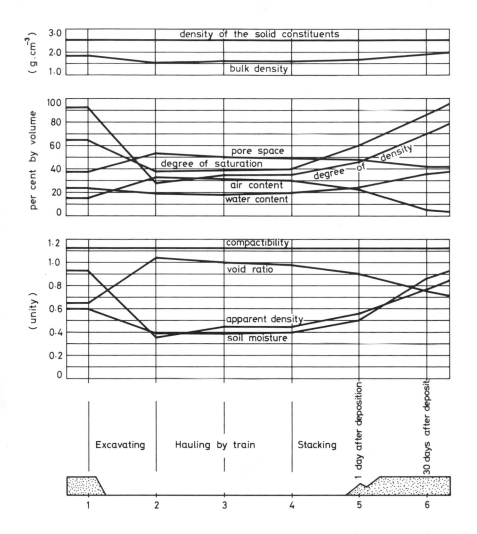

Fig. No. 7, Reversible variations of geomechanic properties in a glauconitic quicksand in a mine. Helmstedt area, Lower Saxony, W. Germany, after(11).

It is readily understood that the attitude of coal and overburden rock to excavating, conveying, and dumping affects to a certain degree the design of the mining machinery. The efficiency of a

bucketwheel excavator increases with the degree
of fill of each bucket, i.e., it depends on the
strength of both coal and overburden rocks.

To ensure the stability of the slopes is one of
the main topics, geomechanics has to deal with
(1o, 11). The steeper the slopes can be cut by the
excavators, the less the volume of waste material
to be removed. Because this volume increases
to between the second and third power of the
overburden thickness, the mining engineer is
interested in keeping the slopes as steep as
possible.

The few examples mentioned above clearly
illustrate the danger to the safety and productivity of an open pit mine, should the geomechanic
properties of the rocks encountered remain unnoticed. As already stated in chapter 3, these
properties have to be evaluated before mining
commences, because they must be considered in
planning the design of the mine.

4.2 Evaluation of Geomechanic Properties

The reactions of unconsolidated rocks on
unloading and reloading as they will occur in a
working mine can be determined by the geomechanical expert, provided that he has available the
essential data. PIERSCHKE (12) subdivides these
properties into four main categories:

a. properties of the solid matter

b. properties of the voids

c. attitude to water,

d. reactions upon changes in load.

Except for general information on the geological
map, it is only the cores and cuttings of the
exploratory bore holes that are available to
determine the indicated properties (see annex II
for definitions of the important parameters).
Standard tests, many of them practicable in field
laboratories, are used to obtain the respective
data.

Drill holes, that are designed to deliver sediment cores should be sited along the future slope and high wall systems.

To run the laboratory tests, there are available

a. disturbed samples from drill cuttings,
b. undisturbed core samples,
c. radioactive logs.

a. Disturbed samples, chiefly of cohesionless rocks, are screened to obtain data on their grain size distribution and unconformity. This data is useful in computing the intrinsic permeabilities of these sediments. Furthermore, the grain size distribution is helpful in evaluating the capillarity of rocks that tend to freeze.

The specific weight of the solids is determined by means of a pycnometer, a value that is necessary in computing porosity, void ratio, bulk density, and grain size distribution of silty-clayey sediments.

It is possible to obtain approximate values on porosity and void ratio by draining vibrated samples. These parameters are useful in calculating the permeability, and, in addition, they are employed to evaluate the results of compression tests.

Degree of density, and compactibility are properties, the values of which are also measured in vibrated samples. These two factors indicate the ratio of natural packing of grains to the tightest packing and therefore, the bearing capacity of a sediment.

The bulk weight of dry, moist, and saturated unconsolidated rocks is determined together with the porosity. In the case of cohesive rocks, however, these values have to be determined by testing undisturbed samples.

b. More geomechanical information is obtained by

investigating cores, especially of cohesive rocks.
The wet mechanical analysis yields the grain size
distribution curve of rocks with effective grain
sizes smaller than 0.1 mm, i.e., silts and clays.

The content of organic matter in cohesive rocks
is a decisive indicator of their water absorption
capacity, and hence, their bearing capacity.

Variations in the water content of cohesive
rocks are responsible for the conspicuous
variations of their strength characteristics.
It is necessary therefore, to determine their
stiffness or consistency parameters. This data
describes the mechanical state of the rock as
being solid or liquid or in between, whereas the
degree of saturation defines the respective water
content. Consistency and degree of saturation
allow computation of the values of the angle of
internal friction, as well as evaluation of
extractibility and compressibility, and last, but
not least, prediction of the behaviour of the
slopes of dams and cuts.

The shear strength of unconsolidated rock is
its resistence against the failure along a
sliding plane. It consists of the two components
internal friction and cohesion. Undisturbed
samples of cohesive rocks are subjected to triaxial tests or Casagrande tests in special shear
apparatuses.

c. In situ measurements of density and moisture
content, conducted in bore holes or shallow test
pits by neutron or γ-γ-probing, give the most
accurate values of these properties at comparatively
low expense.

4.3 Processing of Data

The numerical values of the geomechanic
properties are mainly used to compute the
stabilities of both the working slopes and high
walls and the waste dumps. Describing this
stability by the equation

$$\eta = \frac{\Sigma \text{ stabilizing forces}}{\Sigma \text{ translating forces}}$$

in which η must be larger than 1, the stabilizing forces are expressed by the shear strength, i.e., the sum of cohesion and internal friction of the natural or dumped materials. The sum of translating forces derives from the horizontal component of the earth pressure, that in turn is a function of bulk weight, height of slope, and pore water pressure. Knowing these values, the slope stability is computed by applying internationally standardized methods (13, 14, 15).

If these calculations show that $\eta \leqslant 1$, either the translating forces have to be reduced, say by a reduction of the pore water pressure (a further drawdown of the water table in the strata behind the slope), or gradient and height of the slopes must be decreased. The other possibility is to increase the stabilizing forces, i.e., to apply a drainage blanket.

Of nearly equal importance is evaluation of the suitability of the footwall strata in a mine or the undisturbed ground just outside the mine to bear the load of a dump to prevent base failures in front of their faces.

As shown by these examples, the results of geomechanic investigations enter directly into the process of planning the layout of the mine. This layout must take into consideration the geomechanic characteristics, that in turn depend on the geologic and hydrologic conditions. In this connection, the shear strength is the decisive parameter. Because of its interdependence with the water content, it dictates the required drawdown in the aquifers, the geometric patterns of the slopes, and even constructional details of the mining equipment.

5. CONCLUSIONS

The author hopes to have demonstrated that any exploration of a virgin lignite field must be a multipurpose one, comprising geologic, hydrologic, and geomechanic features of the layers above and below the seams, in addition to the investigation of the coal itself.

The layout of modern open pit mines, working at great depths of several hundred meters, and their technical design are more affected by the hydrologic and geomechanic properties of the top wall and foot wall strata than by depth and thickness of the lignite seams.

6. REFERENCES

1. Leuschner, H.-J., "Entwicklungstendenzen '7o der Tagebautechnik des rheinischen Braunkohlenbergbaus", Braunkohle, Wärme und Energie, Vo.22, 197o, pp. 2-12.

2. Lüttig, G., "Stand und Möglichkeiten der Braunkohlen-Prospektion in der Türkei", Geologisches Jahrbuch, Vo. 85, 1968, pp. 585-6o4.

3. Staesche, U., "Die Geologie des Neogen-Beckens von Elbistan/Türkei und seiner Umrandung", Geologisches Jahrbuch, B 4, 1972, pp. 3-52.

4. Zensus, Th., "Tagebauplanung mit automatischer Datenverarbeitung" Braunkohle, Wärme und Energie Vo. 15, 1963, pp. 253-266.

5. Lohman, S.W., "Ground-Water Hydraulics", US Geological Survey Professional Paper, no. 7o8, 1972, 7o p.

6. Kegel, K., "Bergmännische Wasserwirtschaft", Verlag von Wilhelm Knapp, Halle (Saale), 1938, 277 p.

7. Hagelüken, M., a: "Entwässerung im Braunkohlentiefbau",

   b: "Hangendentwässerung mit Fallfiltern",

   c: "Hangendentwässerung mit Stecfiltern",

   d: "Sicherheitsmaßnahmen bei der Entwässerung",

Taschenbuch für Bergingenieure, Glückauf-Verlag Essen, 1975, pp. 2o1-2o6.

8. Siemon, H. & Paul,R., "Entwicklung und Planung der Entwässerung im Rheinischen Braunkohlenrevier", Braunkohle, Wärme und Energie, Vo. 22, 197o, pp. 26-32.

9. Prickett, T.A. & Lonnquist, C.G., "Selected Digital Computer Techniques for Groundwater Resource Evaluation". Illinois State Water Survey Bulletin, no. 55, 1971, 62 p.

1o. Dermietzel, E., "Die Bedeutung der Gebirgsmechanik für die Betriebsgestaltung im Braunkohlentagebau besonders für den rheinischen Braunkohlenbergbau", Braunkohle, Wärme und Energie, Vo. 17, 1965, pp. 491-497.

11. Wöhlbier, H., "Die Bedeutung der Bodenmechanik für den Braunkohlenbergbau," Braunkohle, Wärme und Energie, Vo. 11, 1959, pp. 487-492.

12. Pierschke, K. "Für den Braunkohlenbergbau wichtige geomechanische Kennwerte", Unpublished Rheinbraun Standards, Rheinbraun Company Köln, 1973, 22 p.

13. Bishop, A.W., "The use of slip circle in the stability analysis of slopes", Géotechnique, Vo, 5, 1955, pp. 7-17.

14. Janbu, N., "Application of composite slip surfaces for stability analysis", Proc. Europ. Conf. on Stability of Earth Slopes, Sweden, 1954, pp. 43-49.

15. Neuber, H., "Untersuchung der Standsicherheit hoher Böschungen nach der sogenannten Streifenmethode", Fortschr. Geol. Rheinld. u. Westf., Vo. 15, 1968, pp. 245-262.

16. Lohman, S.W. and others, "Definitions of selected Ground-Water Terms - Revisions and Conceptual Refinements", US Geological Survey Water-Supply Paper, no. 1988, 1972, 21 p.

17. Thrush, P.W. (editor), "A dictionary of mining, mineral, and related terms", US Bureau of Mines Special Publication, 1968, 1269 p.

ANNEX I

Definitions of hydrologic parameters
      References (5, 16)

Hydraulic conductivity K: The rate of flow of water at the existing kinematic viscosity, in gallons per day, through a cross-sectional area o 1 square foot of the aquifer measured at righ angles to the direction of flow, under a hydraulic gradient of 1 foot per foot. $[L \cdot T^{-1}]$

Transmissivity T: The rate of flow of water at the existing kinematic viscosity, in gallons per day, through a vertical strip of the aquifer 1 foot wide measured at right angles to the direction of flow under a hydraulic gradient of 1 foot per foot $[L \cdot T^{-2}]$

Intrinsic permeability k: A porous medium has an intrinsic permeability of one unit of length squared if it transmits in unit time a unit volume of fluid of unit kinematic viscosity through a cross section of unit area measured at right angles to the direction of flow under a unit potential gradient $[L^2]$
As a property of the medium alone, the intrinsic permeability is usually expressed in square micrometers

or in darcy units.

$$1 \text{ darcy} = 0,987 \ \mu m^2$$

Storage coefficient S: The volume of water the aquifer releases from or takes into storage per unit surface area of the aquifer per unit decline or rise of head.

Specific yield S_y: The specific yield is the ratio of the volume of water which a saturated rock will yield by gravity drainage to the volume of the rock. It is approximately identical with the storage coefficient of a water table aquifer.

ANNEX II

Geomechanic properties pertinent to open pit lignite mining

References: (12, 17)

1. Solid matter

Grain size distribution: Classes of particles of different sizes, indicated by percentages by weight.

Screen analysis: Dry mechanical analysis of unconsolidated rocks with grain sizes between 0.06 and 60 mm ($\sim$0.0025 — $2\frac{1}{2}$ in) on a set of standard screens.

Wet analysis: Mechanical analysis of unconsolidated rocks with effective diameters of particles $\leqq$ 0.06 mm (0.0025 in) by mixing a sample in a measured volume of water and checking its density at intervals with a hydrometer.

Unconformity coefficient: $U = \dfrac{d_{60}}{d_{10}}$

d_{60} = grain size at 60 % undersize

d_{10} = grain size at 10 % undersize

Organic matter g: $g = \dfrac{W_{dry} - W_{ign}}{W_{dry}} \cdot 100 \quad [\%]$

in which W_{dry} = dry weight of sample before ignition

W_{ign} = weight of sample upon ignition

2. Void

Porosity (pore space) $n = \dfrac{V-V_s}{V} \cdot 100 \quad [\%]$

in which

V = total volume of sample

V_s = volume of solid particles of sample

Void ratio $e = \dfrac{V-V_s}{V_s} = \dfrac{n}{1-n} \quad [\text{unity}]$

Degree of density D $D = \dfrac{n_o - n}{n_o - n_d} \quad [\text{unity}]$

in which

n = pore space at natural packing

n_o = pore space at loosest packing

n_d = pore space at tightest packing

Apparent density D_r
$$D_r = \frac{e_o - e}{e_o - e_d} \quad \text{[unity]}$$

in which e, e_o, e_d are the respective void ratios.

Compactibility D_f:
$$D_f = \frac{e_o - e_d}{e_d} \quad \text{[unity]}$$

Density of solids γ_s:
$$\gamma_s = \frac{W_s}{V_s} \quad \left[M \cdot L^{-3}\right]$$

in which

W_s = weight of solid particles of a sample

V_s = volume of solid particles of this sample.

Bulk density γ :
$$\gamma = \frac{W_w}{V} = \gamma_s \cdot \frac{1+w}{1+e} = \gamma_s (1+w)(1-n) \quad \left[M \cdot L^{-3}\right]$$

in which

w = water content

W_w = weight of moist sample

Dry density γ_d :
$$\gamma_d = \gamma_s \cdot \frac{1}{1+e} = \gamma_s (1-n) \quad \left[M \cdot L^{-3}\right]$$

Density of saturated soil γ_r:
$$\gamma_r = \gamma_s \cdot \frac{1+w}{\gamma_w + w \cdot \gamma_s} = (1-n)\gamma_s + n \cdot \gamma_w \left[M \cdot L^{-3}\right]$$

in which

γ_w = density of water

Density of buoyant soil
$$\gamma' = \frac{\gamma_s - 1}{1-e} = (\gamma_s - 1)(1-n) \quad \left[M \cdot L^{-3}\right]$$

3. Attitude to water:

Gravimetric moisture content w:

$$w = \frac{W_w}{W_s} \quad [unity]$$

in which

W_w = weight of water in the sample

Degree of saturation S_f:

$$S_r = \frac{V_w}{V-V_s} = \frac{n_w}{n} \quad [unity]$$

in which

V_w = volume of water in the sample

n_w = fraction of water saturated pores

Consistency of cohesive unconsolidated rocks

a. Liquid limit LL: The water content corresponding to an arbitrary limit between the liquid and plastic states of consistency of the rock [%]

b. Plastic limit PL: The water content corresponding to an arbitrary limit between the plastic and the semisolid states of consistency of the rock. [%]

Plasticity index PI: PI = LL - PL   [%]

Consistency index CI:CI $= \dfrac{LL - w}{PI}$ [unity]

c. Shrinkage limit LS:The moisture content, expressed as a percentage of the weight of the oven-dried soil, at which a further reduction in the moisture content will not cause a decrease in the

volume of the soil mass, but
at which an increase in
moisture content will cause an
increase in its volume.

Intrinsic permeability:

        see annex I

Capillarity: The rise of water in the in-
terstices of a soil or rock,
in result of the action of
capillary forces.

Maximum adsorbed
water W_{max}:   The maximum water content
withheld by the particles of
the unconsolidated rock, the
effective diameters of which
$\leq$ 0,06 mm (0.0025 in)

4. Reactions upon changes in load

Shear strength T_f: The internal resistance of a
soil or rock offered to shear
stress. It is measured both
in saturated and in dewatered
rocks.

$$T_f = C + 6 \cdot \tan\varphi \; (\text{Coulomb's equation}) \; [M \cdot L^{-2}]$$

in which

c = cohesion
φ = angel of internal friction,
    and

$6 \cdot \tan\varphi$ = friction

11

GEOTECHNOLOGY:

AN INTEGRAL PART OF MINE PLANNING

Richard D. Ellison, Ph.D.

Vice President, D'Appolonia Consulting Engineers, Inc.
Pittsburgh, Pennsylvania, U.S.A.

Allen G. Thurman, Ph.D.

Manager, Rocky Mountain Operations
D'Appolonia Consulting Engineers, Inc.
Denver, Colorado, U.S.A.

1.0 INTRODUCTION

This paper summarizes portions of the Geotechnical Investigation and Design Guidelines Manual prepared in cooperation with the Rocky Mountain Energy Company (RMEC) of Denver, Colorado. By incorporating into one document important aspects of geotechnics to mine planning, investigation, design, and development, the Guidelines Manual was designed to serve those persons responsible for the development of safe, economical mines. Efforts were made to relate geotechnical design procedures to other more or less traditional mine evaluation and design disciplines. As appropriate for the 1976 International Coal Exploration Symposium, the discussion concentrates on the exploration and investigation phases of the Guidelines which lead to final premining design.

Basic definitions, schedules and interprofessional relations fundamental to total project development are discussed in Section 2.0, Geotechnical Personnel and Project Planning. The broad scope of important geologic conditions are discussed in Section 3.0, Geotechnical Considerations. These two sections establish the background for the more detailed discussions of geotechnical investigation procedures presented in subsequent sections.

Two basic factors influence the economics of a given mineral reserve: the value of the ore in the ground; and the cost of mining, processing and marketing the ore. The mining costs are almost entirely dependent on the geologic location of the reserve and the associated geotechnical conditions. These conditions warrant examination on a phased basis generally paralleling and often inseparable from mineral exploration.

Important geotechnical conditions which can be evaluated with minimum effort are discussed in Section 4.0, General Site Evaluation. Section 5.0, Geologic Hazard Analyses Concepts, presents an efficient bookkeeping system of geotechnical factors useful during all investigation phases. Mining concepts which influence an effective detailed geotechnical investigation program are discussed in Section 6.0, Preliminary Mining Evaluation. Together, these three sections describe steps which should be made previous to expensive and time-consuming field and laboratory studies. They logically lead to Section 7.0, Geotechnical Investigations, which describes and evaluates field and laboratory investigations important to cost-effective mining design. These investigations should be conducted simultaneously with the mineral exploration program so that a comparison between the value of the reserve and mining costs can be made.

For completeness, Section 8.0, Final Premining Design, and Section 9.0, In Situ Monitoring, briefly discuss geotechnical input beyond the investigation phases.

The authors are indebted to numerous D'APPOLONIA staff members, Messrs. M. J. Coobaugh and Leon Mayhew of RMEC, and Dr. John F. Abel of the Colorado School of Mines for valuable input to the Guidelines from which this paper is extracted.

2.0 GEOTECHNICAL PERSONNEL AND PROJECT PLANNING

Geotechnical engineering is not always included as a key function in the mine planning and design process, even though considerations of soil, rock and groundwater conditions are at least inferred in many major decisions. It is important, therefore, to define the geotechnical role as it applies to the discussion of this paper.

Table I summarizes a realistic range of geotechnical skills applied to mining. Common definitions for areas of specialization are presented in Table II. Since individuals are often

TABLE I

AREAS OF TECHNICAL EXPERTISE IN GEOTECHNICAL INVESTIGATION AND DESIGN

Planning Phase	Geotechnical Input	Typical Principal [1,2] Geotechnical Investigators	Investigators Providing [2] Necessary Supplemental Data
Feasibility Studies	General site evaluation	Mining engineer-geologist structural geologist	Photogeologist geotechnical engineer hydrogeologist geochemist petrologist seismologist
Conceptual Planning / Preliminary Design	Site-specific investigations	Mining engineer-geologist geotechnical engineer hydrogeologist	Geophysicist photogeologist structural geologist geochemist petrologist
Final Design	Geotechnical designs	Geotechnical engineer hydrogeologist	Mining engineer-geologist geophysicist structural geologist engineering geologist

(1) PRINCIPAL INVESTIGATORS MAY VARY DEPENDING UPON SITE CONDITIONS.

(2) AN INVESTIGATOR WITH PROPER EXPERIENCE AND TRAINING MAY PROVIDE NECESSARY EXPERTISE IN MORE THAN ONE AREA.

TABLE II

COMMON DEFINITIONS
OF SPECIALISTS IN GEOTECHNICS (1)

GEOTECHNICAL ENGINEER – An individual trained in civil engineering (majoring in soil and rock mechanics) and geology, and experienced in application of scientific methods and engineering principles and mechanics to analysis and design of structures involving the interaction with the earth crust.

ENGINEERING GEOLOGIST – An individual trained in geology and other geosciences and in civil engineering, and experienced in investigations directed to solving geological problems posed by structures involving the interaction with the earth crust.

MINING GEOLOGIST – An individual trained in mining and geology, and experienced in (1) the study of the nature and occurrence of mineral deposits, and (2) the geologic aspects of mine planning and operation.

HYDROGEOLOGIST – An individual trained in geology, hydrology and groundwater hydraulics, and experienced in the study of groundwater characteristics, occurrence, movement and utilization.

STRUCTURAL GEOLOGIST – An individual trained and experienced in the study of geologic structures who has the ability to determine the form and arrangement of the rocks, and deformations in the structure that occurred either during the deposition or as a result of subsequent tectonic actions.

PHOTOGEOLOGIST – An individual trained in geology and other geosciences, and experienced in geologic interpretation of landforms by means of aerial and satellite imagery, and other types of airborne remote sensing.

GEOPHYSICIST (Engineering Geophysicist) – An individual trained in physics, geology, electronics and related disciplines and experienced in application of quantitative concepts and principles of physics and mathematics in geologic exploration in order to discover the character and changes in properties of the upper portion of the earth crust.

PETROLOGIST – An individual trained in geology and mineralogy, petrography and related disciplines, and experienced with the study of origin, occurrence, history, internal structure and texture and resulting properties of rocks. (2)

GEOCHEMIST – An individual trained in geology, chemistry and related disciplines and experienced in the application of chemical principles to subsurface and surficial geological phenomena such as genesis and nature of ore deposits, weathering of minerals and their effect on natural soil and water composition, etc.

SEISMOLOGIST – An individual trained in geophysics, geology, geography and related disciplines and experienced in study of earthquakes, their frequency, distribution and effects on the earth crust and resulting consequences on structures; by means of the study wave propagation in the earth.

---

(1) Many successful practicing professionals may have backgrounds varying from those stated.

(2) Sometimes the term "petrologist" is used for individuals concerned with only igneous and metamorphic rocks, while an individual with primary concern for sedimentary rocks is terms a "sedimentologist."

expert in more than one area, separate investigators from the ten identified disciplines will seldom be required, but it is unlikely that any one investigator will have in-depth knowledge in all areas. Each project should be staffed to satisfy the specific geotechnical complexity of the particular mine.

Figure 1 illustrates efficient integration of geotechnical engineering within the planning-investigation-design-monitoring functions of a major mining operation. This paper concentrates on the initial geotechnical investigative functions.

The basic mine development roles may be defined as:

- Mineral Evaluation - exploring and identifying the boundaries of the ore body and the ore grade.

- Mining Engineering - analyzing mine economics; optimizing mine layout and materials handling; selecting mining method and equipment; and scheduling operations.

- Operations - developing access shafts, exploratory drifts, experimental mining areas, etc., and extracting and producing ore.

- Geotechnical Engineering - investigating and analyzing geologic conditions; establishing geotechnical design parameters; designing soil and rock-related factors such as slope stability or room and pillar size; and testing and/or monitoring during operations to arrive at optimum recovery ratios with assured safety.

- Environmental Assessment - analyzing environmental conditions, including impacts by mining.

The definition of mining engineering is intentionally restrictive to more clearly illustrate the function of the geotechnical engineer. However, the role identified as geotechnical engineering can be handled effectively by appropriately trained and experienced professionals in civil engineering, engineering geology or mining engineering.

An important factor not shown by this engineering oriented flow diagram is the owner's role in evaluating corporate economics and marketing potentials. The owner must make these major corporate decisions, but he will rely heavily upon the cost and profit schedule generally developed by the mining engineer. The owner's decisions have major

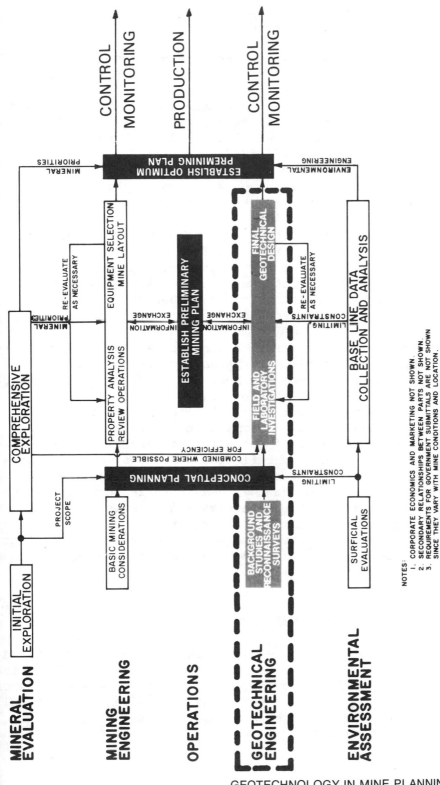

NOTES:
1. CORPORATE ECONOMICS AND MARKETING NOT SHOWN.
2. SECONDARY RELATIONSHIPS BETWEEN PARTS NOT SHOWN.
3. REQUIREMENTS FOR GOVERNMENT SUBMITTALS ARE NOT SHOWN SINCE THEY VARY WITH MINE CONDITIONS AND LOCATION.

FIGURE 1 – FLOW DIAGRAM OF GEOTECHNICAL ENGINEERING IN MINING

importance at each of the three decision points noted by the darkened center areas on Figure 1.

Figure 2 emphasizes mining engineering considerations from a more classical grouping of tasks rather than by geotechnical engineering considerations as discussed in this paper. Geotechnical information provided by field and laboratory investigation is noted by a specific task near the beginning of the flow diagram, but geotechnical analyses associated with the determination of cutoff grade and mine design are only inferred within other mining engineering tasks. The steps where geotechnical input is necessary for optimal design are indicated by heavy line weights on the function block.

Figure 3 adapts the flow diagram to a hypothetical time schedule. The first major topic, Total Mine Planning, pertains to the overall mine development procedure including the mineral evaluation, engineering and operational functions of mine and site development. The second major topic schedules the major control and decision points.

The final topic identifies the basic tasks of the geotechnical program. Typical cost ranges for the tasks are illustrated to demonstrate how the most cost-effective geotechnical program can be planned. For example, General Site Evaluation Studies are very low cost tasks, but they produce very valuable information with major impact on all subsequent steps. The magnitude and range of costs for site-specific investigations are much higher and reflect the wide variety of conditions that may exist for small and large, and routine and complex mine areas. Of course, considerable thought must be devoted to maximizing the investigation benefit-cost ratio for the highest cost tasks. If a study will not lead to significant operating savings or required safety and environmental improvements, it should be discarded or significantly limited.

3.0 GEOTECHNICAL CONSIDERATIONS

Table III introduces the usual mining related geotechnical factors and ranks them for specific mine design consideration This table is used to provide quick, preliminary judgments of the effort appropriate for geotechnical studies for particula mining situations. The geotechnical factors are grouped in five major categories:

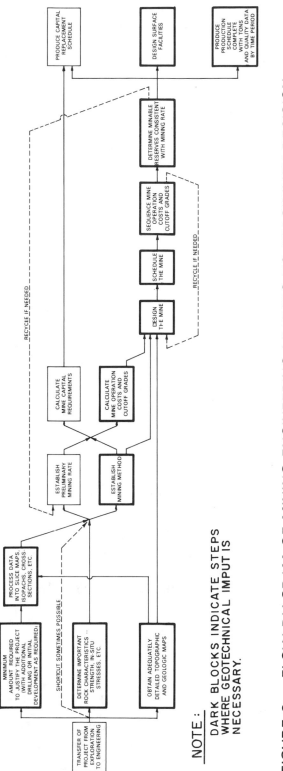

NOTE:

DARK BLOCKS INDICATE STEPS
WHERE GEOTECHNICAL INPUT IS
NECESSARY.

FIGURE 2 — FLOW DIAGRAM OF MINING ENGINEERING LEADING TO FEASIBILITY DECISION

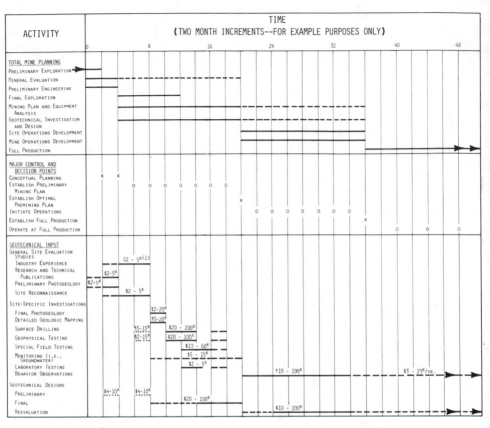

ACTIVITY	TIME (TWO MONTH INCREMENTS--FOR EXAMPLE PURPOSES ONLY)

TOTAL MINE PLANNING
PRELIMINARY EXPLORATION
MINERAL EVALUATION
PRELIMINARY ENGINEERING
FINAL EXPLORATION
MINING PLAN AND EQUIPMENT ANALYSIS
GEOTECHNICAL INVESTIGATION AND DESIGN
SITE OPERATIONS DEVELOPMENT
MINE OPERATIONS DEVELOPMENT
FULL PRODUCTION

MAJOR CONTROL AND DECISION POINTS
CONCEPTUAL PLANNING
ESTABLISH PRELIMINARY MINING PLAN
ESTABLISH OPTIMAL PREMINING PLAN
INITIATE OPERATIONS
ESTABLISH FULL PRODUCTION
OPERATE AT FULL PRODUCTION

GEOTECHNICAL INPUT
GENERAL SITE EVALUATION STUDIES
INDUSTRY EXPERIENCE $2 - 5^{K(1)}
RESEARCH AND TECHNICAL PUBLICATIONS $2-5^K
PRELIMINARY PHOTOGEOLOGY $2-5^K
SITE RECONNAISSANCE $2 - 5^K
SITE-SPECIFIC INVESTIGATIONS
FINAL PHOTOGEOLOGY $2-20^K
DETAILED GEOLOGIC MAPPING $5-20^K
SURFACE DRILLING $5-15^K $20 - 200^K
GEOPHYSICAL TESTING $2-15^K $20 - 100^K
SPECIAL FIELD TESTING $10 - 60^K
MONITORING (I.E., GROUNDWATER) $5 - 15^K
LABORATORY TESTING $2 - 5^K
BEHAVIOR OBSERVATIONS $10 - 100^K $3 - 20^K/YR.
GEOTECHNICAL DESIGNS
PRELIMINARY $4-10^K $4-10^K
FINAL $20 - 100^K
REEVALUATION $10 - 100^K

LEGEND:
• • • • • EVALUATION AND MAJOR INFORMATION EXCHANGE PERIODS LEADING TO CONCEPT DECISIONS.

━━━━━ MAJOR ACTIVITY PERIODS.

━ ━ ━ CONTINUING ACTIVITIES ON AN AS-NEEDED BASIS.

X MAJOR DECISION POINTS, LEADING FROM ONE STAGE OF DEVELOPMENT TO ANOTHER.

O POINTS OF MAJOR CONTROL BETWEEN MINING ENGINEERING AND GEOTECHNICAL ENGINEERING.

(1) THE INDICATED COST RANGES (IN $1,000) ARE REPRESENTATIVE FOR A 5,000 TO 20,000 ACRE, 500-FOOT DEEP UNDERGROUND MINE.

FIGURE 3 – EXAMPLE BAR SCHEDULE OF GEOTECHNICAL ACTIVITIES

TABLE III

THE IMPORTANCE OF GEOTECHNICAL FACTORS TO SPECIFIC MINING CONSIDERATIONS

The table below cross-references geotechnical factors (rows) against specific mining considerations (columns), grouped under **UNDERGROUND MINING**, **SURFACE MINING**, and **SURFACE FACILITIES**. Values follow the legend (1 = High applicability probability of success; 2 = Depends on site conditions; 3 = Low).

Group	Geotechnical Factor	ENTRIES	ROOF CAVING CHARACTERISTICS	OVERALL ROOF SUPPORT	LOCAL ROOF SUPPORT	ROCK BURST POTENTIAL	RIB AND PILLAR STABILITY	SUBSIDENCE	FLOOR HEAVE	WASHOUT SAND CHANNELS ETC	IMPACT ON MULTIPLE SEAM MINING	RIPPABILITY	TRANSPORT SYSTEMS	HANDLING OF WATER	GAS CONDITIONS	MONITORING SYSTEM DESIGN	SPONTANEOUS COMBUSTION	CUT SLOPE STABILITY	FLOOR HEAVE	SPOILS STABILITY	HAUL ROAD DESIGN	HANDLING OF WATER	RIPPABILITY	REHABILITATION	MONITORING SYSTEM DESIGN	SPONTANEOUS COMBUSTION	FACILITIES SITING	STRUCTURES	STORAGE PILES	CUT AND FILLS	IMPOUNDMENTS AND SOLID WASTE DISPOSAL	TRANSPORT SYSTEMS	SPONTANEOUS COMBUSTION
REGIONAL GEOLOGY	GEOMORPHOLOGY	2						2										2		2	2		2	2			2	2	2	2	2		
REGIONAL GEOLOGY	STRUCTURAL GEOLOGY	2	2	2				2					2	2	2		2	2			2	2	2	2	2		2					2	2
REGIONAL GEOLOGY	STRATIGRAPHY AND LITHOLOGY		2	2				2					2	2	2		2	2			2	2	2	2			2					2	2
REGIONAL GEOLOGY	REGIONAL STRESS PATTERNS	2	2	2		2	2	2	2		2				2		2	2	2	2							2				3	3	
LOCAL GEOLOGY	STRUCTURE	1	1	1	1	1	2	1	2	1	1	1	1	1	1			1	1	1	2	1	1	1	1		1	1	1	1	1	1	
LOCAL GEOLOGY	STRATIGRAPHY AND LITHOLOGY	1	1	1	1	1	1	1	1	1	1	1	1	1	1			1	1	1	1	1	1	1	1		1	1	1	1	1		
LOCAL GEOLOGY	IN SITU STRESS	1	1	1	1	1	1	1	1		2	2	2	2	1	1		1	1	1		2	2			1		2	2		2	2	2
LOCAL GEOLOGY	SURFICIAL DEPOSITS	1							1			1	2					1		1	1	1	2	1	2		1	1	1	1	1	1	1
LOCAL GEOLOGY	NATURE OF EROSION	1	2	2	2		2	2	1									2	2	2	1	1	1	1	1		1	1	1	1	1	1	2
LOCAL GEOLOGY	MAN-MADE ALTERATIONS	1	2	2	1	1	1	1	1		2		1	1	2	1	2	1	1	1	1	1	1	1	1	2	1	1	1	1	1	1	2
GROUNDWATER CONDITION	UNCONFINED AQUIFERS	1	2	2	2		1		2		1		1	2		2	2	2	1	1	1	1	1	1	1	2	1	1	1	1	1	1	
GROUNDWATER CONDITION	CONFINED AQUIFERS	1	1	1	1	2	2	1	2	1	2	1	1	1			2	1	1	2	2	1	2	1	1	2	1	2	2	2	1	2	
GROUNDWATER CONDITION	AQUICLUDES	1	1	1	1	2	2	1	2	1	2	1	1	1			2	1	1	2	2	1	2	1	2		1	2	2	2	1	1	
GROUNDWATER CONDITION	EFFECTS OF FRACTURES	1	1	1	1	1	1	1	1	1		1	1	1	1			1	1		1	1	2	1	1		1	1	1	1	1		
SURFACE CHARACTERISTICS	TOPOGRAPHY	1	2	2	2	2		1					2	2				1			1	1	1	1	1		1	1	1	1	1	1	
SURFACE CHARACTERISTICS	SURFACE HYDROLOGY	1	2	2	2	2			2		2		2	1	2			1			1	1	1	1	1	1	1	1	1	1	1	1	2
SURFACE CHARACTERISTICS	EROSION	1	2	2	2	2			2					1	2			1			1	1	1	1	1			1	1	1	1	1	
SURFACE CHARACTERISTICS	TRAFFICABILITY	1		1	2	2	2		2					1		2		2	1		1	1	1	1	1		1		1	1	1	1	
SURFACE CHARACTERISTICS	MAN-MADE ALTERATIONS	1	2	2	2	2		1			2		2	2	1	2		1	1	1	1	1	1	1	1		1	1	1	1	1	1	1
SURFACE CHARACTERISTICS	EXISTING AND PROBABLE DEVELOPMENT	1	2	2	2	1	1	1		1	2		1	2	1			1	2	1	1	1	1	1			1	1	1	1	1	1	
ROCK AND SOIL PROPERTIES	RESISTANCE TO WEATHERING	1	2	2	3	2	2	2		2	1	1		1			2	1	2	1	1	1	1	1	1	2	1	1	1	1	1	1	2
ROCK AND SOIL PROPERTIES	LOAD-DEFORMATION BEHAVIOR	1	2	1	2	2	2	2	2				2	2	1			1	1	1	1		1	1			1	1	1	1	1	1	
ROCK AND SOIL PROPERTIES	STRENGTH	1	2	1	2	2	2	2	2		1	2	2	1	2			1	2	1	1		1	1			1	1	1	1	1	1	
ROCK AND SOIL PROPERTIES	TIME AND ENVIRONMENT DEPENDENCY	1	2	1	2	2	2	2	2		1	1	2	1	1			2	2	1	1		1	1			1	1	1	1	1	1	
ROCK AND SOIL PROPERTIES	EXCAVATION CHARACTERISTICS	1	2		2	1			2	1		2	2					1	2		1		1				1	1	1	1	1	1	
ROCK AND SOIL PROPERTIES	TRAFFICABILITY	1				1			1			1	1			2		1	1	1	1	1	1	1	1		1			1	1	1	
ROCK AND SOIL PROPERTIES	ERODIBILITY	1				2	2					1	2					1		1	1	1	1	2	1	1	1		1	1	1	1	2

LEGEND

APPLICABILITY / PROBABILITY OF SUCCESS	IN ALL AREAS	OFTEN	FREQUENTLY	OCCASIONALLY	VERY SELDOM
HIGH	1				1
DEPENDS ON SITE CONDITIONS	2	2		2	2
LOW				3	3

- Regional Geology
- Local Geology
- Groundwater Conditions
- Surface Characteristics
- Rock and Soil Properties

The mining considerations are grouped in three major categories:

- Underground Mining
- Surface Mining
- Surface Facilities

The subheadings of the major categories include most of the pertinent mine design factors. Particular mining situations may require additional subheadings.

The geotechnical factors and mining considerations are interrelated in two ways to indicate (1) common applicability and (2) estimated reliability of predictions. The style of printing in each square indicates the degree of applicability of the geotechnical factor to the mining consideration. Use of this applicability rating is demonstrated by two examples:

- If overall roof support is thought to be a major con-sideration, reading down the third column from the left identifies and ranks the geotechnical factors which should be considered in the geotechnical investigation program.

- The need to study a particular geotechnical factor can be established by reading across the respective row to identify the mining considerations where the resulting information can be used in design.

The number designations 1 through 3 indicate the probability of successfully relating a geotechnical factor to any mining consideration, with 1 being most probable. By such confidenc level considerations, the planner of a geotechnical study may

- Evaluate the appropriate staged or total effort to be devoted to studying a particular technical subject.

- Determine situations where it is important or valuable to use different procedures supplying redundant information.

- Establish the degree of conservatism to be assigned to particular design parameters.

- Determine areas where field monitoring may be required to supplement premining design analyses.

The details of Table II are complex and not easily followed without considerable study. Once understood, however, this presentation replaces the need for a voluminous text to identify the appropriate scope of a geotechnical investigation.

Closely related to the Probability of Success aspect of Table III are ways in which geotechnical engineers predict the behavior of various mining considerations. Table IV summarizes the bases and general reliability of predicting mine behavior. An understanding of predictive procedures and limitations is essential for planning proper and efficient geotechnical investigation and monitoring programs.

Subsequent sections identify procedures for logical geotechnical investigation programs. The most common investigation techniques are introduced in Table V for the same mining considerations used in Tables III and IV.

Table V ranks the importance of different investigation techniques for different mining considerations, allowing quick, preliminary judgment of the types and number of investigations appropriate for a particular planned operation. Again, appreciation of the probable success of using any investigation technique is essential for (1) obtaining required data on time, (2) avoiding excessive investigation costs, and (3) generally maintaining credibility of the geotechnical role in mine planning and design.

4.0 GENERAL SITE EVALUATION

The first and often most cost-effective step in the geotechnical investigation is the General Site Evaluation. Usually, four tasks are included:

- Review of available geologic literature and related documents.

- Study of available satellite imagery and aerial photography.

- Ground confirmation by fly-over and drive-through reconnaissance.

TABLE IV

QUALITATIVE EVALUATION OF PRESENT PREDICTABILITY OF MINING CONSIDERATIONS BEHAVIOR

MINING CONSIDERATIONS (BASIS FOR PREDICTING BEHAVIOR)	TRANSFER OF TECHNOLOGY (1)	FIELD INVESTIGATIONS (1)	ANALYSIS (1)	MODEL TESTS (1)	EMPIRICAL RELATIONS (1)	OVERALL PREDICTABILITY (2)	REQUIREMENT FOR MONITORING (3)
UNDERGROUND MINING							
ENTRIES	2	3	2	4	3	3	5
ROOF CAVING CHARACTERISTICS	4	3	3	5	3	3	3
OVERALL ROOF SUPPORT	4	3	2	3	2	2	4
LOCAL ROOF SUPPORT	4	4	3	4	3	2	2
ROCK BURST POTENTIAL	5	5	4	5	3	4	4
RIB & PILLAR STABILITY	3	3	2	4	3	2	4
SUBSIDENCE	5	3	3	3	3	3	3
FLOOR HEAVE	4	4	3	5	3	3	4
WASHOUT, SAND CHANNELS, ETC.		3	5		4	3	
IMPACT ON MULTIPLE SEAM MINING	4	4	3	4	4	3	2
RIPPABILITY	2	2	2		—	—	
TRANSPORT SYSTEMS	4	2	3		3	2	5
HANDLING OF WATER	3	3	3		4	3	2
GAS CONDITIONS	5	4	4		4	4	—
MONITORING SYSTEM DESIGN	2	3	2	4	3	2	
SPONTANEOUS COMBUSTION	2	3	3	4	2	2	—
SURFACE MINING							
CUT SLOPE STABILITY	2	3	2	5	2	2	2
FLOOR HEAVE	3	3	2	5	4	2	3
SPOILS STABILITY	1	4	1	5	3	—	5
HAUL ROAD DESIGN	2	3	1		—	5	5
HANDLING OF WATER	2	3	2		4	2	2
RIPPABILITY	2	2	2		—	—	
REHABILITATION	2	2	2		2	2	—
MONITORING SYSTEM DESIGN	2	3	2	4	3	2	
SPONTANEOUS COMBUSTION	2	3	3	5	2	3	—
SURFACE FACILITIES							
FACILITIES SITING	1	2	1	5	5	—	5
STRUCTURES	1	2	2	5	5	5	5
STORAGE PILES	1	2	1	3	3	5	5
CUTS AND FILLS	1	2	1	3	2	—	4
TRANSPORT SYSTEMS	2	2	2	3	3	2	5
IMPOUNDMENTS AND SOLID WASTE DISPOSAL	1	—	—		3	—	—
SPONTANEOUS COMBUSTION	2	4	3	5	2	3	—

NOTES:

(1) NUMBER DESIGNATIONS FOR PREDICTING BEHAVIOR VARY FROM:
1 = PREDICTIVE TECHNIQUE IS APPLICABLE IN PRACTICALLY ALL CASES.
5 = TECHNIQUES CAN BE APPLIED, BUT CONFIDENCE LEVEL WILL BE LOW.

(2) NUMBER DESIGNATIONS FOR OVERALL PREDICTABILITY VARY FROM:
1 = VERY PREDICTABLE IN ALL CASES.
5 = TECHNOLOGY AND EXPERIENCE FOR PREDICTION DO NOT EXIST.

(3) NUMBER DESIGNATIONS FOR REQUIREMENT FOR MONITORING VARY FROM:
1 = FORMAL MONITORING PROGRAM REQUIRED IN ALL CASES BECAUSE OF IMPORTANCE, REGARDLESS OF PREDICTABILITY.
5 = MONITORING CAN BE LIMITED TO OCCASIONAL OBSERVATIONS.

(4) DARKER NUMBERS INDICATE HIGHEST PREDICTIVE BASIS FOR EACH MINING CONSIDERATION.

TABLE V

RELATIONSHIPS OF INVESTIGATION TECHNIQUES TO MINING CONSIDERATIONS

INVESTIGATION TECHNIQUES / MINING CONSIDERATIONS	ENTRIES	ROOF CAVING CHARACTERISTICS	OVERALL ROOF SUPPORT	LOCAL ROOF SUPPORT	ROCK BURST POTENTIAL	RIB AND PILLAR STABILITY	SUBSIDENCE	FLOOR HEAVE	WASHOUTS (SAND CHANNELS, ETC.)	IMPACT ON MULTIPLE SEAM MINING	RIPPABILITY	TRANSPORT SYSTEMS	HANDLING OF WATER	GAS CONDITIONS	MONITORING SYSTEM DESIGN	SPONTANEOUS COMBUSTION	CUT SLOPE STABILITY	FLOOR HEAVE	SPOILS STABILITY	HAUL ROAD DESIGN	HANDLING OF WATER	RIPPABILITY	REHABILITATION	MONITORING SYSTEM DESIGN	SPONTANEOUS COMBUSTION	FACILITIES SITING	STRUCTURES	STORAGE PILES	CUT AND FILLS	IMPOUNDMENTS AND SOLID WASTE DISPOSAL	TRANSPORT SYSTEMS	SPONTANEOUS COMBUSTION IN WASTE PILES
BACKGROUND STUDIES																																
LITERATURE SEARCH	**2**	3	3	3	3	3	2	3	3	3	3	3	3	3	3	3	3	3	3	3	3	3	2	3	2	2	2	2	3	2	2	2
MINING HISTORY	**2**	**2**	**2**	3	2	2	2	2	2	3	2	3	2	2	3	2	2	2	2	3	3	2	2	3	2	3	3	3	3	3	3	2
PHOTOGEOLOGY	1	3	3	3	3	3	**2**	3	2	3	3	-	-	-	3	3	2	2	2	2	3	3	2	3	2	2	2	2	2	2	2	3
SURFACE MAPPING																																
PHOTOGEOLOGY AND REMOTE SENSING	**1**	2	2	3	3	3	**2**	-	2	2	3	-	3	3	3	3	2	2	2	1	2	3	2	3	2	**1**	**1**	**1**	2	**1**	**1**	3
DETAILED GEOLOGIC MAPPING	**1**	2	2	2	2	2	**1**	2	2	3	2	-	2	2	3	2	2	2	3	2	3	2	2	2	**2**	1	1	1	2	2	2	3
SUBSURFACE INVESTIGATION																																
ROTARY DRILLING	**2**	2	2	3	2	2	2	2	2	2	2	3	2	**1**	**1**	3	3	3	3	3	2	2	3	2	2	2	2	2	2	3	2	2
CORE DRILLING	**1**	**1**	**1**	3	2	**1**	**1**	**1**	3	**1**	**1**	3	2	2	2	2	2	2	2	2	**1**	2	3	2		**1**	**1**	**1**	**1**	2	**1**	2
GEOPHYSICAL LOGGING	**2**	**2**	**2**	3	2	2	**1**	2	3	2	2	3	**1**	2	2		3	3	3	3	2	2	3	3	**2**	2	2	2	2	3	2	**2**
GEOPHYSICAL INVESTIGATION																																
SEISMIC REFLECTION	3	3	3	3	3	3	3	3	**2**	3	3	3	2	3	3	3	3	3	3	3	3	3	3	3	3	3	3	3	1	2	2	3
SEISMIC REFRACTION	2	2	2	3	2	2	2	2	3	2	**1**	3	2	2	3	3	2	3	3	3	3	2	3	3	3	2	2	3	3	3	3	3
SEISMIC CROSS HOLE	**2**	2	2	3	2	2	2	2	2	2	**1**	3	2	3	**2**	3	2	3	3	3	3	2	3	2	3	3	2	3	3	3	3	3
ELECTRICAL METHODS	3	3	3	3	3	3	3	3	3	3	3	3	2	3	2	3	3	3	3	3	3	3	3	3	3	2	2	3	3	3	3	3
GRAVITY METHODS	3	3	3	3	3	3	3	3	3	3	3	3	3	3	3	3	3	3	3	3	3	3	3	3	3	2	3	3	2	3	3	3
MAGNETIC METHODS	3	3	3	3	3	3	3	3	3	3	3	3	3	3	3	3	3	3	3	3	3	3	3	3	3	3	3	3	2	3	3	3
FIELD TESTING AND INSTRUMENTATION																																
IN-SITU STRESS	**2**	2	2	2	2	2	2	-	2	2	2	2	3	3	3	3	**2**	2	2	3	3	2	2	3	3	2	2	2	2	2	2	2
MODULUS OF ELASTICITY	2	**2**	1	2	2	1	1	2	-	2	**1**	2	3	3	3	3	**1**	2	2	3	3	**2**	3	3	3	2	2	3	2	3	2	3
GROUNDWATER AND GAS TESTING	**2**	2	2	2	2	2	2	2	2	3	3	2	2	2	2	3	**1**	**1**	1	2	**1**	2	**1**	**1**	3	**1**	**1**	2	**1**	2	2	3
VIBRATION	**2**	2	2	2	2	2	2	-	-	-	2	2	-	-	1	-	**2**	-	1	2	-	2	-	1	-	1	1	1	2	1	1	-
LABORATORY TESTING																																
STRESS–STRAIN	**2**	1	1	2	2	1	1	2	-	2	-	3	3	3	2	2	**1**	2	2	3	2	2	2	2	2	2	2	2	1	2	1	2
CREEP	2	2	1	2	2	1	1	2	-	2	-	3	2	2	1	2	2	3	2	2	3	2	2	2	2	3	2	3	2	3	2	3
STRENGTH	2	1	1	2	2	1	1	2	-	**2**	2	3	2	3	1	2	**1**	2	2	3	2	2	2	2	3	2	2	2	2	2	2	3
WEATHERING AFTER EXCAVATION	2	1	1	2	2	1	**2**	2	-	2	3	2	2	2	2	2	**1**	**1**	1	1	2	3	**1**	2	2	2	1	1	**1**	**1**	2	2
RIPPABILITY	2	1	1	2	2	1	2	2	-	2	**1**	2	2	2	2	3	**1**	1	3	1	3	**1**	2	3	3	2	2	3	2	**1**	2	3
GEOPHYSICAL	3	**2**	2	2	2	2	2	3	3	2	2	3	2	2	2	2	2	2	3	3	2	3	3	3	2	2	3	3	3	3	3	3
GEOCHEMICAL	3	3	3	3	3	3	3	**2**	-	3	2	3	**2**	**2**	**2**	2	2	**1**	2	2	2	2	2	3	2	2	2	2	2	2	2	2
SOILS TESTING	**2**	-	-	-	-	-	-	-	-	-	-	-	3	-	3	3	**1**	3	**1**	1	2	-	**1**	-	-	**1**	**1**	**1**	**1**	1	**1**	2
WATER QUALITY	1	3	3	3	3	3	**2**	2	-	3	3	3	**1**	**1**	**1**	2	2	2	2	2	**1**	3	**1**	1	2	**2**	**2**	1	2	2	2	2

LEGEND

APPLICABILITY / PROBABILITY OF SUCCESS	IN ALL AREAS	OFTEN	FREQUENTLY	OCCASIONALLY	VERY SELDOM
HIGH	**1**	1	1	1	1
DEPENDS ON SITE CONDITIONS	**2**	2	2	2	2
LOW	**3**	3	3	3	3

- Evaluation of regional mining experience and correlation with the general site conditions.

The example bar schedule on Figure 3 coordinates these tasks with other mine development activities. The entire General Site Evaluation can normally be completed in one or two working months, generally, at a cost of $5,000 to $20,000. Variations depend upon the size and geologic complexity of the site, the quality and extent of information available, and the planned mining method.

The following discussion outlines types and sources of information most commonly important to the successful completion of this initial investigation step.

Review of Available Geologic Literature and Related Documents

Valuable information that can be economically gained from a literature search includes:

- General to specific information of the regional and local geologic and groundwater conditions.

- Identification of major geologic anomalies.

- Specific geologic information from nearby operations or major construction activities.

- General knowledge of actual or potential conditions in the ore body and adjacent strata that should be investigated.

- Sensitive environmental factors that will have to be resolved.

- Seismic data, if appropriate.

Important sources of information include:

- Local and national government agencies.

- Universities.

- Open files for nearby projects.

- Closely related publicly available documents, such as Safety or Environmental Reports.

Study of Available Satellite Imagery and Aerial Photography

The science of photogeology and the use of aerial photographs can be applied very quickly and economically when publicly available material is used. For many parts of the world, several types of imagery and photography are available from government agencies. These include:

- LANDSAT 1 and 2 (formerly ERTS-1) satellite imagery obtained from an altitude of about 570 miles, including coverage of the same area during different seasons.

- Skylab black and white, color, and color infrared photographs taken during 1973 and early 1974 from an altitude of 270 miles are also available from the EROS Data Center.

- Color and color infrared high altitude aerial photographs obtained from NASA aircraft, flown at altitudes of about 65,000 feet. These photographs provide transitional data between the small-scale satellite imagery and large-scale conventional photographs.

- Low altitude (conventional) black and white aerial photographs are readily available for most of the United States through several government agencies. Often they are all that will initially be necessary to identify vegetation units, apprise existing developments, and construct an accurate preliminary geologic map of the area of interest.

The importance of expert interpretation of imagery and photographs cannot be overemphasized. Although obvious geologic features can be recognized by many geotechnical professionals, a great deal of experience is required to complete an accurate and full geologic analysis.

Major cost benefit advantages that can be gained from proper implementation of this background study task include:

- Photo interpretation provides the only practical means for locating many significant anomalies that otherwise could go undetected throughout an entire geotechnical investigation.

- The identification of geologic trends over a larger mine area can be valuable for planning the most efficient site-specific investigative program.

- Even if no unusual or particularly interesting geologic features are identified, the cost is so low that the "negative" result offers a high benefit cost ratio by avoiding subsequent, more costly field investigation searching for potential anomalies that do not exist.

- There is no quicker way to obtain a general assessment or "feel" for a new and unknown area than by reconnaissance photo interpretation.

Ground Confirmation by Fly-Over and Drive-Through Reconnaissance

Every general site evaluation study should include field observations by the _principal_ geotechnical investigator to supplement and confirm conclusions drawn from the literature or photogeology tasks. The site reconnaissance program may vary from as little as two days to as much as two weeks. When possible, the survey should be undertaken by two geotechnical experts of different but complementary backgrounds to allow for the exchange and critique of ideas as observations are made. Reconnaissance tasks usually include:

- A fly-over aerial survey, using a light aircraft or helicopter, to provide an overall view of site conditions and to identify areas of greatest importance for surface observations.

- A drive-through/walk-through site survey to observe slope conditions, stream conditions, surficial deposit characteristics, rock outcrops, and other major features judged by other studies to potentially be important, particularly during photogeologic study.

- Detailed collection and evaluation of rock samples are generally not part of the initial site reconnaissance program. However, most investigators generally confirm the rock characteristics at the site, and at complex sites, it may be prudent to spend at least several days obtaining and evaluating selected outcrop, stream sediment, and soil samples.

Evaluation of Regional Mining Experience and Correlation with the General Site Conditions

When available, information gained by evaluating experiences at nearby mines is particularly valuable for identifying geologic conditions which may influence mine design. Care must be taken to assure that the geologic conditions are similar, or if different, that the differences are fully considered. Also the primary characteristics of the experience must be categorized: did geologic conditions dictate the experience, or was the experience precipitated by the mining procedures used?

The information gained from the General Site Evaluation Study is the basis for early identification of major limiting factors and planning a subsequent, more detailed investigation. Recommendations at this stage must be clearly presented for decisions to be made by the mining engineer and management. The Geologic Hazard Analysis Technique, a bookkeeping tool for visually presenting the data, is introduced in the next section.

5.0 GEOLOGIC HAZARD ANALYSIS CONCEPTS

The way a mine opening behaves is seldom due to only one geologic condition, but rather behavior changes with areal extent because of changes in combinations of factors. For example, a thin shale roof alone may not result in serious rock falls, but a severe problem occurs when a thin shale roof lies beneath an aquifer with high water pressure. It is very logical then to use a visual bookkeeping system to keep track of the various combinations that exist at any mine.

An effective procedure for this purpose, termed "Geologic Hazard Analysis," is introduced in this section. It has applicability at all phases of investigation from initial exploration through final design and operations monitoring. For geotechnical investigations, a preliminary version should be established during the General Site Exaluation Phase, discussed in Section 5.0, and then updated with improved information with each subsequent more detailed investigation.

The procedure consists of (1) mapping onto separate maps those conditions known to have design significance and

(2) overlaying each of the maps to establish where combinations of "good conditions" and "bad conditions" cumulatively occur (Figure 9), requiring major consideration in design.

As a further example, and again considering the roof problem, it is generally accepted that the roof control and support conditions depend on:

- The lithology above the roof.

- The frequency and characteristics of bedding planes.

- The frequency and characteristics of jointing.

- Planes of major weaknesses, including fracture zones and/or faults.

- Water seepage and its effect on rock properties.

- Gas and water pressures in strata overlying the roof stratum.

- Future chemical modifications such as the oxidation of pyrite materials.

- The thickness and conditions of draw slate or rider coal seams that may be left in place.

- In situ stresses.

- The magnitude of stress relief that may have occurred due to surface erosion and stream channel or valley development.

- Buried ancient stream channels above the mining horizon

- Orientation of entries with respect to the cleat.

If all of these factors were known in detail for every location, the prediction of entry roof control requirements would be highly accurate. Figures 4 through 9 illustrate how these data are used in a cumulative manner to complete a Geologic Hazard Analysis:

- Figure 4 is the base map for a proposed mining area, showing the surface drainage features, the coal structu contours and the location of preliminary borings.

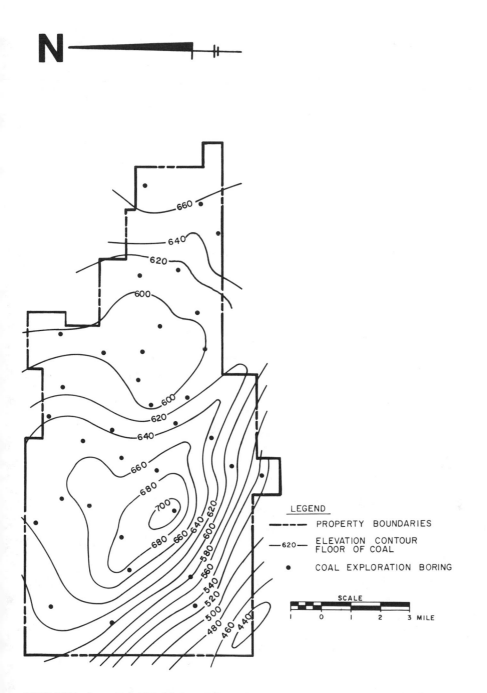

N

LEGEND

- - - - PROPERTY BOUNDARIES

—620— ELEVATION CONTOUR
FLOOR OF COAL

• COAL EXPLORATION BORING

SCALE

1   0   1   2   3 MILE

FIGURE 4 — BASE MAP FOR GEOLOGIC HAZARD ANALYSIS

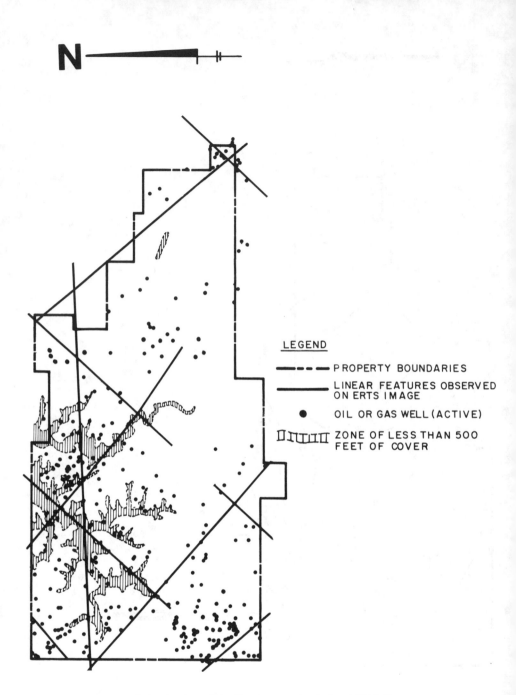

N

LEGEND

PROPERTY BOUNDARIES

LINEAR FEATURES OBSERVED ON ERTS IMAGE

OIL OR GAS WELL (ACTIVE)

ZONE OF LESS THAN 500 FEET OF COVER

FIGURE 5 — LOCATION OF GAS AND OIL WELLS, MAJOR
LINEAMENTS AND AREAS OF POTENTIAL
STRESS RELIEF

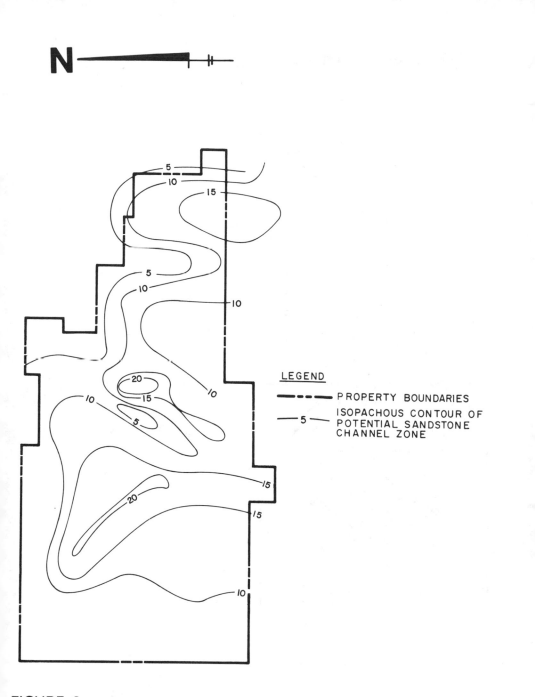

N

FIGURE 6 — ISOPACHOUS CONTOURS OF POTENTIAL
SANDSTONE CHANNEL ZONES

LEGEND

▬ ▬ ▬ ▬ PROPERTY BOUNDARIES

⎯ 5 ⎯ ISOPACHOUS CONTOUR OF
POTENTIAL SANDSTONE
CHANNEL ZONE

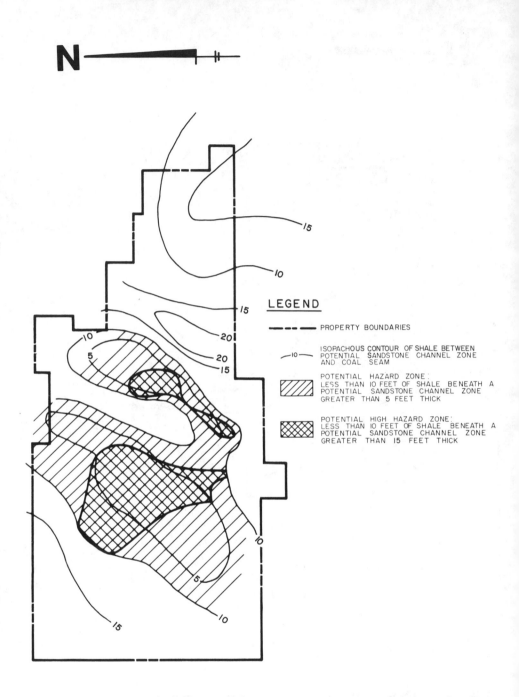

N

LEGEND

— — — — — PROPERTY BOUNDARIES

⌢10⌢ ISOPACHOUS CONTOUR OF SHALE BETWEEN POTENTIAL SANDSTONE CHANNEL ZONE AND COAL SEAM

POTENTIAL HAZARD ZONE: LESS THAN 10 FEET OF SHALE BENEATH A POTENTIAL SANDSTONE CHANNEL ZONE GREATER THAN 5 FEET THICK

POTENTIAL HIGH HAZARD ZONE: LESS THAN 10 FEET OF SHALE BENEATH A POTENTIAL SANDSTONE CHANNEL ZONE GREATER THAN 15 FEET THICK

FIGURE 7– POTENTIAL HAZARD ZONES DUE TO THIN SHALE ROOF BENEATH SANDSTONE ZONE

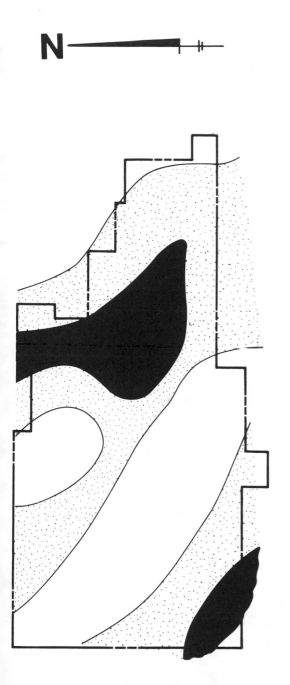

N

LEGEND

- - - PROPERTY BOUNDARIES

ZONE OF POTENTIALLY
HIGH PORE PRESSURE IN
SANDSTONE

ZONE OF POTENTIALLY
VERY HIGH PORE PRESSURE
IN SANDSTONE

FIGURE 8 — AREAS OF POTENTIALLY HIGH WATER HEAD
ABOVE SHALE ROOF

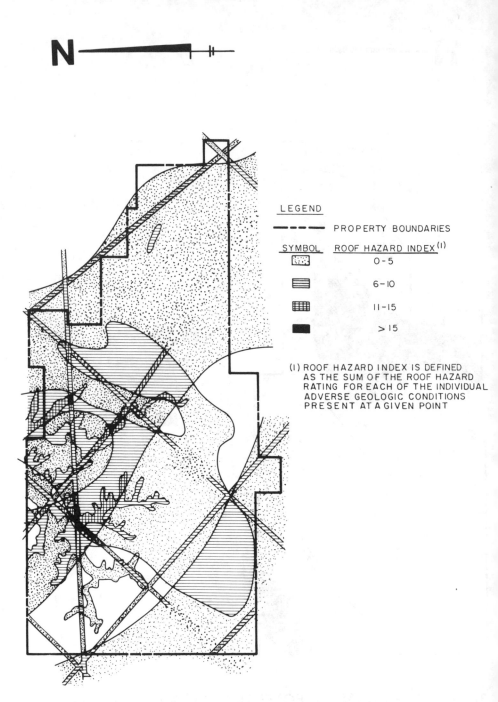

LEGEND

— — — — PROPERTY BOUNDARIES

SYMBOL   ROOF HAZARD INDEX[1]

0 - 5

6 - 10

11 - 15

> 15

(1) ROOF HAZARD INDEX IS DEFINED
AS THE SUM OF THE ROOF HAZARD
RATING FOR EACH OF THE INDIVIDUAL
ADVERSE GEOLOGIC CONDITIONS
PRESENT AT A GIVEN POINT

FIGURE 9 — COMPOSITE GEOLOGIC HAZARD MAP

- Figure 5 shows major fracture zones, oil and gas well locations, and areas of maximum stress relief from topographic unloading.

- Figure 6 shows the potential thickness of the sandstone channel zones above the mine roof.

- Figure 7 shows zones of potential thin roof shale, overlain by significant sandstone zones.

- Figure 8 shows areas of expected maximum water pressure in the overlying sandstone.

- Figure 9 is a composite map indicating combinations of these conditions which may contribute to roof control problems--areas are identified in four degrees of roof complexity, varying from normal problem potential to maximum problem potential.

The value of this information in establishing mining limitations and in determining additional geotechnical analyses that should be undertaken is obvious. Supplemented with additional geologic condition maps, the data can be used to establish design procedures leading to control of each important mining stability factor.

The example was the application of geologic hazard mapping to underground mining and roof control. By adapting the conditions to be mapped as overlays, the technique can be equally applied to other mining problems, such as reclamation problems, strip mine limits, etc. Relative costs of mining in one area versus another can then be visually demonstrated. Examples of additional mappings are:

- Coal thickness and quality.
- Methane production.
- Spontaneous combustion potential.
- Sand channel, clay dike, or fault densities.

6.0 PRELIMINARY MINING EVALUATION

The Preliminary Mining Evaluation step identifies unsuitable mining methods and arrangements, and ranks the feasible mining alternatives. The resulting information is factored into investigation and planning so that:

- All necessary data are obtained during the investigation program without undesirable backtracking.

- Costly, unnecessary or redundant studies are eliminated from the program.

The identification of limiting factors and the development of conceptual plans for feasible mining alternatives can only be made by very experienced individuals with broad interdisciplinary backgrounds. Also required is appropriate communication among the minerals evaluation, mining engineering, operations, geotechnical engineering, and environmental investigators.

Geotechnical limiting factors often become evident by evaluating answers to questions applied to the General Site Evaluation data, such as:

- Can discontinuities be anticipated that will restrict the use of some mechanized mining equipment?

- Do the mine roof or floor conditions restrict the use of any types of equipment or support systems?

- Can water inflow quantities be controlled by specifying special mining methods or sequence?

- Can overburden conditions dictate the location and type of entries and surface facilities?

- Are there preferred directions that influence roof or rib stability, and that dictate mine layout or the location of basic haulage routes?

- Are there any other geological features, such as highly fractured zones or major anomalies that will dictate the mining methods or arrangement?

The combination of limiting factors identified by each discipline often clearly defines those mining systems that

will not work. Also, the identification of probable arrangements for feasible mining systems will occur. And the
requirements for a more detailed investigation program will
be identified. Often, the important criteria will be
determined by addressing questions such as:

- What characteristics of the floor, walls and roof will
  be important for designing permanent and temporary
  support systems in an underground mine--and how much
  of the geologic profile should be investigated in
  detail?

- What is the economic impact of various outside slopes
  at a surface mine, considering the equipment and ore
  recovery--and what are the important parameters necessary
  to design the optimum slope?

- If an unconventional system, such as top slicing of a
  thick seam appears attractive, what is the probable
  mining configuration and what special parameters are
  important for design?

Conceptual plans developed from this process are used to
establish the number and type of geotechnical borings; the
manner in which exploratory and geotechnical borings can be
combined; areas where geophysical methods can supplement or
eliminate borings; the types and number of field and laboratory tests; and an indication of risk conditions that cannot
be investigated in sufficient detail--requiring that a mining
method be discarded or that in-mine monitoring be increased.

7.0   GEOTECHNICAL INVESTIGATIONS

The primary goals of the detailed geotechnical investigation is to (1) outline an accurate three-dimensional picture
of the geology including structure, lithology, and hydrology;
(2) predict the interaction of the geological components as
related to mining; and (3) design the best configuration to
suit mining methods to be used.

The cost effectiveness of the investigation will increase
when:

- All testing and evaluation contributes information
  directly related to the issue.

- The investigation is arranged in phases to permit modification as data become available.

- The investigation employs techniques offering multiple use of particular efforts--for example, borings required for mineral exploration may concurrently provide geotechnical data.

The most valuable geotechnical investigation must be developed by a very experienced engineer who understands the design use of the resulting data; the limitations of the techniques; and the general requirements of the nongeotechnical facets of the mining development.

Table VI lists basic investigative techniques applicable to mining and ranks them in degree of applicability and reliability. Used with Tables III and V, this summary provides a basis for planning a geotechnical investigation. Figure 10 indicates interrelationships between various facet of the investigation.

Fundamental to an optimum geotechnical investigation is selection of the best techniques for the given site conditio Table VII, Alternate Investigation Techniques, is presented to assist the planning of the final details of a geotechnica investigation. By considering alternate investigation techniques, full consideration can be given to such factors as availability of equipment and manpower, special site or mining conditions, and coordination with other investigation

In all cases, the actual investigation must be supervised and periodically reviewed by experts to assure applicability of the data and to make necessary changes. The Geologic Hazard Analyses technique, discussed in Section 5.0, should be continuously updated as the investigation program progresses. The resulting "pictorial" view provides a measure of data completeness.

The following discussions summarize the purpose and types of most frequently used geotechnical investigations.

Photogeology and Remote Sensing

Low altitude black and white photographs for topographic mapping are normally adequate. However, it may also be

TABLE VI

THE APPLICABILITY OF INVESTIGATION TECHNIQUES FOR VARIOUS GEOTECHNICAL FACTORS

	Investigation Technique → / Geotechnical Factors ↓	Photogeology and Remote Sensing	Detailed Geologic Mapping	Rotary Drilling	Core Drilling	Geophysical Logging	Reflection	Refraction	Seismic Cross Hole	Electrical Methods	Gravity Methods	Magnetic Methods	In-Situ Stress	Modulus of Elasticity	Groundwater Monitoring	Vibration	Stress-Strain	Creep	Strength	Weathering After Excavation	Rippability	Geophysical	Geochemical	Soils Testing
REGIONAL GEOLOGY	Geomorphology	1	1																					2
	Structural Geology	1	2	2	1	2	2	2	2	2	2	2	3	3	2	3	2	3	2	2	3	3	3	2
	Stratigraphy and Lithology	2	2	2	1	2	2	2	2	2	2	2	3	3	2	2	2	3	2	2	3	3	2	2
LOCAL GEOLOGY	Structure	1	1	2	1	2	2	2	2	2	2	2	2	3	2	3	2	2	2	2	3	3	3	2
	Stratigraphy and Lithology	2	1	2	1	2	2	2	2	2	2	2	3	3	2	2	2	2	2	2	3	2	2	2
	In-Situ Stress	3	2	3	2	2	3	3	2	3	3	3	1	2		2	1	2	2	2	3	3		3
	Surficial Deposits	1	1	2	2	2	3	3	3	2	3	3			2	3	2	2	2			3	3	1
	Nature of Erosion	1	1	2	2	3	3	3	3	3	3	3				3	3	2	2	1	2	3	2	2
	Man-Made Alterations	1	1	2	2	3	3	3	3	3	3	3			2	2	2	2	2	2	3	2	3	2
GROUNDWATER CONDITIONS	Unconfined Aquifers	2	2	2	1	2	2	2	2	2	2	3			1	3	2	2	2	2	3	2	2	2
	Confined Aquifers	3	2	2	1	2	2	2	2	2	2	3			1	3	2	2		2	3	2	2	3
	Aquicludes	3	2	2	1	2	2	2	2	2	2	3			1	3	2	2	3	2	3	2	2	3
	Effects of Fractures	2	2	2	1	2	2	2	2	2	2	3	2	3	1	2	2	2	2	2	2	2	3	3
SURFACE CHARACTERISTICS	Topography	1	1																	2	3			1
	Surface Hydrology	2	1												2					2	3		3	2
	Erosion	1	1														3	2	2	2	2	2	2	2
	Trafficability	2	1													3	2	2	2	1	1	2	2	1
	Man-Made Alterations	1	1													2	2	2	2	2	2	3	3	1
	Existing and Probable Devel.	1	1												2	2	2	2	2	2	2	3	2	1
ROCK & SOIL PROPERTIES	Resistance to Weathering	2	2	3	1	2	2	2	2	2	3	3	2	2	2	2	3	3	3	1	3	3	2	1
	Load Deformation Behavior	3	1	2	1	2	3	3	2	2	3	3	2	2		2	1	2	1	2	2	3	3	1
	Strength	3	1	2	1	2	3	3	2	3	3	3	2	2		2	1	2	1	2	2	2	2	1
	Time & Environment Dependency	3	1	3	2	3	3	3	3	3	3	3	2	2	2	3	2	3	3	3	3	2	2	1
	Excavation Characteristics	3	2	2	2	2	3	3	2	2	3	3	2	2	1	2	2	2	2	2	1	2	2	1
	Trafficability	3	2	3	2	2	3	3	3		3	3		2	2	2	3	3	2	1	3	3	3	1
	Erodibility	2	2	2	2	2	2	3	2	2	3	3		2	2	2		2	2	2	2	3	2	1
	Potential for Spontaneous Combustion	2		2	1	2	2	3		2	3	3		2							1			

LEGEND

Applicability / Probability of Success	In All Areas	Often	Frequently	Occasionally	Very Seldom
HIGH	1	1	1	1	1
DEPENDS ON SITE CONDITIONS	2	2	2	2	2
LOW	3	3	3	3	3

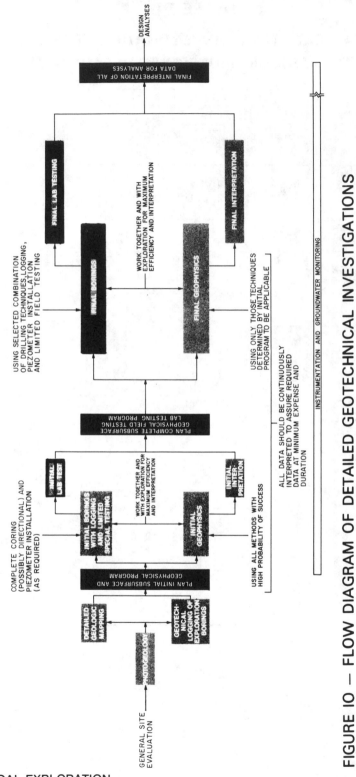

FIGURE 10 — FLOW DIAGRAM OF DETAILED GEOTECHNICAL INVESTIGATIONS

TABLE VII

ALTERNATE INVESTIGATION TECHNIQUES

Basic Investigation Technique	Photogeology and Remote Sensing	Detailed Geologic Mapping	Rotary Drilling	Core Drilling	Geophysical Logging	Reflection	Refraction	Seismic Cross Hole	Electrical Methods	Gravity Methods	Magnetic Methods	In-Situ Stress	Modulus of Elasticity	Groundwater Monitoring	Vibration	Stress-Strain	Creep	Strength	Weathering After Excavation	Rippability	Geophysical	Geochemical	Soils Testing
SURFACE MAPPING																							
Photogeology and Remote Sensing		b/2																					
Detailed Geologic Mapping	b/1																						
SUBSURFACE INVESTIGATION																							
Rotary Drilling				a/2		b/1	b/1	b/1	b/1	b/1	b/1												
Core Drilling			b/1		b/1																		
Geophysical Logging				b/2		b/2	b/2	b/2	b/2	b/2	b/2												
GEOPHYSICAL INVESTIGATION																							
Reflection			a/2	a/2	a/1			b/2	b/2	b/1	b/1												
Refraction			a/2	a/2	a/1			b/2	b/2	b/1	b/1												
Seismic Cross Hole												b/1				b/1							
Electrical Methods			a/2	a/2	a/2	b/1	b/1			b/1	b/1												
Gravity Methods			a/2	a/2	a/2	b/2	b/2	b/2	b/2		b/1												
Magnetic Methods			a/2	a/2	a/2	b/2	b/2	b/2	b/2	b/2													
FIELD TESTING AND INSTRUMENTATION																							
In-Situ Stress																							
Modulus of Elasticity								b/2								b/1							
Groundwater Monitoring									b/1														
Vibration						b/2	b/2	b/2															
LABORATORY TESTING																							
Stress-Strain			b/2					b/2				b/2	a/2										
Creep																							
Strength			b/2																				
Weathering After Excavation																							
Rippability								b/2															
Geophysical					b/2	b/2	b/2	b/2	b/2	b/2	b/2												
Geochemical									b/2	b/2	b/2												
Soils Testing																							

NOTE:

THE USER OF THIS TABLE MUST RECOGNIZE LIMITATIONS IN COMPARING TECHNIQUES FOR GENERALIZED SITUATIONS. THIS TABLE IS A PLANNING GUIDE ONLY.

LEGEND

a – INDICATES THAT THE ALTERNATE TECHNIQUE CAN PRACTICALLY ALWAYS BE USED IN PLACE OF THE BASIC TECHNIQUE.

b – INDICATES THAT THE ALTERNATE TECHNIQUE CAN SOMETIMES BE USED IN PLACE OF THE BASIC TECHNIQUE.

1 – INDICATES THAT THE ALTERNATE TECHNIQUE SHOULD COST LESS THAN THE BASIC TECHNIQUE.

2 – INDICATES THAT THE ALTERNATE TECHNIQUE USUALLY COSTS MORE THAN THE BASIC TECHNIQUE.

appropriate to obtain low altitude color and color infrared photography, or occasionally, remote sensing data such as thermal infrared imagery. When the aerial photography contractor has multiple camera capabilities, the additional cost for multiple photos is small.

Major uses of photogeology in geotechnical investigations include:

- Mapping of stratigraphy (bedrock units).

- Mapping of rock structure, attitude (strike and dip), displacement, zones of weakness, or areas of potential water inflow such as faults and fractures.

- Mapping of surficial materials, alluvial deposits, unusual soil or water conditions, buried geologic features, landslide areas, subsidence and other natural or man-made disturbed areas.

Detailed Geologic Mapping

The next least costly geotechnical investigative step is the detailed field mapping based on observations and measurements of surface features. A typical detailed field mapping study consists of:

- Examination of topography and its relationship to geologic conditions.

- Identification of lithologies and measurement of rock structural parameters.

- Verification of faulted and fractured zones observed in photogeology.

- Observation of erosion characteristics and valley sediments.

- Observation and field testing of major springs and seepage areas.

- Excavation of test pits in areas planned for constructed facilities.

- Research of past mining or exploration activity.

Subsurface Investigations

In many respects, borehole drilling represents the least efficient investigation tool because:

- Significant costs are encountered for the entire length of the borehole, when pertinent information is isolated to only one portion of the depth.

- A borehole provides information at only one spatial location and normally allows limited prediction about formational variations between borings, particularly if the spacing is large.

- It is very difficult to obtain important information about geologic anomalies from borings.

Despite these inefficiencies, borehole drilling is always fundamental because it is the only means of actually observing conditions that do exist beneath the ground surface. Considerations which will minimize this apparent conflict include:

- Coordination of mineral exploration drilling with the geotechnical investigation.

- The less costly investigative techniques (photogeology, remote sensing, geophysical investigations and detailed geologic mapping) should be used extensively to identify the best locations for borings and to provide a method for correlating data from separate borings.

- Expert interpretation should be made of information resulting from each investigative technique as it is completed, and the investigation plan should be regu- larly reevaluated so that unnecessary borings are eliminated and necessary borings are made in optimal locations.

- The total borehole program should include a balance of core and rotary drilling. General guidelines for selecting drilling methods are summarized in Table VIII.

TABLE VIII

APPLICATION AND REQUIREMENTS FOR DRILLING METHODS

GEOTECHNICAL AND ECONOMIC FACTORS / DRILLING TECHNOLOGY	SAMPLING				LOGGING		TESTING					ECONOMIC FACTORS				
	Soil	Rock	Mineral Body	Water	Geophysical	Drilling Log	In Situ Strength	Vibrations	Shear Strength	Pumping	Permeability	Equipment Cost	Drilling Speed	Drilling Cost	Approximate Cost $/Ft.	Approximate Speed Ft./Hr.
Shallow Power and Hand Drilling	1	3	3	1	3	-	-	-	1	1	1	a	b	a	4	15
Cable Tool Drilling	3	3	3	1	1	3	1	1	-	1	1	a	a	a	4	4
Direct Circulation Rotary Drilling – Water	-	2	3	2	1	1	1	1	-	2	2	b	b	b	5	40
Direct Circulation Rotary Drilling – Mud	-	2	3	3	1	1	2	1	-	3	3	b	b	b	6	40
Direct Circulation Rotary Drilling – Air	-	2	3	1	1	1	1	1	-	1	1	c	c	b	8	60
Reverse Circulating Rotary Drilling Water	-	3	3	2	1	1	1	1	-	2	2	b	b	b	6	40
Diamond Core Drilling	3	1	1	2	2	1	1	1	-	2	2	b	a	c	12	5
Air Percussion (Blast Holes)	-	3	3	1	-	2	-	-	-	-	3	a	c	a	2	80

LEGEND

1 = Very applicable in most cases

2 = Applicable sometimes in conjunction with other tests

3 = Indirectly applicable in some cases

a = Relatively low

b = Moderate cost

c = Relatively high

- When possible, geophysical logging should be used in place of coring to save costs, while still providing necessary data for evaluating:

  - Rock porosity and permeability.
  - Rock density and moduli.
  - Rock stratum.
  - Fractured or less competent rock zones.
  - Aquifer locations.
  - Mineral body location.

Geophysical logging applications are summarized in Table IX.

- Early consideration should be given to obtaining oriented cores at selected borings where the direction of original sedimentation or cleat and joint weakness is important to mine planning.

- Appropriate exploration and/or geotechnical borings should be used for required field testing and instrumentation (discussed later in this section) to minimize drilling footage and field rig time.

Geophysical Surveys

Although geophysics has proven valuable for oil and mineral exploration for many years, its acceptance as an engineering tool has only been recent. Its successful use at many projects suggest that this technique has great practical importance to mine site investigations. Of particular note:

- Valuable information can be obtained on a total area basis for a large site--a condition that is not possible using any number of boreholes.

- Faults and fracture planes can be located and evaluated-- they do not lend themselves to identification and analysis using borehole techniques.

- The cost per acre of investigated areas is much less than for borehole techniques.

TABLE IX

APPLICATION AND REQUIREMENTS OF BOREHOLE GEOPHYSICAL LOGGING

GEOPHYSICAL METHOD / GEOTECHNICAL AND ECONOMIC FACTORS	Soil-Rock Interface	Weathered Zone	Lithologic Changes	In Situ Strength	Rippability	Porosity	Groundwater	Density	Fractured Zones	Vibrations	Modulus of Elasticity	Geochemical	Equipment Cost	Operation Speed	Operation Cost	Approximate Cost $/Ft.	Approximate Speed Ft./Hr.
Resistivity	1	2	1	3	-	1	1	3	3	-	3	1	a	a	a	.50*	100
Spontaneous Polarization	1	2	1	3	-	3	2	3	3	-	3	1	a	a	a	.50*	100
Natural Gamma Radiation	2	2	1	3	-	3	3	3	3	-	3	3	a	a	a	1*	100
Neutron Log	2	2	2	3	-	1	1	1	3	-	3	3	b	a	b	2*	100
Gamma-Gamma Log	2	2	1	1	-	2	3	1	3	-	1	3	b	a	b	2*	100
Thermal Log	-	3	-	-	-	-	1	-	3	-	-	-	a	b	b	.50	50
Caliper Log	2	2	3	-	-	-	2	-	1	-	-	-	b	b	b	1	50
Sonic Log	1	2	1	1	2	1	2	1	1	1	1	-	c	b	c	2	50
Borehole Camera (Televiewer)	2	1	3	-	-	-	1	-	1	-	-	-	c	c	c	2	20

* When conducted simultaneously, the total cost will be $2.00 to $3.00/Ft.

LEGEND

1 = Very applicable in most cases

2 = Applicable sometimes in conjunction with other tests

3 = Indirectly applicable in some cases

a = Relatively low

b = Moderate cost

c = Relatively high

Design information that can be obtained from engineering geophysics includes:

- Strata thickness and attitude.

- Material competency (rippability).

- Rock stress-strain properties (including the effects of fracturing and bedding).

- Structural faults.

- Lithology and mineral changes.

- Major aquifer locations.

Table X offers assistance in determining which geophysical techniques should be considered and to what extent the resulting data may be applicable.

To be accepted as a valuable investigation tool, geophysical programs should be planned and implemented considering the following points:

- Geophysics must be regarded as a supplementary or complementary exploration tool to be used in conjunction with other investigative techniques—it is not an absolute substitute.

- The planning of an efficient geophysical investigation must incorporate the thoughts of specialists who will eventually interpret and use the data.

- Geophysical investigations should normally be initiated on a phased basis with continuing evaluation to assure that pertinent, interpretable data do result.

- The geophysical investigation must have the close supervision of a geophysicist knowledgeable in procedures, equipment limitations and data analysis.

Field Testing and Instrumentation

Field testing and instrumentation are necessary when existing conditions or material characteristics cannot be determined through geologic studies or laboratory testing.

TABLE X

APPLICATION AND REQUIREMENTS FOR GEOPHYSICAL SURVEYS

GEOTECHNICAL AND ECONOMIC FACTORS / GEOPHYSICAL METHOD		Soil-Rock Interface	Weathered Zone	Lithologic Changes	In Situ Strength	Rippability	Porosity	Groundwater	Density	Fractured Zones	Vibrations	Modulus of Elasticity	Geochemical	Equipment Cost	Operation Speed	Operation Cost	Range of Unit Costs
		GEOTECHNICAL FACTORS												ECONOMIC FACTORS			
SEISMIC	Refraction	1	1	1	-	1	2	2	2	2	1	2	-	b	b	b	
	Reflection	3	3	2	-	3	3	3	3	3	2	2	-	c	c	c	
	Cross-Hole	3	3	2	3	3	2	3	2	3	1	1	-	b	b	c	
ELECTRIC	Resistivity	1	1	2	-	-	2	1	3	3	3	3	2	b	b	b	
	Self Potential	1	1	2	-	-	2	1	3	3	3	3	2	b	a	b	
	Induced Polarization	1	1	2	-	-	2	1	3	3	3	3	2	b	a	c	
OTHERS	Gravity	2	2	2	-	-	2	2	1	2	-	3	-	a	a	b	
	Magnetic	3	3	2	-	3	3	3	3	2	-	3	-	a	a	a	
	Radiometric	-	-	3	-	-	3	3	3	3	-	-	-	b	b	b	

LEGEND

1 = Very applicable in most cases

2 = Applicable sometimes in conjunction with other tests

3 = Indirectly applicable in some cases

a = Relatively low

b = Moderate cost

c = Relatively high

However, all field testing is expensive and must be carefully planned as a supplement to other procedures. The field testing and instrumentation procedures most applicable to mine development are:

- Measurement and monitoring of groundwater and gas emission conditions.

- Measurement and evaluation of ground vibrations resulting from blasting.

- Measurement of stress-strain characteristics of material that cannot be adequately measured by laboratory testing.

- Measurement of in situ stress conditions that may have major impact on rock mechanics analyses.

Of these four, groundwater level and gas monitoring are most frequently used and should be established in coordination with exploration drilling. The measurement of vibrations due to blasting is probably the least complex field testing program. However, its importance is limited to sites where significant blasting is expected.

The measurement of the stress-strain characteristic of rock in boreholes is not complex, but the usefulness of the resulting data for geotechnical analyses is often questionable. The costs and applicability of the data must always be weighed against alternative procedures such as cross-hole geophysical testing of the rock mass or laboratory testing of recovered samples, down hole acoustic logging, or simply the selection of approximate values based on experience and judgment.

The question of in situ stress conditions can be of major concern when evaluating minability--if high horizontal stresses do exist, they can significantly reduce stability and contribute greatly to rock burst problems. However, most regional geologic studies will show that stress levels are not high or that the direction of stress will not greatly influence mining. Therefore, in situ stress testing is usually not a major investigation factor. Even when high stresses are determined to be important, testing should be planned on a staged basis so that it can be terminated if results are not reliable. The available testing procedures are very costly, require a very high level of technical sophistication, and have relatively low probability of being successful.

Laboratory Testing

The most common laboratory tests for rock are:

- Stress-strain or elastic modulus.
- Creep or plastic flow.
- Compressive strength.
- Tensile strength.
- Triaxial and shear strength.
- Drillability or rippability.
- Resistance to weathering.
- Geochemical and mineralogical characteristics.

Although accurate rock properties information is desirable for design, the investigator must recognize the limitations of laboratory testing and provide for a proper balance with field testing and general geologic interpretation. Limitations and the requirement for expert planning and interpretation include:

- Laboratory tests can be run only on small intact core samples, while the behavior of the rock mass may be more dependent upon discontinuities or anomalies, such as bedding planes, fractures, etc.

- The core samples suitable for testing often represent the best rock encountered, and result in unconservative quantitative data.

- It is very difficult to simulate in situ conditions during laboratory testing of rock materials.

- Rock conditions can vary significantly throughout the mine area and many costly laboratory tests would have to be conducted to obtain statistically reliable, yet representative information.

In general, the testing of the geotechnical properties of rock is best conducted as a carefully planned combination of field and laboratory testing. For example, the modulus of elasticity of rock can be determined: (1) locally by borehole load deformation gages, (2) regionally by crosshole geophysics, or (3) it can be estimated based on available core recovery correlations. Laboratory testing for this property should be restricted to a reasonable number of confirmation tests.

Several rock tests have lesser restrictions and are
commonly required. These include tests to determine the
drillability or rippability, the geochemical and mineralogical
composition, and sensitivity to weathering under varying
moisture and atmospheric conditions. Testing for these
properties is not complicated, and sufficient experience is
available for accurate interpretation of the resulting data.

Discussion of soil testing procedures is not included in
the paper because of limited applicability to mine design and
because procedures for testing and analyzing soils are well-
defined. In the orderly development of a mine, however, soils
testing is necessary for the construction of processing, preparation and storage structures; the construction of transport
systems; mineral storage stability; overburden or spoil material
stability; and where applicable, the design of water or slurry
disposal impoundments.

3.0 FINAL PREMINING DESIGN

Although beyond the basic scope of this paper, brief
discussion of geotechnical design procedures is presented
for completeness.

Geotechnical design differs from other types of engineering design. For example, in the design of steel or concrete
structures, the material properties and loading conditions
are generally well known, allowing accurate analyses by well
established scientific methods. Further, the material
properties and other parameters are generally constant with
time or the changes are well defined. Under such conditions,
the requirements for engineering judgment are limited. In
contrast, the material properties and loading conditions of
geologic materials in a mining environment cannot be exactly
predicted for any point in time or space. Accordingly,
geotechnical design must rely heavily upon the experience and
judgment of the engineer as well as upon scientific analyses.

With the above cautionary note, it is emphasized that
mine design based on predictive geotechnical criteria is
routinely accomplished. Often the level of design confidence
will be high enough to allow mine development to move directly
to full production. In complex situations, the premining
design may have to be confirmed by monitoring during an
initial exploratory mining operation. In either case, the
premining geotechnical investigation and design is the only

way to optimize the mining configurations and foresee potential difficulties which could otherwise cause unexpected and expensive interruptions in mining.

Procedures for Final Premining Geotechnical Design can be summarized in the following five categories:

- Geologic Hazard Analysis (Section 5.0) provide the basis for the designers to use all information resulting from the geotechnical investigation and to subdivide the mine area into design domains. The domains indicate basic limitations on operations, areas of potential problems, probable solutions for these problems, and the magnitude and scope of design necessary to initiate mining.

- Rock Mechanics Analyses make use of various equations from elasticity and elasticity-plasticity theory to quantitatively predict the behavior of various mining configurations; set anticipated dimensions such as pillar size or roof span; evaluate alternative support systems; determine optimum surface mine slopes; and establish extraction ratios. The scope of rock mechanic analyses can vary from very inexpensive (linear elastic theory) to very expensive (nonlinear, three-dimensional finite element). The required analyses must be determined for each site to be consistent with the accuracy of available parameters and the importance (cost or safety) of the results.

- Physical Model Testing seldom is required for single mine design studies, but when appropriate, they can be used to predict and evaluate the relative behavior of alternative configurations or support systems.

- Hydrologic Analyses predict the magnitude of surface and/or groundwater that must be controlled and the impact of regional and local changes due to mining. Analyses for surface flow are well defined and standard. Analyses of groundwater are less straightforward and can vary from simple empirical estimates to complex computer predictions. The scope of analyses must be tailored to the requirements of each mining situation, and the adequacy of available geologic parameters.

The geotechnical designer can contribute in a parallel
role with regard to gas emission. Analyses techniques
for water and gas flow are similar.

- Soil Mechanics and Foundation Design Analyses are
  required for most surface construction and site development work, including highwall and spoil slopes for
  surface mines, the disposal of waste material and the
  construction of impoundments.

9.0 IN SITU MONITORING

The technology, experience and professional personnel are
available to predict the geotechnical behavior of typical
mining operations with a high degree of assurance that the
mine will be workable, safe, and cost effective. In other
cases the geologic conditions may be unusually complex, or the
mining method may be unusually sensitive to geologic conditions. In such cases, adjustment of the mining plan should
be anticipated at the time operations are initiated.

The prudent geotechnical engineer will anticipate discrepancies, between conclusions of the premining investigation
and the actual conditions. To allow for discrepancies,
particularly at small mines, the best geotechnical design
may incorporate conservative features, recognizing that they
may impose unnecessary limitations on mining. More generally,
good premining design will include procedures for monitoring
in-mine conditions to verify design assumptions, or to
identify conditions that dictate design changes for improved
safety or decreased operating costs. In any event, some
planned monitoring program should be implemented to assure
that mining is not continued under conditions varying by
more than preestablished limits--surprises, caused by
unrecognized rapid change in geologic conditions as the mine
is advanced, should not occur.

Each monitoring program should be carried out in two stages:

- In situ testing and performance evaluation to be scheduled and completed while operations are being initiated.
  The emphasis of such monitoring is to supplement the
  premining investigation to more accurately define
  geotechnical factors which may affect the basic mine
  design. Although the premining plan is not always

modified as a result of such monitoring, it is not considered to be the final plan until the monitoring program has been completed.

- Control monitoring which is scheduled to be maintained after full production operations have been established. The emphasis of such monitoring is to assure that geologic conditions do not change beyond that which has been considered in the geotechnical design. This generally means that the mining procedure does not change, but it may also mean that the mining procedure changes in a preestablished manner as changing geologic conditions are encountered.

Common techniques for either stage of monitoring may be divided into four categories:

- General Observations in Monitoring
- Deformation Monitoring
- Stress Monitoring
- Water, Gas and Environmental Monitoring

As is the case for all aspects of geotechnical investigation and design, the monitoring program should be regularly evaluated for its effectiveness: is it producing the data desired? does it relate to the geologic conditions encountered and the mining methods used? are the data being interpreted soon enough to properly influence mining decision and are the data being summarized and stored in a condition useful to the design of future mines?

10.0 CONCLUDING REMARKS

It is worth repeating that the general behavior of any mine is a direct result of the surrounding geologic condition The ability of geological scientists and engineers to define these conditions and predict how they will behave is much higher than commonly recognized. Proof of this conclusion is documented by the many successful foundation and underground construction projects completed every day.

Maximum benefit of geotechnology can be realized only if it is incorporated as an integral and continuous part of the mine investigation, planning and design process. The intent of this paper has been to identify this role and to outline many of the available techniques to make the program successf

Section Five Discussion

MR. WALTON: (Walton and Coppach) I would like to ask both the speakers one or two questions about digability in surface mine operations. Firstly to the two men from d'Appolonia, I very much liked your plan showing roof quality which seemed to be on a large scale, very similar to the sort of thing that is being done in South Africa, and which Barton in the MGI have been doing on tunnelling conditions. Have you done a similar exercise, and how would you go about it, on digability in open-cast operations or for advising clients on directions of advance or even equipment? Secondly, to Dr. Voigt, I would like to enquire of him precisely what sort of testing would he recommend for the selection of bucketwheel excavators or assessing the suitability of overburden for bucketwheel excavator operations? At present, the only real technique seems to be to send cubes off to Germany to get them tested. Surely there are recognised geo-technical tests which can be done?

R.D. ELLISON: The geologic hazard analysis technique has many different applications. It can be used and is used, for surface mining. In most instances it is less exacting on a surface mine because the problems are not as complex and therefore, you have less parameters to consider, but certainly if you had a fractured system, a faulted system, if your joint pattern or your depth was going to have major impact on slope, you would use the same type of technique to identify zones where the slope should be steeper or shallower and how you would progress from one zone to another zone in the manner that is going to create the least number of problems. Does that answer your question?

Mr. WALTON: Yes, apart from digability.

R.D. ELLISON: Digability meaning how hard is it to remove the material and what type of equipment? Yes, certainly there it has direct application in that if you have your layers of rock very significantly different and you have some hard rock that peters out, either due to topography or just because it peters out in the system, you would use the identical overlay system and you may base it on percentage of hard rock versus soft rock to identify where a bucket-wheel may be used or where a shovel is required, and how these would interface.

DR. R. VOIGT: I am afraid I can't answer your question completely, because the bucket-wheel excavator technique has been applied so far, only in soft rock and tertiary strata consisting of unconvoluted sands, clays, gravels, silt and materials of this kind and wherever bucket-wheel excavating is now going on, Germany, Poland, Greece, Yugoslavia, Spain - it is in tertiary soft rock areas . As far as I know right now, there is an investigation underway in Queensland, Australia, where semiconsolidated rocks, I think sort of limestone or marlstone, are investigated to check if they, after pre-blasting, can be excavated by bucket-wheels. Does it answer your question?

MR. WALTON: Not entirely. There are a number of tests which I understand are done on specific cutting forces. The fact that bucket-wheels are considered for a large range of relatively recent strata doesn't necessarily mean that all relatively recent strata associated with coal seams, can be bucket-wheeled, and I think there are a number of us who would be very interested to have general guidelines for testing and testing methods that would be appropriate for the selection of bucket-wheel equipment.

DR. VOIGT: I forgot something. In Townsend, the northwest of Canada, bucket-wheels have been applied, but I don't know that these tests have been successful. At least the operations have been going on for at least five years, but I am sorry I cannot go farther in this detail because I don't know.

R.D. ELLISON: Have I mentioned that the U.S. Bureau of Mines in Denver are undertaking a number of studies and some are going to turn into demonstration projects as I understand it, to determine the extent that the bucket-wheel can be used particularly for the thick coals in the west and you might write to the Denver Research Centre of the Bureau of Mines and ask - I don't have the details on it.

PROF. HADSELL: I don't know whether I can speak for the School of Mines, Paris and the Royal School of Mines as well as the Colorado School of Mines, but I think they feel that they have been turning out geo-technologists for a number of years and been calling them mining engineers. I would appreciate it if you would comment on this subject.

MR. ELLISON: Sure, our staff consists of mining engineers, mining geologists, engineering geologists, geophysicts, geologists, and I have no pride of authorship. I think we are all professional individuals and we all fit into the picture, so a geo-technical engineer can be a civil engineer, can be a mining engineer, can be an engineering geologist, and if he is capable, his particular title doesn't bother me at all.

S. GAZANFER: (TKI - Turkish Coal Enterprises): Dr. Voigt mentioned the overburden/coal ratio which is the yardstick for operators but one feels that one must include also other parameters, such as the amount of water pumped per ton of coal produced, and I wonder if Dr. Voigt has got any typical figures, on any mines in the Rheinbraun area, on how much water has to be pumped per ton of coal produced. That is my first question. The other one is the pumping wells. I agree that they go beyond the depth of the open pit so it must be around 200 meters in depth. Do they have any caving problems when they drill these and instal the turbine pumps? My next question is about the safe working slopes of the Rheinbraun open pit mines, what do they regard as 'as safe as possible'. And can they quote the values for cohesion and the internal pressure margins for the overburden which is I believe composed of clay and sand?

DR. VOIGT: It is nearly impossible to indicate any figures which fit, which give an overall picture of the water to coal ratio. We have now five mines running at different depths and hydrologically at different locations. The minimum is in our western-most mine - Zukunft mine near Aachen, where the ratio of ton coal to ton water is approximately 1 to 2, (Two tons of water per ton of coal), but in the Fortuna mine, pictures of which I have shown, the ratio is about 12 tons of water per ton of mined coal, but this figure is only valid for a certain time because we are still lowering the water table. About by the end of 1978 we will have steady state conditions, and only the annual recharge will be discharged and then we expect the water to coal ratio to about between 6 or 7 to 1, but it is usually impossible to indicate any figures for the future.

Your second question with regard to cavings. We have had serious problems in certain clay layers which have water molecules in their crystal lattice. These clays swell and may collapse. We drill wells at the maximum diameter of 2 metres and by caliper measurements we have found that the caving extends to 6 metres in diameter. Now we have overcome the problems by adding mud stabilisers to the drilling mud - usually we drill only with pure water - and this appears to have been the right solution. Caving has ceased to be a serious problem. As to your last question: The average slope as we call it, the general slope of a mine is approximately 1-3 and locally in certain clay sediments of course it is flatter, we reduce it to 1-5 and locally, especially in the working benches it may be steep, nearly 90°, and these steep slopes may last for several weeks. We don't consider them to be a problem, it is only the permanent slopes or high walls we are concerned with and these we usually keep at 1-3 as I told you before. Right now I am unable to quote you any figures as to values of cohesion and internal friction.

SECTION 6

12

ESTIMATING THE POTENTIAL OF A COAL BASIN

R.G. Wilson

Consultant Geologist
CRA Exploration Pty. Limited
Melbourne, Australia

INTRODUCTION

Estimating the potential of a coal basin is a very
broad topic encompassing numerous economic and geological
factors. Essentially the task breaks down into four
major facets, viz.

1) Regional geological appraisal

2) Engineering and reverse economic studies

3) Detailed evaluation of coal deposits

4) Feasibility studies

Facets (1) and (3) involve mainly geological input and
are generally conducted by the exploration department of
a coal company whilst facets (2) and (4) involve engineers
and economists to a large degree with only minor input by
geologists.

Any single one of these facets merits a whole paper being
devoted to it at a conference such as this and because the
total scope is so wide, I have chosen deliberately to
confine my contribution to a discussion of the first

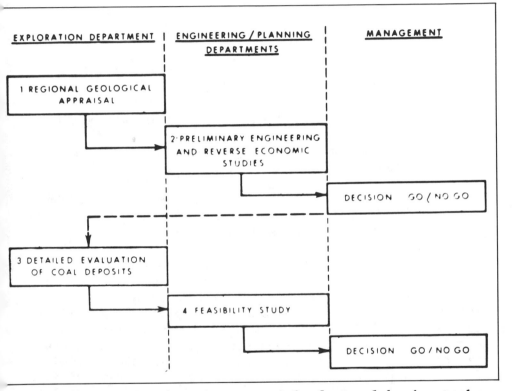

Figure 1. Estimating the potential of a coal basin--tasks and responsibilities.

facet, the regional geological appraisal of a coal basin.

This regional appraisal involves the following individual tasks :

- Determine the extent and boundaries of the selected basin.

- Determine the vertical distribution of time stratigraphic units within the basin.

- Determine the lateral distribution of lithofacies within the basin.

- Isolate the potential coal bearing formations in the stratigraphy.

- Determine the geological history of the basin.

- Reconstruct the palaeogeography of the potential coal bearing formations.

- Collect data from existing oil wells and water bores and test any coal or carbonaceous horizons for rank and type.

- Fill in any gaps by field mapping and sampling.

The approach described in this paper uses a basic picture of coal as a sedimentary rock, a knowledge of the geological factors which influence coal properties and an understanding of the depositional environments of coal bearing strata in order to construct a conceptual model of a coal depository. Such a model is reliably predictive within certain limits and its use enables the coal explorationist to systematically collate information to the point where he or she can reach an objective decision as to the likely extent and nature of any coal deposit present.

The ultimate objective of such a regional appraisal is to provide an assessment of the following :

i)   Likely maximum and minimum in situ tonnages.

ii)   Likely coal quality and its variation.

iii)   Broad variations expected in seam thickness.

iv)   Structure in terms of maximum, minimum and

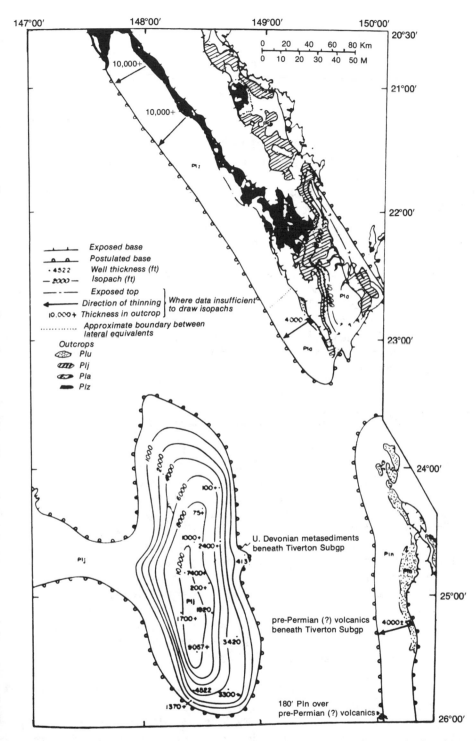

Figure 2. Distribution and outcrops of Camboon Andesite, Reids Dome Beds, Carmila Beds, and Lizzie Creek Volcanics.

average dips, and degree of faulting.

v)   Range of depths at which coal may occur and
         some broad idea as to the proportions of
         coal tonnage mineable by open cut and under-
         ground methods.

From this information engineers and economists can
commence a preliminary appraisal of the potential viability
of any coal deposit at the locality in question. Capital,
operating and infrastructure costs can be estimated and
set against the likely realisation to give some idea of
the size, and quality of the target that needs to be
found in order to establish an economic operation. The
coal geologist can then reassess the situation based on
this preliminary economic study and decide whether the
chances of finding the required target merit continued
exploration. The geologist's judgement at this time is
critical for management could be called upon to commit
large sums of money at the next stage which is the
detailed evaluation of the coal deposits by an intensive
drilling programme.

COAL TYPE AND THE GEOLOGICAL FACTORS INFLUENCING IT

Coal is a combustible sedimentary rock formed from
the accumulation of plant debris which have been subsequently modified by a combination of biological, chemical
and physical processes. Microscopically coal consists of
of number of basic constituents known as macerals which
were first described in 1935 by Stopes[14]. These macerals
in coal may be regarded as analagous to minerals in other
sedimentary rocks and in the same way as mineral assembl-
ages largely determine the properties of rocks, so the
proportions of different macerals present has a profound
influence on the properties of a coal seam. For convenience in description macerals are often grouped by their
petrographic characteristics into three groups known as
vitrinite, exinite and inertinite[4].

Vitrinite is the major constituent of bright coal and
it plays an important role in the ability of a coal to
form coke on carbonisation. Exinite is the major constituent of dull coal; it is hydrogen rich and yields tar on
carbonisation. Inertinite, as its name implies, is non
reactive on carbonisation.

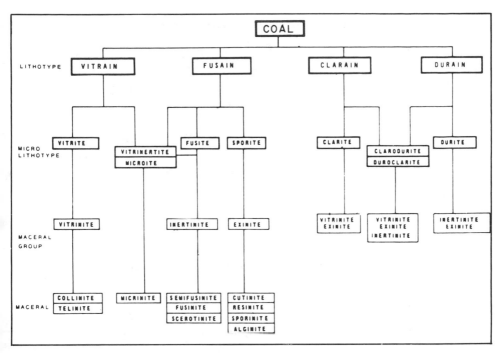

Figure 3. Nomenclature used in coal petrology.

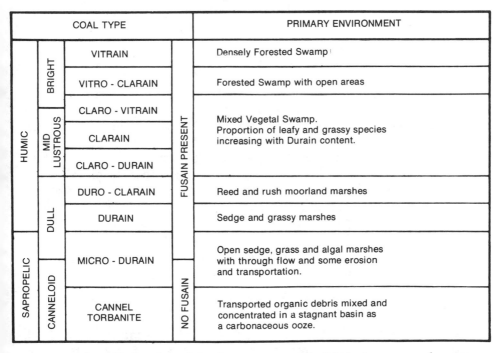

COAL TYPE			PRIMARY ENVIRONMENT
HUMIC	BRIGHT	VITRAIN	Densely Forested Swamp
		VITRO - CLARAIN	Forested Swamp with open areas
	MID LUSTROUS	CLARO - VITRAIN	Mixed Vegetal Swamp. Proportion of leafy and grassy species increasing with Durain content.
		CLARAIN	
		CLARO - DURAIN	
	DULL	DURO - CLARAIN	Reed and rush moorland marshes
		DURAIN	Sedge and grassy marshes
SAPROPELIC	CANNELOID	MICRO - DURAIN	Open sedge, grass and algal marshes with through flow and some erosion and transportation.
		CANNEL TORBANITE	Transported organic debris mixed and concentrated in a stagnant basin as a carbonaceous ooze.

Figure 4. Relationship between coal lithotypes and peat swamp environment.

Vitrinite and inertinite are both thought to be derived from woody plant tissue, the former in a reducing environment where rapid burial in a moist peaty soil led to preservation, and the latter in an oxidising environment. Exinite on the other hand is thought to be derived from spores, leaf and herbaceous tissues[7,13].

Heavily forested swamps in present day conditions are dry swamps i.e. dense tropical forests in which the water table lies a few centimetres below the surface of a generally spongy humic soil. Herbaceous swamps on the other hand are usually wet with abundant free standing surface water or of the temperate highland plateau type - where the prevailing climate precludes the growth of trees.

In tropical to sub-tropical conditions where vegetation proliferates, dry forested swamps are found in extensive fluvial basins e.g. the Amazon Basin; on deltas and the flood plains leading to them e.g. Mississippi delta; and along extensive coastal plains parallel to linear shorelines e.g. Malay Peninsula. Wet herbaceous swamps on the other hand, are found in flooded embayments, lagoons, intermontane land locked basins and on temperate highland plateaux. Often the dry forested swamps of a major fluvial regime are separated by only a short distance from a land locked wet swamp environment e.g. the Midland Peat Province of the Everglades as it relates to the forested paralic plain of the Gulf Coast in Florida, U.S.A.

Coal type is ultimately determined by the relative proportions of the three basic maceral groups in the make-up of a coal seam, and the genesis of these maceral groups can be broadly related to the swamp environment in which the coal was formed. Furthermore, as certain swamp environments are preferentially developed in certain physiogeographic settings, it follows that if the explorationist is able to unravel the geological history and reconstruct the palaeogeography of a coal depository, he or she will be able to predict (within certain limits) the type of coal most likely to be present within it.

COAL RANK AND THE GEOLOGICAL FACTORS INFLUENCING IT

The properties which govern the most suitable end utilisation of a coal depend not only on coal type but also upon coal rank. Rank is an arbitrary measure of the maturity or degree of coalification of the coal and its

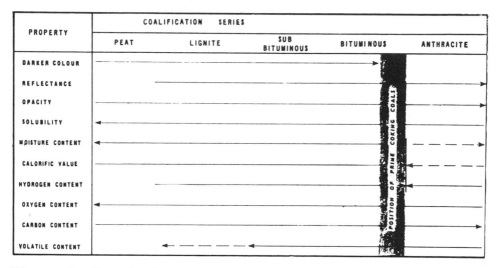

Figure 5. Trend of some coal properties with increasing rank.

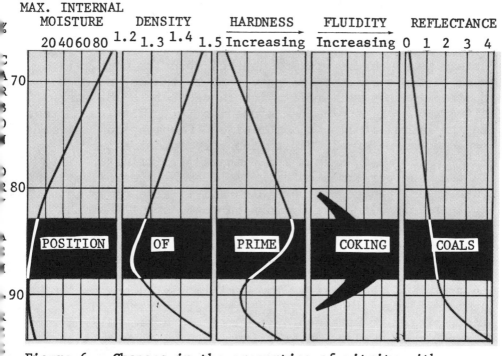

Figure 6. Changes in the properties of vitrite with coalification.

existence was first proposed in 1883 by Von Gumbel.

M. and R. Teichmuller[17] have suggested that there are
two stages to coalification. The first involves mainly
biological activity and leads to the formation of lignite.
This stage is likened to diagenesis in other sedimentary
rocks. The second stage leads to the formation of bituminous and anthracitic coals and involves a combination of
chemical and physical processes. This stage can be likened
to metamorphism in other sedimentary rocks.

Differences in the rank of various coals were first
noted by Hilt in 1873 during his studies of the coals of
the Aachen region in Germany. From these studies Hilt
proposed a relationship between the depth of burial of a
coal seam and its fixed carbon and volatile contents. The
principle that within a single shaft or borehole the rank
of coal increases with depth, is now known as Hilts Law
and it has been shown to hold good for all coal basins
except those which have been subjected to local abnormal
geothermal conditions.

The ultimate rank of a coal can be measured by chemical
parameters such as volatile, carbon, hydrogen and moisture
content or by physical parameters such as the mean maximum
reflectance of vitrinite. The reliability of these parameters as a measure of rank depends upon the comparative
rank of the coal being measured. For example, moisture
content is only reliable in the range peat to low rank
bituminous coal and carbon content or vitrinite reflectance
in the range lignite to anthracite. Perhaps the best all
round measure for all except the very lowest ranks of
coal is the mean maximum reflectance of vitrinite. This
test can be performed on very small quantities of material
such as chippings from boreholes and it also gives intelli-
gible readings for exploratory purposes on all but the
most severely weathered material.

There has been much discussion in the literature
concerning the causes of increased coalification both by
chemists and geologists. The decisive influence of temperature is now widely accepted by most workers and
increasing rank with depth of burial is universally acknowledged to be a function of rising crustal temperatures at
depth (11,19,21). Variations in temperature with depth
of burial (the geothermal gradient) are not constant
however, and they can vary considerably from place to
place. Thus in areas of low geothermal gradient relatively

low rank coals can be expected even where these have been buried to considerable depths during their geological history. This situation contrasts with that in orogenic domains with high geothermal gradients in which geologically young coals which have never been deeply buried can nevertheless attain relatively high rank.

The period of time over which the coal has been subjected to increased temperature is another important factor in the attainment of rank. Karweil[5] discussed this in some detail and he determined that a coal of 19% volatile matter could have been formed by exposure to a temperature of 200°C over 10 million years, 150°C over 50 million years or 100°C over 200 million years. He also demonstrated that a lower temperature of say 50° - 60°C even if applied over 300 million years could never lead to the development of a similar rank of coal.

The more obvious effects of increasing rank on coal properties are as follows :

i)   Colour darkens.

ii)   Reflectance of vitrinite increases.

iii)   Opacity increases.

iv)   Solubility decreases.

v)   Moisture content decreases to high rank bituminous then increases slightly in anthracites.

vi)   Calorific value increases to high rank bituminous then may decrease slightly in anthracites.

vii)   Hydrogen content increases to high rank bituminous then decreases.

viii)   Oxygen content decreases.

ix)   Carbon content increases.

x)   Volatile content decreases in the range lignite to anthracite.

The point in the coalification series where moisture

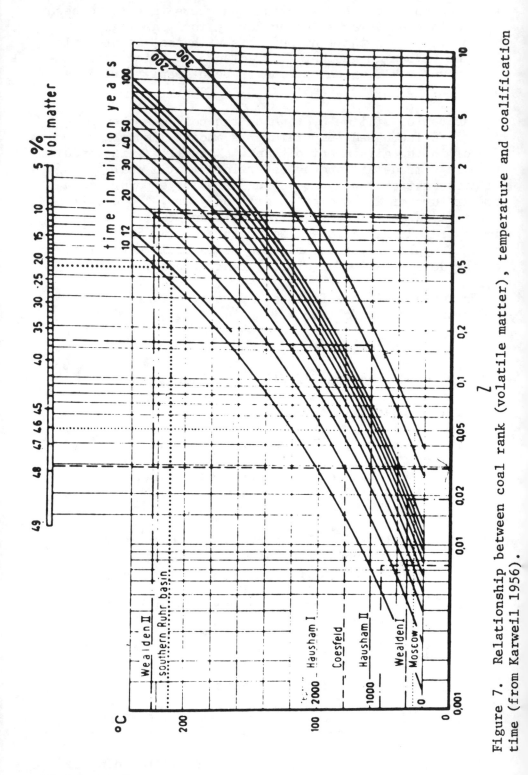

Figure 7. Relationship between coal rank (volatile matter), temperature and coalification time (from Karweil 1956).

content, calorific value and hydrogen content trends undergo a reversal is known as the coalification break. This break occurs at 29.5% volatile content of vitrinite and forms the boundary between gas coal and fat coal. As the point is approached there is a considerable build up of methane in coal seams and this is expelled on subsequent coalification. This factor can be of great importance in assessing the likely underground mining environment of coals of certain rank.

The Bowen Basin in Queensland, Australia, serves to illustrate the influence of an orogenic setting on the rank of coal. During the time of maximum coal deposition (Upper Permian) the Basin was generally open to the south east and the coal measures were deposited on the foredeep plain of a western hinterland with the major orogenic axis in the east. The climax of the Hunter Bowen orogeny was the uplift and crumpling of the main orogenic axis in the Upper Triassic. This resulted in the formation of a deformed block in the east central part of the basin which is known as the Dawson Tectonic Zone[1]. This zone marks the area of highest geothermal gradient in Permian and Triassic times and the effect of this on the rank of coal in the Bowen Basin is marked. Semi-anthracites are present within the Dawson Tectonic Zone and these grade through low volatile coking to high volatile bituminous and even sub-bituminous coals away from it.

As rank variations are dependent to a large degree on geothermal gradients, they show great variance from one basin to another. During 1971 and 1972 CRA Exploration conducted an Australia wide study with the objective of rating sedimentary basins in order of priority for potential coking coal occurrence. Reflectance determinations were carried out on coal and coaly material from government subsidised oil wells, State core libraries, company core stores and outcrop samples taken from a number of different sedimentary basins. The intracontinental basins of Western Australia which developed as rifts or depressions in the Pre Cambrian shield showed only a slight increase in rank with depth and nowhere was there definitive evidence of the presence of high rank bituminous coal. In the orogenic basins of Eastern Australia however, there was a marked increase in rank with depth which reaches a climax in the development of semi-anthracites.

In addition to basin wide geothermal gradients, more local hot spots can be demonstrated in many coal basins

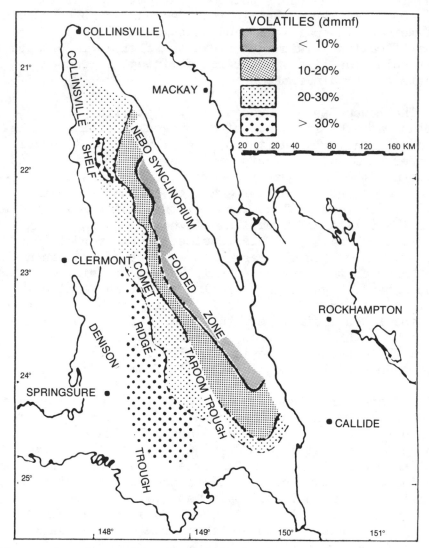

Figure 8. Generalized map of the Bowen Basin showing
variations in coal volatile content (or inferred rank)
with respect to structural elements.

within orogenic domains. The devolatilisation and amelioration of coal by igneous intrusions is mostly a very local phenomenon with the effects noticeable over areas of up to hundreds of metres away from the coal/intrusive contact. In certain circumstances, where the intrusive bodies are extensive laccoliths or plutons, these effects can be more widespread and lead to the formation of workable pockets of high rank coal. Examples of this are the Bukit Assam coalfield in South Sumatera, Indonesia, the Erkelenz area of West Germany, and parts of the Capella District in the Bowen Basin, Queensland, Australia.

In this latter case one restricted area of coal which is in close proximity to a basement high of Devonian granite is markedly devolatilised. The average rank of the Lower Permian coals in the area is medium volatile (30% - 36% d.a.f.) bituminous. Within an area of a few square kilometres however the average rank is low volatile (18% - 20% d.a.f.) bituminous and the only explanation yet offered for this is reactivation of the Devonian granite during the Upper Permian volcanism which was widespread in this part of Queensland[23].

SEDIMENTARY AND TECTONIC ENVIRONMENT OF COAL MEASURE SEQUENCES AND THEIR EFFECT ON THE NATURE OF A COAL DEPOSIT

In the same way that the properties of a coal are related to its depositional environment and geological history by being dependent upon coal type and coal rank, so the mineability of a seam in terms of its present inclination, degree of dislocation, continuity and thickness variation is also determined by the nature of the depository and its subsequent tectonic history.

Strakhov[15] proposed a four fold classification of coal environments based on sedimentation and tectonics as follows :

    i)   Orogenic paralic

    ii)   Orogenic intracontinental

    iii) Epeirogenic paralic

    iv)   Epeirogenic intracontinental

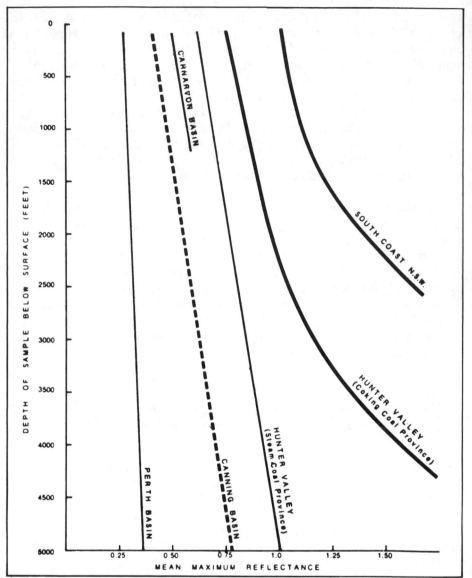

Figure 9. Some Australian coal basins variation in reflectance of vitrinite with depth.

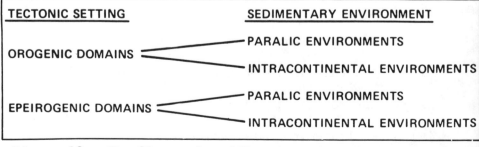

Figure 10. Strakhovs classification of coal environments.

This classification incorporates two distinctive sedimentary environments favourable for coal deposition, paralic and intracontinental. A second dimension is added as the nature of the sedimentation and the lithofacies deposited in both these environments will largely depend upon the tectonic setting of the depositional basin.

In orogenic domains, such as the foredeep areas of active island arcs, rapid differential uplift and subsidence often leads to the development of growth structures within the sedimentary pile and these contemporaneous features can materially affect the lateral continuity and thickness of coal seams. Such orogenic settings are also susceptible to higher than normal geothermal gradients and are thus the favoured loci for the development of high rank coals. As orogenesis inevitably closes with a period of intense uplift, this tectonic setting is also the most likely to result in a deeply buried sequence containing higher rank coal being rapidly returned to mineable depths.

Epeirogenic domains such as the stable margins of cratonic blocks and depressions in the surface of ancient shields, on the other hand, are characterised by gentle, almost imperceptible, subsidence and sedimentation over a long period of geological time. Under these extremely stable conditions where there was never any significant relief in the hinterland, the lithofacies are mature oligomict clastics and the sequence deposited often contains only few but unusually thick coal seams. Such a setting gives little opportunity for deep burial, metamorphism and subsequent uplift of the coal sequence, and in consequence, epeirogenic coals are often of low rank. The stability of such environments can be indicated by the presence of numerous unconformities and fossil weathering horizons such as bauxites which occur within the coal bearing sequence.

As well as the effects of the tectonic setting on the nature of coal seams and their associated strata, the sedimentary processes within the actual depositories also have a marked influence. Paralic environments include lobate (deltaic) or linear shorelines and have as an essential feature a plain inclined towards the sea and lying in front of a rising mountain chain or high relief hinterland. The fluvial and deltaic processes by which sedimentation occurs in such a setting involve the

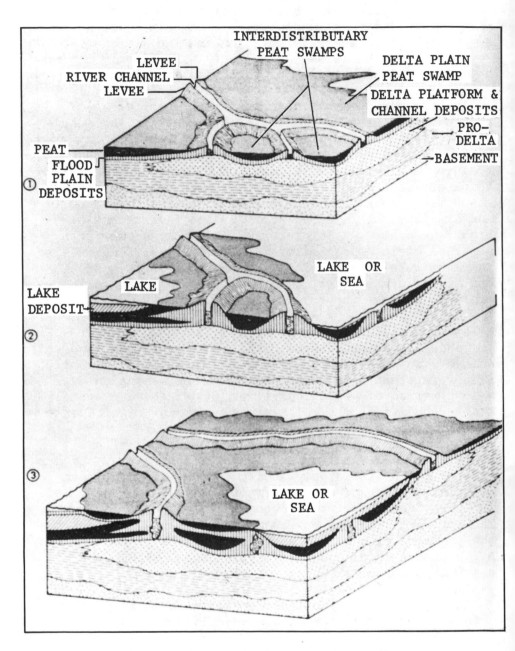

Figure 11. Development of a deltaic coal sequence.

deposition of shifting channel conglomerates and alternating sands, silts and muds which interfinger and result
in multiseam successions of coal measures in which individual coal seams are subject to random washout and rapid
unpredictable splitting and coalescence [2,3,20].

Intracontinental environments are more diverse and
sometimes they are related to previous paralic situations
by virtue of regression or uplift. More often they are
intracratonic depressions or intermontane basins depending
on whether their tectonic setting was epeirogenic or
orogenic. In contrast to paralic situations, marine
transgressions seldom reach these intracontinental basins
and as marine influences have been demonstrated conclusively
by Suggate[16] to influence the sulphur content of coal
seams, intracontinental coals can generally be expected
to have relatively low sulphur contents in contrast to
the more variable sulphur contents of paralic coal seams.

Using the environmental classification of Strakhov[15]
and a knowledge of coal type and rank, it is possible to
generalise to the extent that bright coals can be mainly
expected in paralic environments or embayments therefrom
where they will show a tendency to occur in multiseam
sequences which show considerable lateral variation.
Dull coals on the other hand can be mainly expected in
intracontinental and intermontane basins in thinner coal
measure sequences containing few but thick and more
uniform coal seams. Furthermore, high rank coals can be
mainly expected in orogenic domains whilst low rank coals
will usually be found in epeirogenic domains.

Within any single basin the effects of the palaeogeography at the time of deposition on the nature of the
coal seams is also marked. In areas of unusually rapid
subsidence, (e.g. close to major hinge lines) thick
accumulations of coal can be expected, but due to the
frequent relative changes in relief between basin and
hinterland leading to renewed periods of clastic deposition,
such coals are often highly banded with numerous fine
partings of dirt. Cleaner coals are thus to be expected
away from the faulted margins of major basins and preferably
where the underlying basement forms a stable shelf or rise.
A good example of this situation is the Bowen Basin of
Queensland, Australia where there is a general positive
correlation of areas of maximum subsidence (troughs) with
dirtier banded coals and areas of stability (shelves and

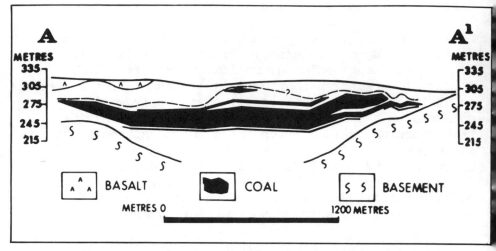

Figure 12.   Sketch section across Blair Athol coal deposit.

	PARALIC	INTRACONTINENTAL
OROGENIC	1. Thick variable coal measure sequence 2. Numerous individual coal seams 3. Rapid lateral facies variation 4. Marine transgressions common 5. Coal seams subject to rapid local splitting and washout 6. Coals mixed but mainly bright 7. Ranks from sub bituminous to anthracite	1. Relatively thin but variable coal measure sequence 2. Fewer individual coal seams 3. Broad facies variation 4. Marine transgressions rare 5. Coal seams subject to regional splitting 6. Coals mixed but mainly dull 7. Broad rank spectrum sub-bituminous − bituminous
EPEIROGENIC	1. Thin monotonous coal measure sequence 2. Relatively few coal seams 3. Monotonous oligomictic sediments 4. Marine transgressions common 5. Uniform coal seams 6. Coals mixed but mainly bright 7. Low to medium rank	1. Very thin monotonous coal measure sequence 2. Very few but thick coal seams 3. Monotonous oligomictic sediments 4. Marine transgressions rare 5. Lenticular coal seams 6. Coals predominantly dull 7. Low to medium rank

Figure 13.   Effect of depositional environment and tectonics on coal measure characteristics.

rises) with cleaner coals in any given coal bearing formation.

REGIONAL EXPLORATION METHODS

The major tool in the regional exploration for coal is the detailed basin study. In this respect and at this level of application coal exploration has much in common with petroleum exploration and the bulk of the data collected by oil companies is pertinent to the search for coal. Using the concepts outlined in this paper and accepting their limitations, it is possible to establish a model of a coal depository in any basin from a knowledge of published geological and geophysical data supplemented by a certain amount of field work.

After selecting an area for study the first problem is to establish the boundaries of the sedimentary basin concerned. Intracontinental basins often have their boundaries clearly defined by surrounding basement but paralic basins are more likely to overlap one another in time and in consequence some of their boundaries are more obscure. Geological mapping and a combination of seismic, gravity and magnetic surveys are the accepted tools for accurately delineating basin margins. An example of the use of a detailed gravity survey in this situation is the exercise carried out by the Australian Bureau of Mineral Resources in 1959 at Blair Athol in Queensland[8].

The Blair Athol basin lies some 30 kilometres to the west of the western margin of the Bowen Basin and probably formed as an intermontane basin within a drainage system which entered the Bowen Basin at a major delta[9]. It contains a typical intermontane coal measure sequence which is relatively thin (200 metres) and contains a clean dull coal seam of unusual thickness (up to 33 metres). Present day relief in the area is slight and there is extensive Quarternary sand cover masking the outcrop of the Permain sequence. Gravity readings were taken at quarter mile centres along lines half a mile apart over a total area of about 26 square miles. The results show a close correlation between the extent of the basin and a pronounced gravity low.

The second major task of the basin study is to determine the time stratigraphy of the basin and the lithofacies distribution within the interesting formations.

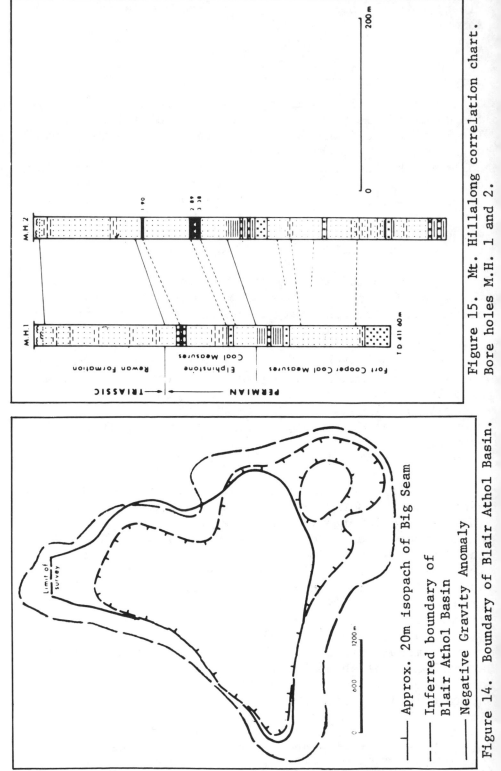

Figure 15. Mt. Hillalong correlation chart. Bore holes M.H. 1 and 2.

Figure 14. Boundary of Blair Athol Basin.

Approx. 20m isopach of Big Seam

Inferred boundary of Blair Athol Basin

Negative Gravity Anomaly

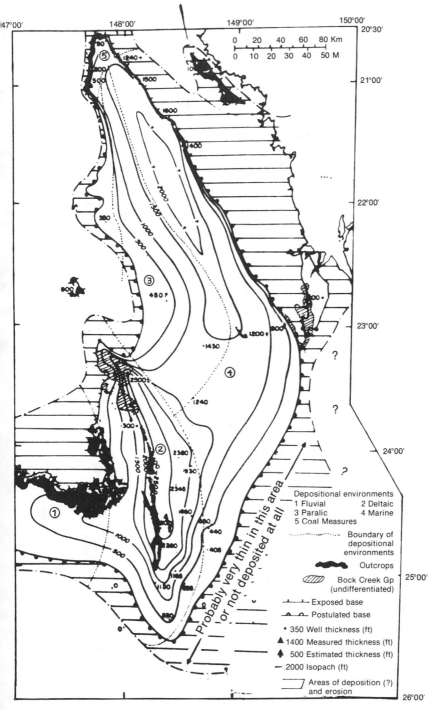

Figure 16. Distribution and outcrops of Gebbie Subgroup and its equivalents.

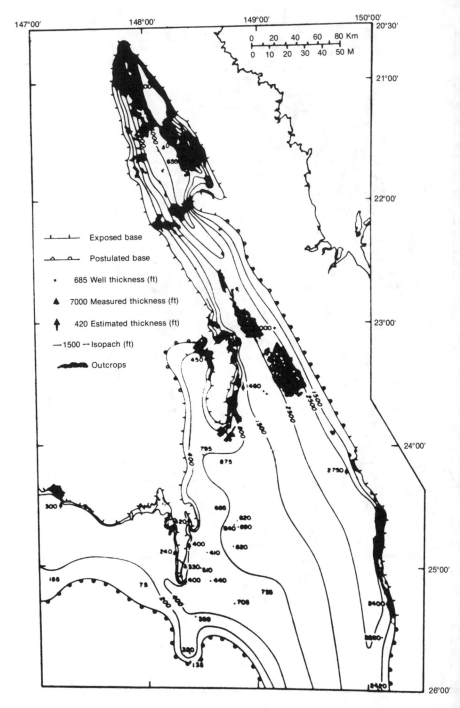

Figure 17. Distribution and outcrops of Blackwater Group.

In many cases this information can be built up from published geological data and the records of previous oil and water bores. In certain circumstances however, where there is no previous subsurface information, it may be necessary to undertake a limited programme of stratigraphic drilling to complement field mapping and air photo interpretation. Any such stratigraphic drillholes should be carefully sited by the field mapping party and the programme should have the objective of intersecting as complete a succession as possible. This implies the siting of the holes to allow for a recognisable overlap in stratigraphy. The results should be carefully correlated borehole to borehole and against field data using lithology and palaeontology. Absolute ages can be obtained by micropalaeontology and palynology and samples for such studies should be taken at regular intervals in drillholes in order to accurately determine the presence of any disconformities or major breaks in the sequence.

During the course of field work or drilling programmes any coaly material encountered should be sampled and, depending upon the judgement of the geologist concerned, submitted for proximate analysis, sulphur determination, petrographic description and reflectance measurement. By relating such results to the stratigraphy and vice-versa, meaningful conclusions can be drawn as to the potential of different formations for various types and rank of coal.

As the basin study progresses it will be possible to produce geological maps, subsurface cross sections, structure contours and isopachs of formations, coal rank and quality maps and palaeogeographic reconstructions sufficient to provide data for a preliminary engineering study of a potential coalfield and to outline areas of interest for intensive exploration should this be warranted.

1. **DETERMINE EXTENT AND BOUNDARIES OF THE SELECTED BASIN**

2. **DETERMINE VERTICAL DISTRIBUTION OF TIME STRATIGRAPHIC UNITS WITHIN THE BASIN**

3. **DETERMINE LATERAL DISTRIBUTION OF LITHOFACIES**

4. **ISOLATE POTENTIAL COAL BEARING FORMATIONS**

5. **DETERMINE GEOLOGICAL HISTORY OF BASIN**

6. **RECONSTRUCT PALAEOGEOGRAPHY OF POTENTIAL COAL BEARING FORMATIONS**

7. **PRODUCE CONTOUR AND ISOPACH MAPS OF FORMATIONS AND COAL PROPERTIES USING OIL WELLS, WATER BORES AND EXISTING DATA.**

8. **FILL IN GAPS BY FIELD MAPPING AND STRATIGRAPHIC DRILLING**

Figure 18.   Tasks involved in the regional geological appraisal of a potential coal basin.

REFERENCES

1. Dickins, J.M. and Malone, E.J., "Geology of Bowen Basin, Queensland", Bull. 130 Aust. B.M.R., 1973

2. Elliott, R.E., "Deltaic processes and episodes: The interpretation of the productive coal measures in the East Midlands, U.K.", Mercian Geologist pp. 3,2, 1969.

3. Hacquebard, P.A. and Donaldson, J.R., "Carboniferous coal deposition associated with flood plain and limnic environments in Nova Scotia. In environments of Coal Deposition". Geol. Soc. Amer. Sp. Pap. No. 114 , 1964

4. International Committee for Coal Petrography. International Handbook of Coal Petrography, Cenres, Paris, 1963.

5. Karweil, J. "Inkohlung pyrolyse und primare migration des erdols", Brennst. Chcm. pp. 17, 161, 1966

6. Krevelen, D.W. Van. Coal Science, Elsevier, Amsterdam, 1958.

7. Krevelen, D.W. Van. Coal, Elsevier, Amsterdam, 1961

8. Neumann, F.J.G. "Blair Athol gravity survey, Queensland 1959", Report No. 94, Australia, Bureau Mineral Resources. 1965.

9. Osman, A.H. and Wilson, R.G., "Blair Athol Coalfield, Queensland", In Economic Geology of Australia and Papua New Guines, Vol. 2, Aust. Inst. Min. and Met. 1975

10. Raistrick, A and Marshall, C.E. The nature and origin of coal seams, Eng. Univ. Press. 1939.

11. Roberts, J. "Thermodynamics of Hilts Law", Colliery Guardian No. 180, 1950.

12. Selley, R.C. Ancient Sedimentary Environments, Chapman and Hall, London, 1970.

13. Stach, E. "Basic principles of coal petrology: Macerals, microlithotypes and some aspects of coalification." In Coal and Coal Bearing Strata, Eds. D. Murchison and T.S. Westoll. Oliver and Boyd. 1968.

14. Stopes, M.C. "On the petrology of banded bituminous coal". Fuel, London, pp. 14,4. 1935.

15. Strakhov, N.M. Principles of Lithogenesis, Vol. 2 Oliver and Boyd. 1962.

16. Suggate, R.P. "New Zealand coals their geological setting and its influence on their properties". N.Z. Dept. Sci.Ind. Res. Bulletin 134. 1959.

17. Teichmuller, M. and Teichmuller, R. "Geological aspects of coal metamorphism". In Coal and Coal Bearing Strata. Eds. D. Murchison and T.S. Westoll, Oliver and Boyd. 1968.

18. Trotter, F.M. "The devolatilisation of coal seams in South Wales". Proc. Geol. Soc. 1442. 1947.

19. Trotter, F.M. "The genesis of high rank coals". Proc. Yorks. Geol. Soc. pp. 29,4. 1954

20. Wanless, H.R. et al. "Conditions of deposition of Pennsylvanian Coal Beds. In Environments of Coal Deposition". Geol. Soc. Amer. Sp. Pap. No. 114. 1964.

21. White, D. "Progressive regional carbonisation of coals". Trans Amer. Inst. Min. Met. 71. 1925.

22. Wilson, R.G. "Coal and Coal Environments". Confid. Rpt. to CRA Exploration. 1972.

23. Wilson, R.G. "Capella District". In Economic Geology of Australia and Papual New Guinea Vol. 2. Aust. Inst. Min. and Met. 1975.

13

COAL PETROGRAPHY AS AN

EXPLORATION AID IN THE WEST CIRCUM-PACIFIC

Peter G. Strauss*

Chief Coal Geologist

Nigel J. Russell**

Coal Petrologist

Allan J.R. Bennett***

Coal Petrologist

C. Michael Atkinson*

Senior Coal Geologist

ABSTRACT

The advantages of utilising microscopy techniques on coal samples in the field are often not sufficiently appreciated. Normally petrographic results are valid for weathered samples in contrast to more conventional chemical analyses.

The techniques involved in field-sampling and in the determination of the rank and type of coal are briefly reviewed. The interpretation of the petrographic results assists in predicting usage parameters and in providing a preliminary geological and economic evaluation of a coal deposit.

An attempt is made to summarise the coal-type provinces throughout the west Circum-Pacific region and examples are presented taken from initial exploration surveys in the region including Australia, New Zealand and Indonesia.

*   Robertson Research (Australia) Pty. Limited, Sydney.
**   Robertson Research (Singapore) Private Limited.
*** Commonwealth Scientific and Industrial Research Organisation, Division of Mineralogy, Sydney.

1. INTRODUCTION

The advantages of utilising coal petrography as an aid in exploration for coal and the basic principles involved, are still insufficiently appreciated by many organisations. The aim of this paper is to outline its application with emphasis on the search for, and development of coal deposits in the west Circum-Pacific region.

Initially, work on coal petrography was undertaken mainly in Europe and the U.S.A. and certain differences in terminology and interpretation still persist (1) (2) (3). However, basic terms are defined in Table No. I.

Over the years research was also carried out in Russia, Canada and in the west Pacific region in Japan and Australia. In particular, various Japanese organisations have systematically collected much data on coals worldwide and related the information to the blending and utilisation, especially to their coking coal requirements. It is partially due to Japanese insistence that many companies throughout the world are now obtaining, during exploration, petrographic data on their potential as export coals.

2. COAL PETROGRAPHIC TECHNIQUES

One of the main macroscopic features of coal is its distinctive banded appearance. In sections perpendicular to the bedding, the bands or layers are bright, dull or finely banded and have in general, clearcut boundaries, varying from a fraction to many inches in thickness. These characteristic bands also differ markedly in their chemical and physical properties, so that a variation in the relative proportion of these bands, as developed in a seam, alters its overall properties. Therefore, a careful assessment is made of the relative proportion of the bright and dull bands present in a coal seam, particular reference being made microscopically to the type and distribution of characteristic components of botanic origin ("macerals") which constitute these distinctive bands.

The present day extensive application of petrographic analysis of coal has developed from pioneer work on coal components (4) (5) (6).

TABLE I

SUMMARY OF NOMENCLATURE IN COAL PETROLOGY

Lithotypes	Macerals		Microlithotypes	
	Maceral	Maceral group and symbol	Microlithotype group	Principal groups of constituent macerals in the microlithotypes
Vitrain (Bright Coal)	Collinite Tellinite	Vitrinite (V)	Vitrite	V >95%
			Vitrinertite	V + I >95%
Fusain	Macrinite Micrinite Semifusinite Fusinite Sclerotinite	Inertinite (I)	Inertite	I >95%
	Cutinite Resinite Sporinite Alginite	Exinite (E)	Liptite	E >95%
Clarain Durain (Dull Coal)	Includes all Macerals		Clarite Durite Duroclarite(V I) Clarodurite(I V)	V + E >95% I + E >95% V + E + I >5% I + E + V >5%

2.1. Preparation of Coal Samples

The method of preparation depends on the procedure used to obtain the coal sample. A bulk sample, for instance, could be crushed if necessary to minus 3/4 inches size with a jaw crusher and at least 90 pounds extracted by sample sub-division. The top size of this fraction should be reduced to 3/16 inch and at least 9 pounds extracted by sample subdivision. This material should then be crushed with a coffee mill to give a top size of 1/16 inch with a minimum of fines. A representative 4 ounces of this material can be extracted by increments and placed in a container for petrographic analysis.

Often only a little material is available from sub-sections of a bore core, and then all material is reduced to below 1/16 inch before extracting any samples for analysis Proximate and petrographic analyses can be carried out on samples weighing only three ounces, but when microscopic analysis alone is required, for instance on samples extracted from oil well ditch cuttings, a mere 1/5 ounce is sufficient. The recommended maximum grain size for petrographic analysis is 1/16 inch but with difficulty, maceral and reflectance analysis can be undertaken on grains as small as .0008 inches

The preparation of polished particulate coal mounts for microscope analysis requires some manual skill. The crushed coal sample is placed in a one inch cube-shaped aluminium container with cold setting polyester resin. When the resin is set, the surface is ground on water-proof carborundum papers using paper grades 220, 400 and 600. Polishing is carried out using chromium oxide and water on a rotating linen-covered lap, followed by magnesium oxide and water on a lap with napped cloth such as Selvyt.

2.2. Petrographic Analysis

2.2.1. Maceral Analysis. Macerals are identified micro-scopically by their form and reflectivity, and are divisible into three groups on appearance and physical characteristics (7) (Table No. I). A minimum of five hundred macerals including mineral matter constituents, are counted when determining the maceral composition of a coal sample.

The three main groups - Vitrinite (or Huminite in sub-bituminous coals), Exinite (or Liptinite) and Inertinite - are each subdivided on the basis of the original and form

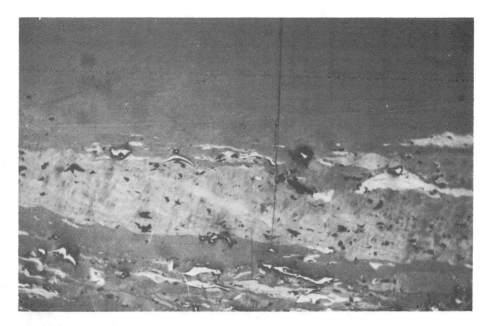

Figure No. 1 - Bulli Seam, Appin Colliery, New South Wales (bituminous coal). Vitrinite grey grading into white semi-fusinite (photomicrograph x 290).

Figure No. 2 - Thailand (sub-bituminous coal). Huminite dark grey with white pyrite inclusions (photomicrograph x 290).

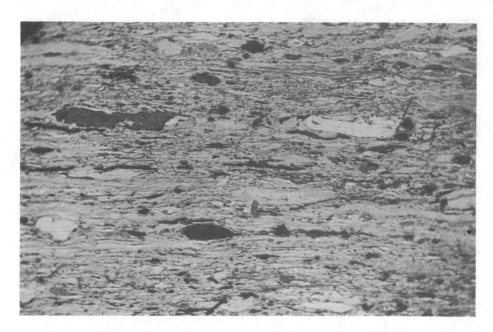

Figure No. 3 – Higashi-Horonai, Japan (sub-bituminous). Liptinite black and white lenticular inertinite in grey huminite (photomicrograph x 290).

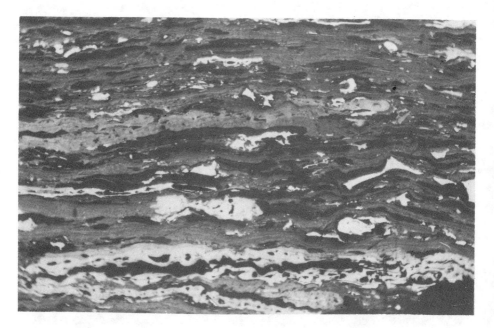

Figure No. 4 – Greta Seam, Aberdare Colliery, New South Wales (bituminous coal). Sporinite black associated with white semi-fusinite in grey vitrinite (photomicrograph x 290

Figure No. 5 – Strongman Seam, Greymouth Coalfield, New Zealand (bituminous coal). Micrinite white angular fragments and black lenses of exinite in grey vitrinite (photomicrograph x 290).

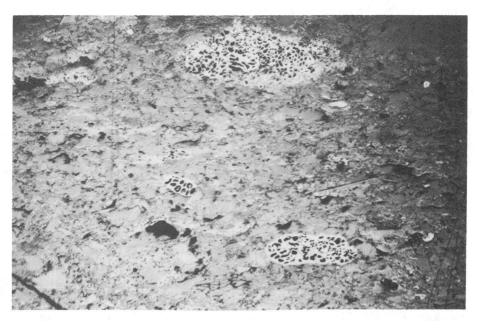

Figure No. 6 – Alison Colliery, Waikota District, New Zealand (sub-bituminous coal). Sclerotinite white cellular in grey vitrinite (photomicrograph x 290).

of the maceral. Examples of various types are given in
Figures No. 1, 2, 3, 4, 5 and 6.

2.2.2. <u>Microlithotype Analysis</u>. In bituminous coals the
maceral groups tend to occur in identifiable associations as
micro-bands about fifty to several hundred microns thick.
As an example, the most frequent grouping comprises a roughly
constant ratio of exinite to inertinite associated with a
wide variation in vitrinite content. Seyler (3) suggested
that the term microlithotype should be used for these maceral
associations (Table No. I).

2.2.3. <u>Measurement of Reflectance</u>. A polished crushed-
coal mount, prepared for maceral and microlithotype examination, can also be used for reflectivity determination. The
mean maximum reflectance ($\bar{R}_o$ max) is the average of seventy
to one hundred readings of the percentage of incident light
reflected vertically from a uniform band of vitrinite under
standard specifications (2).

As the reflectance value increases with coal rank along
with an increase in ultimate carbon content, reflectivity
will rapidly and accurately indicate the carbon content of
a coal seam and its coal rank class. Figure No. 7 presents
a simple coal classification system based on vitrinite and
reflectance values with carbon values superimposed. Other
chemical or rank class properties could similarly be superimposed.

This application has advantages in that the measurement
of rank is based on one homogenous coal constituent (viz.
vitrinite) and is independent of other properties, the results are reproducible and the method is rapid, non-destruct-
ive, uses only very small quantities and can be carried out
on weathered samples.

3. <u>ASSESSMENT OF PROPERTIES FOR COAL UTILISATION</u>

A classification based specifically on both rank and
type was first evolved for Australian coals by Commonwealth
Scientific Industrial Research Organisation in 1970 (8).
The system uses a "type" parameter based on petrographic
analysis (vitrinite and exinite), and a "rank" parameter
determined by reflectance. One advantage of this system is
that, for the west Pacific region, both coal characteristics
from the various coalfields and variations of seams within
the same coalfield can be differentiated (Figure No. 8).

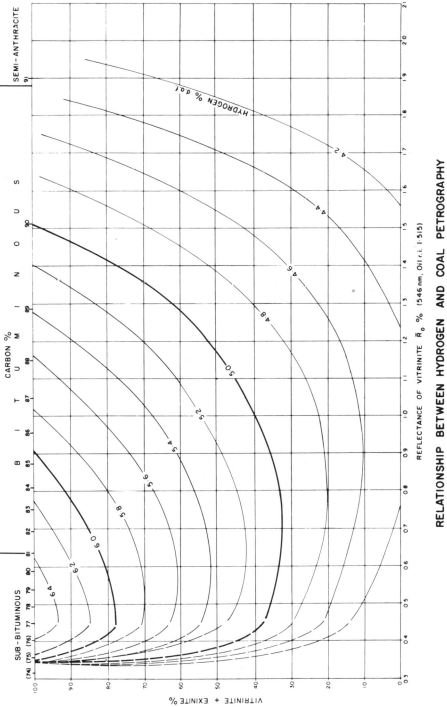

RELATIONSHIP BETWEEN HYDROGEN AND COAL PETROGRAPHY

FIGURE No. 7

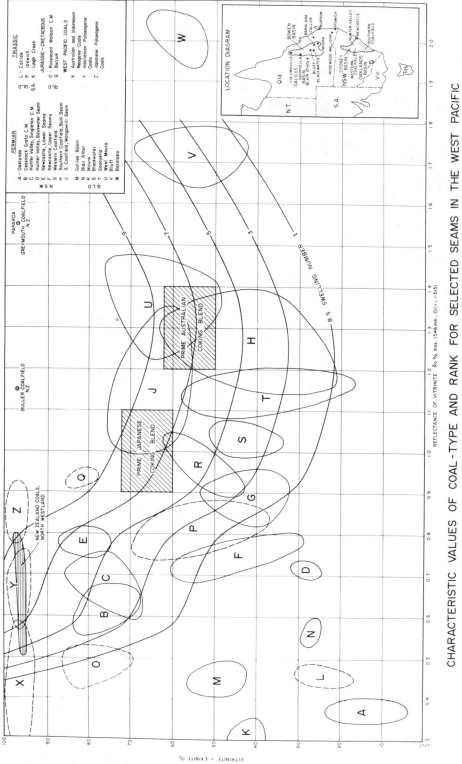

CHARACTERISTIC VALUES OF COAL-TYPE AND RANK FOR SELECTED SEAMS IN THE WEST PACIFIC

FIGURE No. 8

Graphs relating such "use" parameters as swelling index, volatile matter and calorific value to both rank and type simplify the task of assessing the characteristics of a particular seam and comparing them with those of other seams, as well as considering its utlisation potential. For example, for a particular coking coal such as the Goonyella seam of Queensland, the range of relevant "use" parameters can easily be established from the computed mean lines for crucible swelling index (Figure No. 8) and volatile matter (Figure No. 9). Similarly, other "use" parameters can be predicted for a steam-raising coal, such as for the Oaklands Coalfield in New South Wales by transposing the relevant petrographic data field (Figure No. 8) onto the appropriate diagram e.g. Figure No. 10 - calorific value. In the case of conversion to gaseous or liquid hydrocarbons, such as for the Neogene coals of Indonesia (Figure No. 8) it is useful to establish the range of lines for hydrogen values (Figure No. 7).

Current knowledge (Figure No. 8) suggests that an optimum coal blend for Australian conventional coke ovens should contain between 50 and 62% vitrinite plus exinite with reflectance values between 1.2 to 1.4%. For Japanese coke production, however, the optimum values should be between 60 to 72% vitrinite plus exinite with reflectivity values between 0.9 to 1.1%. A great deal of research is still in progress for predicting the behaviour of coals in a blend suitable for making metallurgical coke (9) (10) (11) (12) (13).

During washing coal can be separated from shale by flotation in liquid of specific gravity 1.6, but by using a lower specific gravity bright coal with a better coking potential can be isolated from the more inert dull coal, as the latter frequently contains more mineral matter.

4. PETROGRAPHIC PROVINCES

It is pertinent to outline some regional variations in coal as this assists in establishing exploration targets. Particular reference is made to recent contributions by Cook (14) and Shibaoka and Smyth (15).

4.1. Coal Sequences in the West Pacific Region

Coal-forming conditions were developed in different areas during various periods (Figure No. 11). Basically, Australia, and other continents bordering the west Pacific, can be considered as being formed by the progressive outward accretion of successive orogens. The authors believe that significant

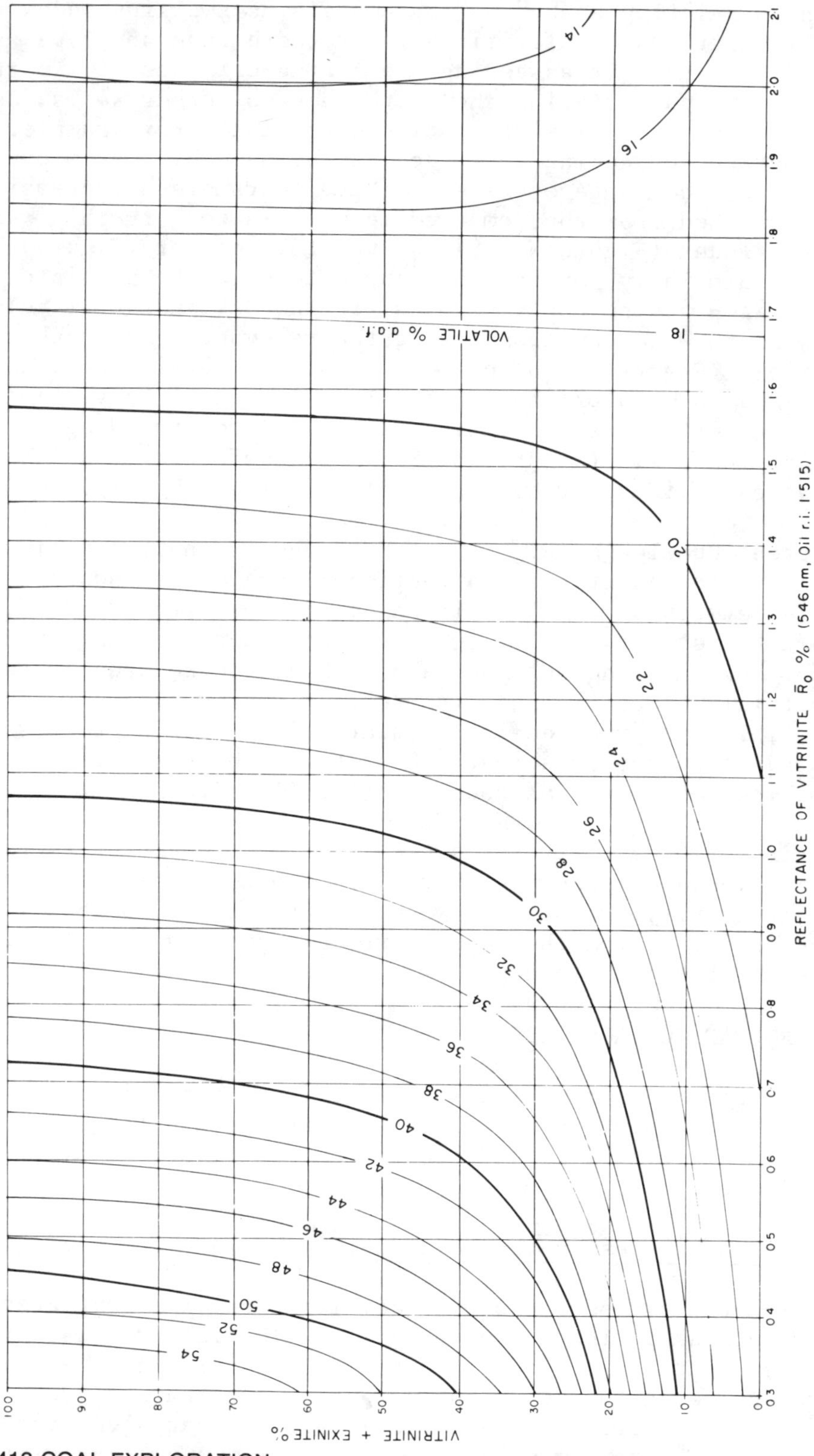

RELATIONSHIP BETWEEN VOLATILE MATTER AND COAL PETROGRAPHY

FIGURE No. 9

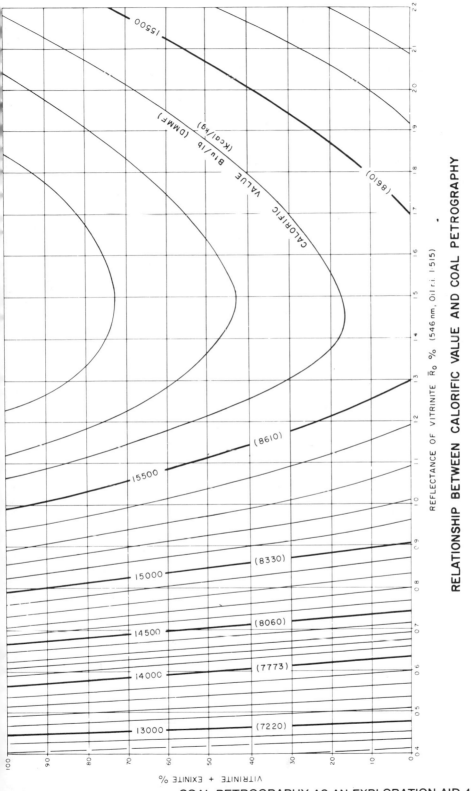

RELATIONSHIP BETWEEN CALORIFIC VALUE AND COAL PETROGRAPHY

FIGURE No. 10

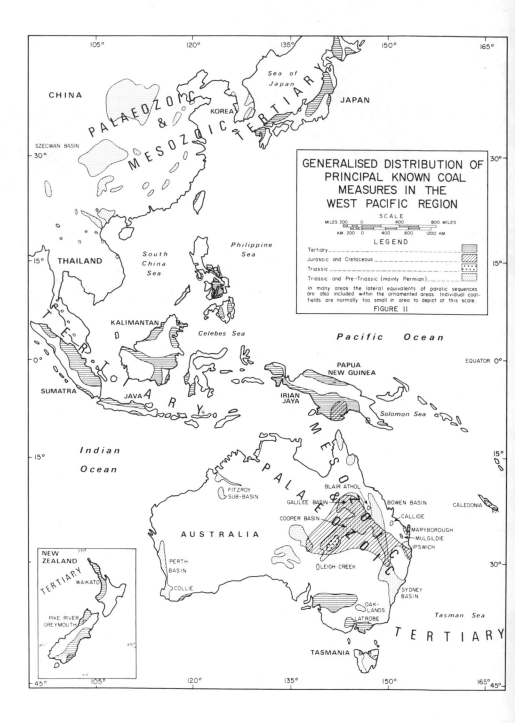

GENERALISED DISTRIBUTION OF
PRINCIPAL KNOWN COAL
MEASURES IN THE
WEST PACIFIC REGION

SCALE

MILES 200   0   400   800 MILES

KM 200 0   400   800   1200 KM

LEGEND

Tertiary
Jurassic and Cretaceous
Triassic
Triassic and Pre-Triassic (mainly Permian)

In many areas the lateral equivalents of paralic sequences
are also included within the ornamented areas. Individual coalfields are normally too small in area to depict at this scale.

FIGURE II

414 COAL EXPLORATION

Palaeozoic coal deposits only occur within these continental margins.

In the Carboniferous significant coal seams are only known from the northern hemisphere e.g. in China.

During the Permian period important coal basins were formed throughout Gondwanaland. In Australia, the most significant reserves occur in the Bowen and Sydney Basins which were formed in a "foreland basin" situation whilst similar type of coals also occur in the Indian sub-continent. Important Permian coals are also developed in China and adjacent areas.

Between the margins of the continents mentioned above, some non-commercial traces of coal of Permian-Carboniferous age have been reported such as in the island chain extending from Irian Jaya through Sumatra towards Burma.

Triassic coals were deposited within continents in smaller and fewer basins than in the Permian, e.g. to the east of the Bowen Basin and in Tasmania and in China.

Jurassic and Cretaceous coal deposits were deposited further east around the Australian continental margins, although, in places the Mesozoic transgressed west across Palaeozoic successions forming intracratonic basins. Coal seams were also formed within parts of the island-arcs of the west Pacific, for instance, in New Zealand, and, of lesser significance, in New Caledonia, Papua New Guinea and Japan.

Tertiary coals are developed throughout the island-arcs of the west Pacific region, including New Zealand, Papua New Guinea, Philippines, Japan as well as in the Indonesian Archipelago. Thick brown coals also occur in Australia and Southeast Asia.

4.2. Coal-type Provinces

4.2.1. Suite of Coal-types. Differences between the coal types characteristic of the Permian-Triassic and those in the Mesozoic-Tertiary deposits within the west Pacific region are demonstrated in Figure Nos. 12 and 13. However, this distinction may not apply to all Southeast Asia, as petrographic data on many coals in this region is scarce.

Figure No. 12 is a preliminary indication of the nature of the principal coal-type provinces, based on maceral anal-

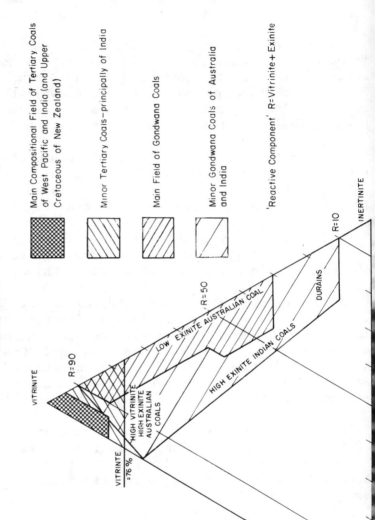

COMPARATIVE PETROGRAPHY OF TERTIARY, CRETACEOUS AND GONDWANA COALS OF WEST PACIFIC AND INDIA

MACERAL ANALYSES
(Mineral matter-free-basis)

FIGURE No. 12

Main Compositional Field of Tertiary Coals of West Pacific and India (and Upper Cretaceous of New Zealand)

Minor Tertiary Coals-principally of India

Main Field of Gondwana Coals

Minor Gondwana Coals of Australia and India

'Reactive Component' R=Vitrinite+Exinite

VITRINITE

R=90

VITRINTE =76%

HIGH VITRINITE HIGH EXINITE AUSTRALIAN COALS

LOW EXINITE AUSTRALIAN COAL

R=50

HIGH EXINITE INDIAN COALS

DURAINS

R=10

INERTINITE

EXINITE

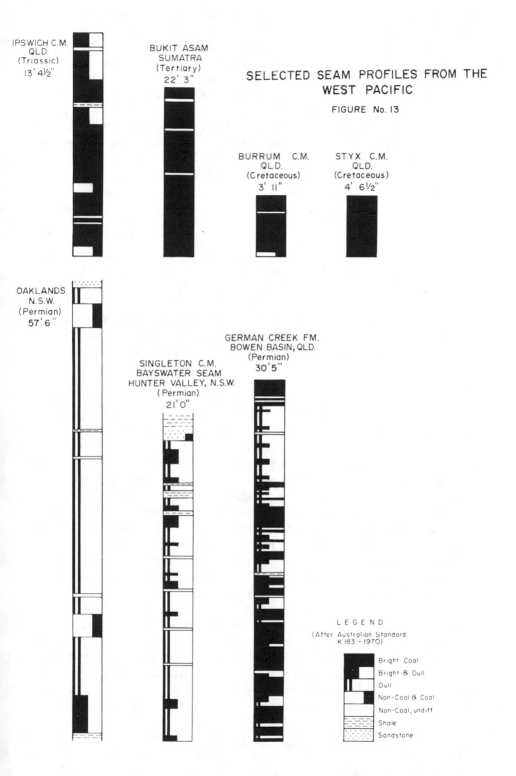

SELECTED SEAM PROFILES FROM THE
WEST PACIFIC

FIGURE No. 13

IPSWICH C.M.
QLD.
(Triassic)
13' 4½"

BUKIT ASAM
SUMATRA
(Tertiary)
22' 3"

BURRUM C.M.
QLD.
(Cretaceous)
3' 11"

STYX C.M.
QLD.
(Cretaceous)
4' 6½"

OAKLANDS
N.S.W.
(Permian)
57' 6"

SINGLETON C.M.
BAYSWATER SEAM
HUNTER VALLEY, N.S.W.
(Permian)
21' 0"

GERMAN CREEK FM.
BOWEN BASIN, QLD.
(Permian)
30' 5"

L E G E N D
(After Australian Standard
K 183 - 1970)

Bright Coal
Bright & Dull
Dull
Non-Coal & Coal
Non-Coal, undiff
Shale
Sandstone

COAL PETROGRAPHY AS AN EXPLORATION AID 417

yses only, the results of microlithotype analyses showing
similar features. The spread of data relating to Figure No.
12 is not claimed to be complete, but maceral analyses of
Tertiary coals occupy a restricted field near the vitrinite
apex, where vitrinite is greater than 76%. A few Indian
Palaeogene coals fall outside the main compositional field.
The Gondwana coals occupy a large field, with the majority
plotting adjacent to the vitrinite-inertinite boundary, the
vitrinite content in a range from 10 to 90% and the exinite
content is typically less than 5 to 10%. There is little
overlap between the separate compositional fields of the
Tertiary and Gondwana coals.

Laurasian Carboniferous coals of Europe and North America
yield maceral analyses that plot within the main compositional field of the Gondwana coals with a tendency to fall
within high vitrinite, relatively high exinite portion of
that field (16).

Figure No. 13 presents selected seam profiles from the
west Pacific region and illustrates the characteristic differences in lithotype between the Permian-Triassic and younger
coals.

4.2.2. <u>Possible Origin of Coal-types</u>. Detailed studies
in Britain have shown that the ultimate coal-type is related
to water table effects in the original peat forming environments. In Australia (17) (14) (15), authors have considered
tectonic stablity as one of the major controlling factors
in the development of the principal coal-types. Character-
istically the seam profile commences with a bright coal phas
and ascends to a dull coal phase as the peat swamp grew upwards and the groundwater level dropped. Generally, seams
comprising many cycles indicate successive periods of more
rapid subsidence and have a higher vitrinite content than
those made-up of one cycle; also, the more dirt bands a seam
contains the higher is the vitrinite content. Thus, in a
rapidly subsiding sequence and a relatively unstable environment, relatively fewer dull coal bands and streaks tend to
occur in a bright coal seam (with or without dirt bands).
At the other extreme, dull coal may form the major component
of a seam profile, deposited in a relatively more stable
environment. This aspect is of importance in locating, expl
oring and developing particular types of coal deposits.

4.2.3. <u>Depositional Environments</u>. Some Permian and Triassic deposits occur in limnic, intramontane or intracratoni
basins which frequently formed at a relatively high topo-

graphical level; in these coal measures formed with a persistent low water table. The seams temd to be of relatively small areal extent, lenticular, very thick and consist mainly of dull coal. Figure No. 8 indicates the petrographic composition of several examples from Australia. Examples include the **Blair Athol Coalfield** (18), the **Oaklands Basin** (19) (**Fig. 14**), the **Callide Coalfield**(20), and the **Leigh Creek Coalfield**.

Paralic-deltaic sequences associated with geosynclines also contain Permian and Triassic coals. However, the environments vary considerably and the generalisations presented here have to be taken as such.

The Bowen and Sydney Basins (Figure Nos. 8 and 11) contain four to five main Permian coal measure sequences (21). Vitrinite-poor coals tend to occur in relatively more stable situations though the vitrinite content is not normally as low as in the examples cited for limnic basins. Firstly, they occur at the basin margins and in particular along the western margins of the Sydney and Bowen Basins and examples are present in the Western Coalfield. Secondly, they are found at the top of some cycles in coal measure basins as they become infilled with sediments. Examples are the coals near the top of Permian succession in the Bowen Basin, the topmost (Bulli) seam in the Southern Coalfield, and possibly the Bayswater Seam in the Hunter Valley in the northwest of the Sydney Basin (Figure No. 13).

Vitrinite-rich sequences occur in more unstable situations, for example commonly in the Singleton Coal Measures of the Hunter Valley, where considerable lateral and vertical seam variations occur; the locally vitrinite-rich seams are thicker than 15 feet (Figure No. 13). Another example is the Wongawilli Seam in the Southern Coalfield which was formed before the Bulli Seam.

Many Mesozoic and Tertiary coals are very rich in vitrinite, exinite, volatile matter and hydrogen content. Figure No. 8 presents some petrological data from the west Pacific region. In Australia good examples of this type are the Jurassic Walloon Coal Measures and the Cretaceous Burrum Coalfield both of southern Queensland. However, exceptionally inertinite-rich Tertiary coals occur in continental regions such as locally in parts of the Indian sub-continent (Assam and Jammu) which strictly fall outside west Pacific region under consideration.

Probably there is no simple relation between tectonic

Figure No. 14 - Oaklands Coalfield, New South Wales (sub-
bituminous coal). Inertinite white fragments with isolated
grey vitrinite bands (photomicrograph x 290).

stability and coal-type for the Mesozoic-Tertiary coals (15) but further geological knowledge is required. Undoubtedly extensive areas containing Mesozoic and Tertiary coals were formed in a paralic environment. Throughout much of the west Pacific region coal is developed in the Palaeocene to Oligocene and in the Mio-Pliocene, the sequences frequently being diachronous and were formed in numerous relatively small, frequently interconnected, fault-controlled sub-basins.

The Mio-Pliocene of south and central Sumatra contain relatively few but thick seams, the principal ones being three or four in number and extending over large areas. The relatively thin sequence of sediments were formed mainly in a low energy environment. By contrast, the Mio-Pliocene sequence of east Kalimantan were deposited in a deltaic, high energy environment with frequent marine incursions. The coal measures are extremely thick and most seams are of variable thickness. Although the Sumatran and Kalimantan coals were deposited in different conditions, no major variations in maceral content have so far been noted. The Indonesian coals in common with other Tertiary coals often contain both finely divided and coarse lumps of resin.

5. RANK VARIATIONS

This section briefly considers the regional causes and variations of coal rank (degree of coalification or maturity of a coal) and not those due to the effects of local thermal metamorphism by igneous intrusions.

Among the major factors which have been considered are depth of burial, heat and pressure engendered by tectonism, nature and depth of basement, time and thermal conductivity (22) (23) (24) (25). However, the authors consider that, in general the geothermal gradient causing coalification is principally due to the nature and depth to basement. Regional occurrences of higher rank coals may be associated with "mobile" zones, especially granitic intrusions which may pre-date or post-date coal deposition and are not necessarily aligned along the depositional axis. The existence of specific "mobile" zones may be subject to speculation. In the Bowen Basin, the maximum deformation occurred in the centre of a "mobile" zone along the axis of the basin and the geothermal gradient in this area is believed to be related both to the depth of burial and to the type of basement. However, nearer the margins of the basin, where the coal measures were deposited on a more stable shelf, effects of radiogenic base-

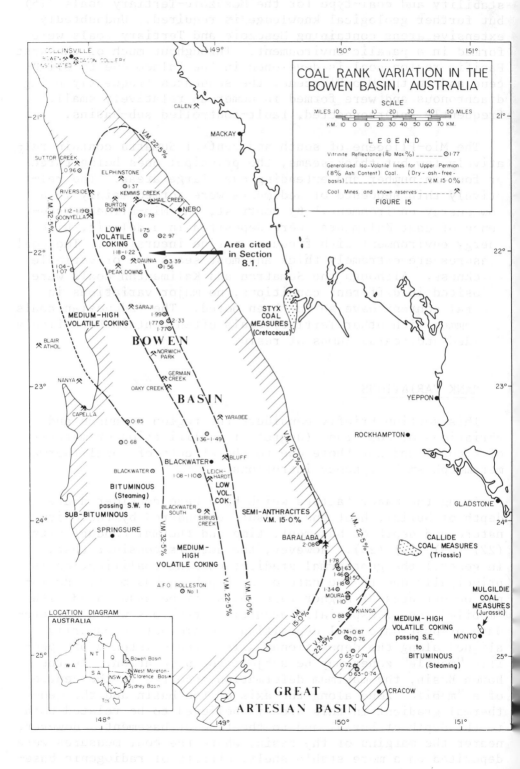

COAL RANK VARIATION IN THE
BOWEN BASIN, AUSTRALIA

SCALE

MILES 10   0   10   20   30   40   50 MILES
KM. 10   0   10   20   30   40   50   60   70 KM.

L E G E N D

Vitrinite Reflectance (Ro Max %) _____ ⊙1·77

Generalised Iso-Volatile lines for Upper Permian
(8% Ash Content) Coal. (Dry-ash-free-
basis) _____ V.M. 15·0%

Coal Mines and known reserves _____ ⚒

FIGURE 15

Area cited
in Section
8.1.

LOCATION DIAGRAM
AUSTRALIA

ment may locally be significant (23).

Mr. J.W. Beeston of the Geological Survey of Queensland (personal communication) has prepared a preliminary iso-reflectance map of the uppermost Permian of the Bowen Basin. His initial results broadly compare with a highly generalised iso-volatile map presented in Figure No. 15 together with the related distribution of the main coal-types for the Upper Permian seams of this basin. Reflectance measurements at selected locations are also given for comparison. The distribution of the iso-volatile lines shows that the rank decreases from the centre to the margins of the basin. Thus, the anthracitic coals with volatile matter of under 15%, occur in the axial zone of the basin; a tight series of folds is a common feature of this zone. To the west and southeast occur progressively a low volatile coking coal zone, medium to high volatile coking coal zone followed by bituminous "steaming" coals.

The geothermal gradient of the Bowen Basin is conveniently shown by depth-reflectance curves based on borehole data (Figure No. 16). These show that coals of specific rank can occur at several localities at differing depths. One explanation is that after coalification (a process which is irreversible) subsequent uplift and erosion took place. Thus, on this basis, the anthracitic coal at Baralaba, which may originally have reached this rank at a depth of at least 10,000 feet, as indicated by comparison with the Burunga No. 1 well, was subsequently uplifted and now occurs close to the surface.

Other good examples have been described from New Zealand (26).

6. ADVANTAGES OF COAL PETROGRAPHY AND SAMPLING METHODS FOR EXPLORATION PURPOSES

6.1. Advantages

In the initial stages of exploration the bulk of coal samples are likely to be collected from outcrops, with some samples possibly derived from sub-surface exposures in disused mines, old boreholes and private collections. All these samples are liable to be subject to varying degrees of weathering or oxidation. Many of the physical and chemical properties such as moisture of low rank coals, coking properties, calorific value, volatile matter, fluidity charact-

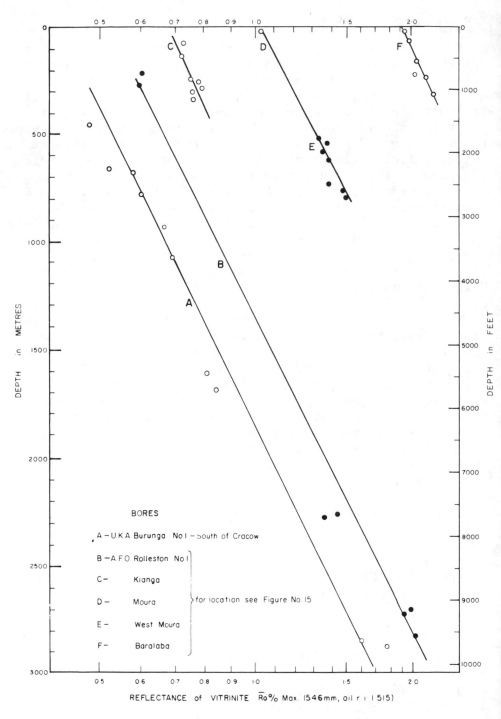

DEPTH−REFLECTANCE CURVES IN CENTRAL BOWEN BASIN, QUEENSLAND

FIGURE No. 16

424 COAL EXPLORATION

eristics, crucible swelling index, etc. are extremely sensitive to oxidation and it is frequently not appreciated that
the determinations often yield suspect results. A similar
argument holds for samples obtained from cored borehole sequences, especially for coking coals, which are frequently
analysed following a delay of some two to three months and
over. In the west Pacific region, where samples have frequently to be transported from one country to another to reach
suitable laboratories and often experience considerable delays, it is advisable to wrap samples in plastic sheets or
air-tight containers.

The advantage of petrographic techniques in evaluating
oxidised samples lies in the fact that oxidation exerts little influence on petrographic parameters except in cases
where the alteration has proceeded to a very advanced stage.
In fact, the effects of weathering can frequently be seen in
the vitrinite as enlarged cracks around core of unaltered
material on which the determinations are made. As indicated
previously, the broad correlation between petrographic parameters and physical and chemical properties permits the prediction of certain "use" parameters.

Subject to such factors as the degree of macroscopically
apparent weathering, representativeness of sample, economic
thickness, etc. it is advisable to undertake limited conventional laboratory analyses, as well as petrographic determinations. A comparison between observed and predicted values
for these "use" parameters often provides an indication of
the degree of oxidation of the sample and avoids the possibility of down-grading or misinterpreting the results simply
based on preliminary physical and chemical analyses.

The other advantage, already apparent from previous discussions is that the estimation of rank by means of reflectivity determinations, based on a pure coal substance viz.
vitrinite, is more reliably obtained than undertaking an
ultimate analyses on oxidised samples. Thus, the rank and
the variation within a region can be quickly and objectively
studied.

In other words, by assessing at an early stage in the
exploration programme the rank, quality and type of coal
present, without major analytical programmes including bulk
sampling, coking tests, tar contents etc., quick and reliable
decisions can be reached on many aspects on the potential
economic value of the coal deposit.

Certain basic determinations such as ash and sulphur values cannot be determined by petrographic means although some qualitative indication can be given on the abundance and nature of inherent ash and pyrite sulphur present.

6.2. Comments on Field Sampling

The poor standard of sampling still frequently undertaken merits mention of this subject. The validity and interpretation of the results depend to a high degree on the sampling technique. In the initial exploration, data is often required from the following sample types :

Spot Samples: Examples are "grab" samples, which are frequently the cleanest and brightest handpicked sample from one portion of an in situ exposure and "floats" found on river beds, beaches etc. which are totally unrepresentative. A reflectance determination may be warranted, together with a brief petrographic examination for possible distinctive features to indicate its possible nature and origin. No maceral analyses is normally warranted.

Channel Samples (27): The exposed coal seam section should whenever possible, first be carefully cleaned by removal of the more obviously weathered and contaminated portions. After recording a description, a representative channel, preferably with a minimum horizontal area of 16 square inches, should be taken uniformly from the top to the bottom of the exposed section.

Strip or Ply-by-Ply Samples (27): The procedure involved is similar to that of taking a channel sample except that leaves or plies consisting of macroscopically distinct lithotypes and dirt bands are sampled separately.

Samples from bore cores are treated in a similar manner to that of strip samples. Petrographic determinations are preferably undertaken on "clean" coal composites or on "raw" coal of the presumed mineable section or the total seam in order to be representative. An example exists of recent work when petrographic evaluations were erroneously determined for major seams based on an individual ply of inferior coal. In another recent case, two major seams, split by over 30 feet of measures, were combined into one sample.

Borehole Cuttings: Ditch cuttings or chip samples may be available from water bores, petroleum wells, etc. Samples from petroleum wells (including possible solid and side-wall

cores) are especially useful in determining a depth-reflectance gradient, as they often represent a substantial part of
a stratigraphical succession. Frequently a quick systematic
visual logging of the cuttings coupled with a study of geophysical bore logs, ensures noting sections containing a relatively high proportion of coal. If coal is not present as
discrete seams, determinations can often be made on dispersed
coal particles which can be concentrated by froth flotation.
It is advisable to take several samples spread over a vertical depth of at least one thousand feet to ensure obtaining
statistically meaningful results. It is necessary to bear
the following constraints in mind :

1. Caving of coal seams from high horizons giving rise to
   contaminated samples and anomalous low reflectance results.

2. Additives of lignitous material in the drilling mud,
   also contaminating samples.

3. Samples being dried on a hot plate prior to storage
   thus oxidising samples.

4. Coal fragments in geologically reworked sediments.

7. APPLICATION OF COAL PETROGRAPHY IN EXPLORATION

An appreciation of the use of coal petrography can safe-
guard unnecessary expenditure on exploration programmes. It
may well be advisable to discontinue further work on a specific area, or transfer attention to an adjoining region, on
the basis of a preliminary assessment, once the distribution
of rank and type coal has been established by the early use
of coal petrographic technique.

7.1. Regional Rank Variations

These preliminary regional coal rank variations can often
be established by the use of reflectivity determinations of
samples taken in the field and from boreholes, coupled with
a knowledge of the geological environment. Clearly the amount of data that can be obtained varies from region to region.

For some regions the general rank variation has been inferred, e.g. the Bowen and Sydney Basins and New Zealand. For
a more detailed analysis of rank variations in these areas

or regional mapping in lesser known regions, a reasonable spread of samples has to be collected. In practice the ease of doing so varies from region to region. Thus, in the Bowen Basin the Permian is commonly severely weathered to a depth of 30 feet, so that coal outcrops are relatively few, whilst in Indonesia, coal outcrops can normally be easily located.

In considering exploration for coking coals in little-known regions attention should be focussed on the following objectives :

1. Regional Geological Studies: An attempt should be made to locate coals close to the surface which were deeply buried and are of potentially high rank. Also, it is advisable to study the occurrence of "mobile" zones older, penecontemporaneous or younger than the coal measures of interest, including the presence of granitic intrusions. The possible proximity of a radiogenic basement should also be taken into account.

2. Sampling Programmes: It is advisable to obtain representative samples, if possible both from the surface and subsurface. The necessary fieldwork includes stratigraphical mapping, describing and measuring coal seam exposures etc. together with obtaining a good spread of samples. The latter should preferably include taking samples from previous boreholes drilled in the area.

3. Reflectivity Determinations: This should be undertaken on the collected samples in the laboratory, together with selected maceral analyses. If sufficient data is obtained, iso-reflectance or other iso-rank maps, as well as depth-reflectance graphs, should be compiled. Consideration should then be given to the regional variation in rank both laterally and vertically to provide a guide to the most favourable region for locating coking coals.

Similar programmes can be undertaken to locate coals of other ranks or simply to establish the rank configuration of a region of interest.

7.2. Variations in Coal-type

7.2.1. Selecting Exploration Targets. It has been shown that generally the Permian-Triassic and Mesozoic-Tertiary deposits of the west Pacific region form distinct coal-type provinces. The former contain a higher proportion of inertinite with limnic basins normally containing the highest.

The Mesozoic-Tertiary coals, on the other hand, are dominated by the reactive group of macerals.

Thus, in selecting exploration targets for prime coking coals, ideally for Australian purposes containing 50 to 62% reactives, it is necessary to turn to the Permian-Triassic coals, with the appropriate ranks. However, if blending of coals for conventional coke production is being considered, it is feasible to turn towards the more reactive Permian-Triassic as well as Mesozoic-Tertiary coals, again subject to an acceptable range in rank; in such cases it may be possible also to utilise inertinite-rich coals, including those of anthracitic rank. This allows for the conservation of prime coking coal resources, and the utilisation of a wider range of coals.

Coals suitable for power generation purposes are potentially abundant in the west Pacific region, always assuming the presence of favourable economic aspects, ash and sulphur considerations, etc. These include the low rank inertinite-rich, limnic basins of the Permian-Triassic and low rank vitrinite-rich coals from much of the west Pacific region. It is sensible not to exploit coals for power generation purposes which can be utilised, say, for coking coal blends and demand a higher price.

Coals suitable, in the longer term, for liquifaction and gasification processes are subject to the requirements of the processes involved or being developed. The high-volatile, hydrogen-rich Mesozoic-Tertiary coals, frequently rich in resin, are most amenable for hydrogenation purposes.

7.2.2. Exploration Sampling Programmes. As outlined above, representative samples for maceral and/or microlithotype determinations should be collected to obtain detailed petrographic data.

Frequently, specific coal localities have been repeatedly examined and reported on, with negative economic results. Often such localities are the most accessible and past explorers have not looked beyond these occurrences. Commonly, it is advisable to obtain petrographic data on these coal sections, relate the results to those obtained previously by conventional laboratory techniques, and then utilise the information in exploring for coal deposits in the surrounding region.

It is also frequently important in the early exploration

stages to correlate seam sequences, since seams may have distinct characteristics which require to be established for economic evaluation of the deposit, including preliminary reserve assessments. Detailed lithotype descriptions of seam profiles must be undertaken in the field, followed in selected cases by maceral and microlithotype analyses. The seam profiles have to be drafted on appropriate scales. Comparison of the different seams sampled, allows those with a distinctive lithotype profile to be correlated. Those seams sampled on a ply-by-ply basis allow a better correlation to be undertaken as the results of analyses can also be taken into account.

It is thus argued that an understanding of coal petrography provides guidance in selecting exploration targets for particular types of coal, not readily available by other techniques. Further, coal petrography can be of assistance in assessing exploration results with regard to the ultimate utilisation and marketing aspects.

8. SELECTED EXPLORATION EXAMPLES

A few examples are given to indicate how coal petrography can assist in exploration for coal in the west Pacific region.

8.1. North Bowen Basin, Queensland

Preliminary exploration was undertaken on Authority to Prospect 121C, twelve miles south of Nebo, in the north of the Bowen Basin (28). A sequence of Upper Permian coal measures is developed at the northern limit of a tightly folded zone containing anthracitic coal (Figure No. 15). The structure is also controlled by doming associated with a Cretaceous granodiorite intrusion. The Permian sequence is overlain by both Triassic and Tertiary sediments. An aerial photograph interpretation delineated two areas of shallow dipping sediments. A diamond drill hole was put down in each area. The first bore, in the northwest of the Authority, intersected ten shallow dipping seams up to 11 feet thick. Reflectivity values ($\bar{R}_o$ max) range from 2.97% at a depth of 189 feet to 3.58% at 961 feet, indicating an anthracitic coal and a high reflectivity-depth gradient of 0.91%/1000 feet.

The second bore was drilled in the southwest of the Authority. Only small coal lenses from the Triassic sequence at 472 feet and 986 feet were sampled and gave mean reflectivity values of 1.56% and 1.84% respectively. The reflectance-

depth gradient is 0.56%/1000 feet, indicating that all seams
in the presumed underlying Permian coal measures, although
of lower rank than in the first borehole, would also be anthracitic.

From a consideration of the reflectivity values and the
known regional geology it is concluded that all seams within
the Authority are anthracitic. Further, referring to Figure
No. 15, it can be stated that low volatile coking coals would
be located to the west and north of the Authority.

8.2. Oaklands Basin, New South Wales

A brief geological outline of this Permian basin has been
given above while Driver (19), and Palese (29) have described
more fully the geology of the coalfield. Robertson Research
Australia were requested to provide lithological description,
petrographic analyses and chemical analyses of selected cores
of the principal seam. Table No. II presents a condensed
example of a petrological analysis, based on a raw coal total
seam sample, together with some chemical analyses of one core,
the total seam thickness being 57.5 feet, whilst Figure No.
13 presents a seam profile.

The analytical results indicate that the seam contains
74% inertinite, 15% vitrinite and 6% exinite whilst the reflectance value at 0.29% indicates a very low rank coal (Figure No. 12). Classification based on the American Society
for Testing and Materials (A.S.T.M.) System indicates a Sub-
bituminous B Coal and on the Motts System a Sub-hydrous Lignite B. These results are typical for Permo-Triassic limnic
basin-type of coals.

Theoretically this coal could be utilised locally for power
generation purposes. Following McMaugh (30), it is not considered suitable for hydrogenation purposes due to its low
vitrinite and corresponding hydrogen content. However, it
is of interest to note that an exploration area at Oaklands
was recently successfully granted to Mitsubishi Proprietary
Limited for the purpose of investigating conversion processes.

8.3. Pike River Coalfield, New Zealand

A geological field survey was undertaken by Robertson
Research Australia of this outlier located in the northwest
of South Island, of New Zealand. The following details give
part of a sequence that crops out for four to five miles
along or near the Paparoa escarpment (31) :

PETROGRAPHIC AND CHEMICAL ANALYSIS OF A CORE FROM THE OAKLANDS COALFIELD

Analyses of Total Seam (Raw Coal)	
PETROGRAPHIC PARAMETERS:	
Predominant lithotypes Durain, with subordinate amounts of fusain, minor amounts of vitrain and trace of clarain.	
MACERAL:	VOLUME %
Vitrinite	15
Sporinite	1
Resinite	5
Total exinite	6
Fusinite	13
Semifusinite	24
Micrinite	37
Total inertinite	74
Quartz	1
Clay mineral	4
Limonite	Trace
Total mineral matter	5
Total reactive component (Vitrinite + exinite) (mineral-matter-free)	22
$\overline{R}_o$ max.%	0.29 (0.24 - 0.33)
CHEMICAL PARAMETERS:	
Proximate Analysis % (air-dried)	
Moisture	14.2
Volatile matter	24.4
Fixed carbon	48.7
Ash	12.7
Total sulphur (air-dried) Specific gravity (calculated)	0.20 1.48
CLASSIFICATION	SUB-BITUMINOUS B.
A.S.T.M. based on volatile matter (dry ash-free)	33.4%
and calorific value (moist ash-free)	10410 Btu/lb
MOTT'S CLASSIFICATION	SUB-HYDROUS LIGNITE B.
Based on volatile matter (dry mineral-matter-free) and calorific value (d.m.m.f.)	32.3% 12670 Btu/lb.

Age	Formation	Approximate Thickness (feet)
Upper Eocene	Island Sandstone	50 - 100
Eocene	Brunner Coal Measures	50
Upper Cretaceous	Paparoa Coal Measures	2000 (?)
Cretaceous	Hawk's Crag Breccia	0 -5000
	Unconformity	
Lower Palaeozoic	Granite	

The Brunner Seam, 20 feet thick, was sampled at several locations; it has a "pseudo-canneloid" appearance. Microscopic studies indicate the coal (Table No. III) to comprise vitrinite (91%), much which is detrital; this may explain the hackly rather than conchoidal fracture characteristic of bright coals, as well as the absence of distinct lamination, which, in association with the exinite content (5%), gives the coal a canneloid appearance. This distinctive coal has been vaguely called the "James" type in contrast to the "Strongman" type of the Paparoa seams. The reflectance values (R_o max %) average 0.64%. Although the measured volatile matter and calorific values (dry mineral-matter-free basis) (Table No. IV) are close to the predicted values based on petrographic analyses, the crucible swelling indices on the analysed samples are substantially higher. This is possibly due to the abundant organic sulphur, characteristic of the Brunner Seam throughout the Greymouth and Brunner Coalfields, the sulphur combining with oxygen to form sulphur dioxide and trioxide and the resulting gas causing the coal to swell excessively.

The Paparoa group of seams were also surveyed. Patterson records four coals (see 31), which contain interlaminated vitrain and clarain, the vitrain bands possessing typical conchoidal fracture. The coal contains 96% reactives, mainly as vitrinite (Table No. III) and the mean reflectivity value is 0.95%. The predicted crucible swelling index is very high at nine plus (Table No. IV) and the seam can, in this instance, be regarded as a true coking coal; the determined crucible swelling indices are lower, no doubt due to weathering of the samples.

The reflectance data indicates that there is a substantial difference in rank between the Brunner and Paparoa seams. This is supported by the difference in volatile matter values. The two seam groups are separated by only 300 feet of strata,

TABLE NO. III

PETROGRAPHIC ANALYSES – PIKE RIVER COAL FIELD

MACERAL ANALYSIS (%)		PAPAROA COAL (AVERAGE VALUES)		BRUNNER SEAM	
Vitrinite		94		91	
Exinite		2		5	
	Sporinite		1		1
	Resinite		1		4
	Cutinite		–		–
Inertinite		4		4	
	Fusinite		1		1
	Semi-fusinite		2		Trace
	Sclerotinite		Trace		–
	Micrinite		Trace		–
	Semi-micrinite		1		3
Mineral Matter		–		–	
	Quartz		Trace		Trace
	Clay minerals		–		–
	Carbonate		Trace		Trace
	Sulphide		–		Trace
	Limonitic material		Trace		Trace
Reflectance values ($\bar{R}_o$ max)%		0.95 (0.83 - 1.06)		0.64 (0.56 - 0.75)	

TABLE NO. IV

PIKE RIVER COALFIELD

COMPARISON OF PREDICTED "USE" PARAMETERS

WITH "USE" PARAMETERS DETERMINED EXPERIMENTALLY

	Paparoa Coals	Brunner Seam
Petrographic Parameters		
Vitrinite + exinite % (m.m.f.)*	96	96
Mean maximum reflectance in oil ($\overline{R}_o$ max)%	0.95	0.64
Predicted "Use" Parameters		
Volatile matter % (d.m.m.f.)**	34.5	45.0
Calorific value (d.m.m.f.) BTU/lb.	15,400	14,350
Crucible swelling index	9+	3 - 4
**Experimentally-determined "Use" Parameters		
Volatile matter % (d.m.m.f.)	34.5	44.2
Calorific value (d.m.m.f.) BTU/lb.	15,000	14,800
Crucible swelling index	3	7

** Volatile matter and calorific value calculated on a dry-mineral-matter-free
   basis from air-dried values using Parr's equation.

   (1)   Mineral matter % = 1.08 Ash % + 0.55 total sulphur.

   (2)   "Use" parameter = $\dfrac{\text{("use" parameter (air-dried basis)) x 100}}{\text{(100 - Moisture \% (air-dried basis) - Mineral Matter \%)}}$
        (d.m.m.f.)

*   m.m.f. - mineral-matter-free
** d.m.m.f. - dry mineral-matter-free.

which gives a very high reflectance-depth gradient of 0.9%/ 1000 feet. There is no evidence in the Pike River Coalfield for the presence of local igneous intrusions.

The Geological Survey of New Zealand has reported no unconformity between the Brunner and Paparoa Coal Measures in the Pike River Coalfield but an offlap occurs in the north. It is suggested that an unconformity may occur in the floor of the Brunner Seam. Some supporting evidence is found in the detrital nature of the seam, the occurrence of marine strata in the roof and a persistent quartzose sandstone floor

It is suggested that the Paparoa seams were affected by high temperature gradients due to the proximity of granitic basement and/or the considerable thickness of sub-Brunner strata originally laid down at a greater depth of burial. An unknown thickness of strata was subsequently eroded prior to the deposition of the Brunner Seam.

8.4. Indonesia

Unfortunately it is not possible to present results of detailed exploration surveys as is also the case for many other countries, due to confidentiality considerations. Hence, descriptions given in this section must be presented in a generalised manner.

As stated, there are two principal coal-bearing horizons, of Eocene-Oligocene and Mio-Pliocene age, the former occurring prior to a major marine transgression and the latter during marine regression. Indeed large areas consist of low-lying swamps to the present day.

The characteristic maceral composition is :

Vitrinite	86 - 94%
Exinite	4 - 12%
Inertinite	0 - 6%
Vitrinite + Exinite (mineral-matter-free)	44 -100%

It has been shown that there is little known variation in coal-types (Figure No. 12). Hence the prediction of "use" parameters of a coal is largely dependent on the coal rank.

Several authors commented on the relationship between

increase in rank and the age of Indonesian coals. For instance some of the parameters given by van Bemmelen (32) are
as follows :

	"Eocene" Coal	"Pliocene" Coal
Colour	Black	Blackish-brown or brown
Carbon	79.7%	69.4%
Hydrogen	5.2%	5.2%
Oxygen	13.8%	23.4%
Nitrogen plus sulphur	1.3%	2.0%
Moisture	9.9%	28.7%
Volatile matter	33.8%	41.0%
Fixed carbon	56.3%	30.3%
Calorific value	13,000 BTU/lb.	9,000 BTU/lb.

Hooze (33) suggested the following subdivision for east
Kalimantan :

Age	Moisture (%)
Lower Pliocene coal	30
Upper Miocene coal	19
Middle Miocene coal	14
Lower Eocene coal	3 - 6

In fact, for many years attempts were made to correlate
and subdivide the Tertiary on the basis of moisture content
of coal seams. However, more recently, correlations using
planktonic foraminifera, and nannoplankton, have shown that
coals with the similar moisture content from different areas
are not necessarily of the same age.

The following broad rank ranges have been established to
date, based on reflectivity ($\bar{R}_o$ max %) determinations :

Neogene coals	0.25 - 0.60%
Palaeogene coals	0.55 - 0.75%

These rank ranges refer to coals that have not been affected by local metamorphic effects and the overlap in reflectivity values is presumably due to the interplay of temp-

erature and time, the assumption by earlier workers that
lower-rank coals are Neogene age and higher rank coals Palaeogene age is not necessarily correct.

Considerations of reflectivity measurements of thermally
altered coals (e.g. Stach (34) records values ranging from
0.34 to 1.22% for the Bukit Asam Coalfield, south Sumatra)
suggest that the maximum reflectance value for Indonesian
Neogene coals, unaffected by contact metamorphism, may be
0.65% and for Indonesia Palaeogene coals 0.8%; the latter
value may be yet higher as the reflectivity values of Japanese Palaeogene coals can be as high as 1.0%.

On the basis of the petrographic data presented above,
the following approximate "use" parameters may be predicted
(dry mineral-matter-free basis) :

	Calorific Value (BTU/lb.)	Volatile Matter (%)	Crucible Swelling Index
Neogene coals	12,800 - 14,100	47 - 54	0 - 6
Palaeogene coals	14,000 - 14,900	40 - 47	3 - 6

It is therefore apparent that a search for higher rank,
coking coals should be undertaken with seams of Palaeogene
age. At Ombilin in central Sumatra fresh coal of this age
has swelling values of up to six (35) which compares with
values of less than three from surface samples. As indicate
previously the depth of burial, influence of mobile zones,
depth to basement and basement-type should be taken into
consideration in outlining prospective targets for these
coals.

On the other hand, exploration for coals suitable for
hydrogenation purposes; should be stressed in the Neogene
coals which contain an appreciably higher volatile matter
content than Palaeogene coals.

9. SUMMARY AND CONCLUSIONS

It is necessary to understand the varied occurrence, composition and utilisation potential of coal deposits. An
appreciation of coal petrology assists coal exploration targets to be selected, prospected and evaluated in a more scientific and economical manner.

In the west Pacific region, the Permian-Triassic coals occur within the present-day continental areas and tend to be vitrinite-poor, especially in limnic coal basins. This contrasts with the Mesozoic and Tertiary seams which are typically vitrinite and exinite-rich and occur within the island-arc, as well as along the margins of the continents – transgressing in places into the continental interiors. Regionally, higher rank coals tend to be situated in "mobile" zones, where the coals were subjected to higher thermal gradients whilst the lower rank coals occur in the more stable regions. Thus the regional consideration of both coal type and rank allows preliminary exploration targets to be identified in the search for specific coals to be utilised for either coke production, steam-raising or liquification and gasification processes.

When the initial field work is undertaken, possibly followed by exploratory drilling, it is essential at the outset to obtain detailed lithotype descriptions which are also valuable for seam correlation purposes and in addition to acquire coal seam samples which are as representative as possible, although only very small quantities are required for the petrographic analysis itself.

The results of coal petrography are valid for weathered samples in contrast to conventional chemical and physical analyses. This allows a number of "use" parameters to be predicted including the swelling index, calorific value and the carbon, volatile matter and hydrogen content.

In addition, petrographic data enables more reliable rank variation maps to be produced from preliminary reconnaissance surveys which, when combined with depth reflectance gradients, enable a three dimensional model of coal rank and type variations to be developed.

A consideration of the results of all these aspects, provides a most useful support for a preliminary geological and economic evaluation of a coal deposit.

10. ACKNOWLEDGEMENTS

Full acknowledgements are due to BP Petroleum Development Australia Limited, Geological Survey of Queensland, Magellan Petroleum Australia Limited and R.W. Miller and Company Pty. Limited, who made it possible to present selected data. Grateful thanks are also due to other organisations and colleagues, too numerous to mention individually. The paper is presented by permission of Robertson Research International Limited and the Commonwealth Scientific and Industrial Research Organisation.

11. REFERENCES

1. Stopes, M.C., "On the four visible banded ingredients in banded bituminous coals," Proc. Roy. Soc. London, Series B90 1919, 470.

2. International Commission on Coal Petrology, "International Handbook of Coal Petrology," Second edition, 1963, and "Supplement to Second Edition," 1971, Central National Research Society, Paris.

3. Seyler, C.A., "Letter to the nomenclature sub-committee, International Committee for coal petrology," 1954.

4. Thiessen, R., "Constituents of coal through a microscope Proc. Coal Mining Inst. Amer., 1919, 34-35.

5. Stach, E., "The use of oil immersion in the microscopic examination of coal," Gluckauf, 1937, 73, 330-333.

6. Seyler, C.A., "Petrology and the classification of coal, Proc. South Wales Inst. Engng., 1938, 53, (4), 254-327.

7. Stach, E., "Textbook of coal petrology," Gebrueder Borntraeger, Berlin-Stuttgart, 1975, Second Ed.

8. Bennett, A.J.R., and Taylor, G.H., "A petrographic basis for classifying Australian coals," Proc. Aust. I.M.M., 1970, No. 233, 1-5.

9. Ammosov, I.I., Eremin, I.V., Sukhenko, S.I., and Oshurkova, L.S., "Calculation of coking charges on the basis of petrographic characteristics of coals," Koksikhim., 1957, (12), 9-12.

10. Schapiro, N., Gray, R.J., and Eusner, G., "Recent developments in coal petrography," Proc. A.I.M.E., Blast Furn. Coke Oven, Raw Material, 1961, 20, 89-112.

11. Edwards, G.E., and Cook, A.E., "The design of blends for the production of metallurgical coke with particular reference to long term aspects of using New South Wales coals,"

Proc. Aust. Inst. Min. Met., 1972, No. 244, 1-10.

12. Smith, A.H.V., "Calculations of Micum 40 from petrographic data based on 250 kg test oven results," National Coal Board, Yorkshire Regional Laboratory, 1973 (unpub.).

13. Simonis, W., "Pre-calculation of coking abrasion in high temperature coking etc.," Gluckauf - Forschung, 1968, 4, 205-207.

14. Cook, A.C., "The spatial and temporal variation of the type and rank of Australian coals," IN Cook A.C., (Ed) Australian Black Coal - its occurrence, mining and preparation and use, Australian Institute of Mining and Metallurgy, Illawarra Branch; 1975, 63-84.

15. Shibaoka, M., and Smyth, M., "Coal petrology and the formation of coal seams in some Australian sedimentary basins," Economic Geology, 1975, 70, 1463-1473.

16. Mackowsky, M.Th., "European Carboniferous coalfields and Permian Gondwana coalfields," IN "Coal and coal-bearing strata," 1968, Oliver and Boyd Ltd., 325-345.

17. Britten, R.A., Smyth, M., Bennett, A.J.R., and Shibaoka, M., "Environmental interpretation of Gondwana coal measure sequence in the Sydney Basin of New South Wales," IN "Gondwana Geology," Canberra, 1975, Australian University Press, 233-247.

18. Osman, A.H., and Wilson, R.G., "Blair Athol Coalfield," IN "Economic Geology of Australia and Papua New Guinea - 2 Coal," 1975, Australian Inst. of Min. Metall., Monograph Series No. 6, 376-380.

19. Driver, R.C., "Oaklands - Coorabin Coalfield, N.S.W.," IN "Economic Geology of Australia and Papua New Guinea - 2 Coal," 1975, Australian Inst. of Min. Metall., Monograph Series No. 6, 376-380.

20. Svenson, D., and Hayes, S., "Callide Coal Messures, Q.," IN "Economic Geology of Australia and Papua New Guinea - 2 Coal," 1975, Australian Inst. of Min. Metall., Monograph Series No. 6, 283-287.

21. Dickins, J.M., "Correlation chart for the Permian system of Australia," 1976 Bur. Miner. Resour. Aust., Bull. 156.

22. Teichmuller, M., and Teichmuller, R., "Cainozoic and Mesozoic coal deposits of Germany," IN "Coal and coal-bearing strata," 1968, <u>Oliver and Boyd Ltd.</u>, 347-379.

23. Koppe, W.H., and Anderson, J.C., "The influence of basement type on coal metamorphism on the Collinsville Shelf," <u>Queensland Govt. Min. Jnl.</u>, LXXV, July 1974, 245-248.

24. Dunham, K.C., Dunham, A.C., Hodge, B.L., and Johnson, G.A.L., "Granite beneath Visean sediments, northern Pennines <u>Quart. Jnl. Geol. Soc. Lond.</u>, 1965, No. 483, 121, Pt3, 383-417.

25. Bostick, N.H., "Time as a factor in thermal metamorphism of phytoclasts (coaly particles)," <u>Congres. International de Stratigraphie et de Geologie du Carbonfere</u>, Septieme, Krefel August, 1971, Compte Rendu, 1973, 2, 183-193.

26. Suggate, R.P., "New Zealand Coals - their geological setting and its influence on their properties," <u>New Zealand Dept. of Scientific and Industrial Research</u>, 1959, Bulletin 134, pp 113.

27. Standards Association of Australia, "Taking samples from coal seams in situ," <u>Australian Standard</u>, CK5-1964.

28. Strauss, P.G., (Robertson Research (Australia) Pty. Ltd. for R.W. Miller & Co. Pty. Ltd., "Relinquishment report and results of exploration of A.P.121C: progress report for period ending July, 1974," August, 1974, <u>Department of Mines Queensland</u>, Report No. 5193 (unpub.).

29. Palese, G.W., "Oaklands Basin Coal Drilling Programme," <u>Geological Survey of New South Wales</u>, Report No. 1974/090 (unpub.).

30. McMaugh, M.J., "The assessment of Australian black coals for the production of synthetic crude oil," IN Cook A.C., (Ed), Australian Black coal - its occurrence, mining and preparation and use, <u>Australian Institute of Mining and Metallurgy</u>, 1975, 154-160.

31. Wellman, H.W., "Geology of the Pyke River Coalfield, North Westland," <u>N.Z. Jnl. Science and Technology</u>, 1948, 84-95.

32. Bemmelen, R.W., van, "The geology of Indonesia," 1949 (reprinted 1970), 3 vols., <u>Government Printing Office</u>, The

Hague.

33. Hooze, J.A., Onderzoek naar kolen in de straat Laut en aangrenzende landstreken, Jaarb. Mijnw. Ned. Ind., 1888, 17, Techn. Adm. Ged. Pt. 2, 337-429.

34. Stach, E., "Bericht uber die Untersuchung von Proben aus dem Tagebau Bukit Asam, Sud-Sumatra, 1. Teil: von Bericht uber die Petrographische Stuckschliff Untersuchung," 1953, Amt. Bod/Land. Nord-Westf. Krefeld W. Ger., D. Geol. Bandung A.43/BU, 6p. (unpub.).

35. Marubeni Iida and Kaiser Steel, "Ombilin Coalfield west Sumatra, Indonesia," Technical Report of Investigation and Exploration, Marubeni Iida Co. Ltd., and Kaiser Steel International Mining Corp. (unpub. report to P. Tambang), 1971, pp 67.

Section Six Discussion

P.K. GHOSH: Have you any information on the effect of intrusives on increasing the rank of coal?

R.S. QUINTON: There are Cretaceous intrusions particularly in the northern part of the Bowen basin.

P.K. GHOSH: Has it contributed to the increase of the rank of the coal?   That is my question.

R.S. QUINTON: I think the main trends of distribution of the various rank zones were related to the original depth of burial. The axis of the Bowen basin runs north-south.   The main area of deposition has subsequently been the site of major uplift and the geo-thermal gradient in that region is the highest, consequently having had the greatest depth of burial in the higher geo-thermal gradient the coking coals are associated with that north-south belt. To answer your question therefore, the presence of cretaceous intrusions doesn't appear to have significantly affected that overall picture.

P.K. GHOSH: While discussing the causes of increased coalif tion, decisive influence of temperature and increasing rank with depth of burial have been stated to be universal. Experience in India based on extensive drilling in almost all the coalfields of the country, however, points out conclusively that increase in rank of coal is closely relate to exposure of the coal measures to higher geothermal gradie depth of burial having no effect at all.   Rise in temperatur in basement rock and/or the coal measures itself has been mainly controlled by resurgent tectonism and/or igneous intrusives, the latter being primarily responsible for the increase of the rank of the coal seams in the Peninsular fields.

There are two types of intrusives found in the Indian coal-fields, viz. lamprophyres and dolerites - both of which occur either as dykes and/or sills. It has also been observ that while dolerites have practically no affect on the coal seams, lamprophyres have a profound affect on the coal seams converting them into 'Jhama" (burnt coal) as is locally called.   True 'Jhama' has a volatile content of about 5-8% against the volatile constituent of 25-30% in the unaffected

seam. When such intrusives occur as sills within the seam
or seams normally containing good quality coal, these become
totally spoilt. Intrusive lamprophyres have, however, a
very redeeming feature. When it permeates the basement
and/or the intervening barren zones in the coal measures, it
generates a higher geo-thermal gradient responsible for the
increase in the rank of the coal. This phenomenon is typically
displayed in the fields containing coking and/or other
varieties of coal of higher rank. In the fields where such
intrusives are absent, though geologically of same age and
type, the seams are invariably of lower rank with high
moisture content.

Leaving apart the effect of intrusives, tectonism has also
played an important role in the increase of the rank of the
coal seams as typified in the fields falling within the
Himalayan orogeny so much so that the seams have practically
been converted to anthracitic varieties. In the Peninsular
fields, tectonic effects are of little consequence except in
the highly faulted portion or portions of some of the fields.

All the above relates to the Gondwana fields. The coal
measures within the Gondwanas is of the Permian age (250 m.
years) whereas the lamprophyres are of Lower Cretaceous time
(100 m. years).

Tertiary measures of Eocene and Oligocene age contain lignite
beds and also coking coal. The latter has been formed
mainly due to serious thrusting affecting the coal measures.
The fields containing such coals are the ones where Himalayan
orogeny had been operative as in Makum and Dilli-Jeypore
coalfields in Assam in North Eastern part of India. There
are also other such areas. The same coal measures proved at
a depth 10,000 ft. (3045 m.) in course of drilling for oil
in the adjacent area at Rudrasagar being tectonically least
affected do not indicate any increase in the rank of the
coal seams above that of lignitic type. Similarly, in some
of the large coalfields, as for instance, Raniganj field in
Peninsular India, deep drilling to a depth of about 1218 m.
(4000 ft.) had been carried out. But analysis of the coal
seams at that depth vis-a-vis its constituents at the surface
does not show any difference to justify the theory of "depth
of burial". All these demonstrate that the application of
Hilt's law is dependent on some variants other than depth
alone.

MR. HALL: (National Coal Board): You have suggested that petrographic analyses do not replace chemical analyses and yet they do analyse the same thing in certain cases. What in your opinion would be an ideal analysis programme considering economic conditions for a company to go to?

P.G. STRAUSS: By all means carry on what has been done before, but include petrographic analysis at the same time to check that the results you have are valid. For instance, in West Canada they have been sampling oxidised coal and doing chemical analysis. The results are simply not valid because the volatiles have been reduced, the c.v. has been reduced and the coking properties have been lost. You cannot by a petrographic analysis get all the results you need. You have still to do the ash and the sulphur and so on by chemical analysis, but the petrographic analysis will check whether the chemical analysis is valid and you can go on from there. So I am not saying change your programme, but I am saying do some additional work to make sure your conventional laboratory analysis is correct. Does that answer your question?

MR. HALL: Yes it does, except that most companies will try and do things in the cheapest possible way, especially in the initial exploration stages, so would you say then, to analyse for ash, and sulphur and then run a petrographic analysis, to have an approximate result for the other factor

P.G. STRAUSS: I know I am generalising, but if you have a good seam profile and a good sample, do chemical analysis, but if you have a small or unsatisfactory sample then it may be worth-while only to do petrographic analysis because you haven't got enough material there to do all the other work and in fact, it may often be exorbitantly expensive to do all the other types of work, so it depends very much on the situation. If for instance you were thinking of petroleum wells and getting cuttings from it, often it would be better just to do petrographics but in a normal exploration program I would say do both hand in hand.

R.H. HOARE: (National Coal Board): Many years ago a very interesting but problemmatical feature of reflectance measur ments was the development of anisotropic optical properties in high ranked coal. How does this affect or benefit your reflectance applications?

P.G. STRAUSS: Obviously I kept my talk very simple. You find in practice that anisotropic values increase for all coals from low ranked coals to high rank coals. I found in practice that in any such high anistropic value there is a big difference between the minimum and maximum bireflectance. I found when you have a very high value that it may indicate the presence of pressure. In one example in the Bowen basin, very high reflectance in anthracite had a very high bireflectant anistropic value due to granite intrusion. In another part of the Pacific Islands the presence of these high values did suggest the presence of nearby intrusions.

The investigation of world coal resources has extended to areas so remote as to require helicopters to establish exploration sites. Photo courtesy Longyear Co.

SECTION 7

14.
Computer Evaluation and Classification of Coal Reserves
by William H. Smith

Section Discussion

14

COMPUTER EVALUATION AND CLASSIFICATION OF COAL RESERVES

William H. Smith

Consulting Geologist
Champaign, Illinois, U. S. A.

Computer techniques will be widely used in the future to assemble, evaluate and map geological information relating to the detailed characteristics of coal seams that are to be mined. The utilization of automated data processing systems promises to make possible the production of the necessary maps and charts required to define a coal deposit and to calculate reserves much more quickly and at a lower cost than conventional methods. There is currently considerable interest within the mining industry to utilize computer facilities for coal resources assessments and for mine planning and engineering.

Recent developments have brought about the proliferation of digital integrated circuits and microprocessors. This has reduced the size and cost of computer hardware, which will result in the more widespread use of computers in the coal mining industry. Through the utilization of modern computer-graphics techniques, it is now possible to store in a data bank and later retrieve in map format all of the data concerning a mining operation that is required for land management, geological evaluations, and mine engineering.

The next five to ten years will see greatly expanded use of computers to assist coal mining engineers and geologists in feasibility studies and mine planning. The computerized maps and data will then be available to expedite the designing of the mine, and to monitor and control the coal mining and preparation operations after the mine is in operation.

In the United States, several of the major coal companies are utilizing computer facilities for coal mapping, property management, and engineering evaluations, and many others are moving toward the more widespread use of computers for the evaluation and engineering of mining properties.

The United States Bureau of Mines is engaged in a number of programs to demonstrate the feasibility of computer graphics techniques for engineering and management of coal mines. The United States Bureau of Mines and the Office of Coal Research have sponsored studies that now provide computer programs and documentation that are being made available to the mining industry.[1-2]

The United States Geological Survey is undertaking the extensive use of computers and interactive computer graphics systems for coal resources investigations. The survey has undertaken the development of the computer-based National Coal Resources Data System, which will provide a comprehensive data base for coal resources information in the United States.[3]

Coal Data Base

For coal seam mapping and evaluation, the computer can be used to establish an information storage and retrieval system that will integrate all of the information required to complete mapping and feasibility studies for a proposed mining operation. Additional information, such as that obtained during development drilling, can easily be added to the data base as the mine development proceeds to provide updated maps depicting the geology of the coal seam. These maps can then be used later for engineering the mine. This type of mapping is accomplished by establishing a coal data base that contains all of the specific information that has been obtained during the exploration and evaluation program for a mine.

All of the information in the data base is identified geographically by X and Y coordinates to designate its specific location. The Z dimension defines the information that we wish to analyze or to map. The system permits the entry of a wide variety of Z values, such as ground surface elevation, depth and thickness of coal seams, type of roof and floor strata, analytical data, and many other types of information.

In order to properly identify the information both geographically and geologically as it is entered into the data base, it should be coded to identify its location in terms of county, township, mining property, etc. This has been greatly simplified by the adoption of standard coding systems such as the uniform State and County Codes established by the A. A. P. G. Standard coding for stratigraphic information, such as geological formations, coal seams, and other specific rock units, has also been worked out by the United States Geological Survey and many of the State Geological Surveys.

A block diagram of such a data base is shown in Figure 1. It begins with the geographic and topographic mapping information to which property ownership and management information and any geologic data from previous investigations are added. As property acquisition and geologic exploration progresses, new data are added to the data base.

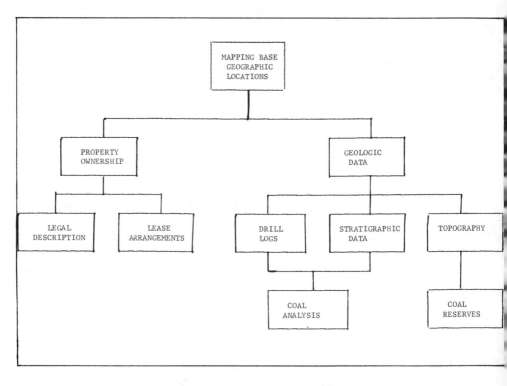

Figure No. 1. Block Diagram of a Coal Data Base

An information storage and retrieval system, which consists of computer software programs, is designed to organize and arrange these data by geographic location based on some type of rectangular coordinate system that ties everything together. The coordinate system that is chosen should be the same as that which will be utilized for mine planning and development. In the United States, a uniform plane coordinate mapping base has been developed for each of the states by the United States Coast and Geodetic Survey, and is being used for this purpose.

A convenient way to structure the data base is to have an access dictionary to store the general information for each drill log, and for all samples taken for quality analysis, etc. Such general information relates to specific locations, sample or drill hole number, date of sampling, etc., and the appropriate address and access numbers for the secondary files.

A series of secondary files are then established where chemical data, for example, can be stored in one file, coal depth and thickness data in another file, etc. Each of these secondary files may have a different format, but each is keyed to the access number and location information so that different files can be interrelated if desired. If an access number and the location for each sample is stored in both the dictionary and in the secondary file, the data in any of the secondary files can easily be used to draw maps at any time directly from the secondary files.

A data storage and retrieval system of this sort begins with the digitized geographic and topographic information and then compiles property ownership information, geological information relating to depth and thickness of the coal seams, coal quality data, etc. The retrieval programs can very rapidly produce a variety of maps by calling for data stored in the data base and formulating it to provide output data from which high speed plotting machines can produce maps at any desired scale.

There are available programs that contain surface graphics capabilities such as the IBM program STAMPEDE that can recall information from the data base and through a series of arithmetic routines produce contour maps, trend surface maps, and compute areas of volumes of coal.[4] Several of the illustrations of computer drawn maps that will be shown as slides were drawn by GEOMAPS, a surface graphics package derived from STAMPEDE, that was developed at the Illinois Geological Survey.[5]

These programs transform the input data into a gridded
numerical surface that represents the input data so that
many lines or data points can be considered at one time for
the solution of complex problems. The numerical surfaces
representing different coal seams or different types of
information on a single seam may be added together, sub-
tracted, or otherwise manipulated in a number of ways to
solve many complex problems. This enables the geologist
or engineer to quickly evaluate a variety of different maps
and combinations of maps to study a problem in ways that
would not be possible if a map of each of the parameters
to be studied had to be compiled by hand in the conventional
manner.[6]

The input data consists of a series of random values
distributed over the map, which may represent, for example,
the elevation of the top of the coal seam, or its thickness.
The program in the computer will construct from this data
a uniform grid having a numerical value at every grid inter-
section and output a contour map of the top of the coal seam,
or an isopach map of the seam thickness. Through the use of
the numerical surface techniques, trend surface maps and
maps of the interval between coal seams can also be provided
to show geological relationships that often are overlooked
because of the difficulty of producing such maps by conventional methods.

If the surface topography has been digitized, it is
possible to compute overburden depth and stripping ratio
for the seam. Using the surface graphics approach, random
variations in the data tend to be smoothed out so that only
the overall relationships are preserved. The grid size can,
however, be varied to provide any desired amount of detail
to be retained.

Coal Reserves

To compute coal reserves by computer, the numerical surface that we have used to produce the isopach of coal thickness is integrated grid by grid for the areas falling within
each category of thickness represented by the isopach lines
to obtain the volume of coal represented. If the coal seam
contains areas previously mined out, or areas not considered
to be minable for other reasons, these can be eliminated
from the estimate by forming a "binary surface" that assigns
zeros wherever mined out or ignored areas appear and ones
elsewhere. When this surface is multiplied grid by grid by

the numerical surface representing the thickness of the coal seam, we obtain a "true" coal surface having zeros wherever the coal has been mined out or for some other reason is to be omitted, and the original coal thickness values elsewhere. This technique can be used to exclude from the reserves estimate coal in areas that for one reason or another should be ignored in computing the coal tonnage from the computer generated maps.

When computing reserves, the geologist may wish to establish a minimum thickness cutoff and to tabulate the reserves by various classes of reliability and seam thickness. Because of the speed with which the computer can compute areas and volumes, the computerized approach to reserves evaluation allows a much greater degree of flexibility in the number of categories into which coal reserves may be classified than does the traditional method of hand planimetering.

In the traditional methods of measuring and computing coal reserves utilizing the polar planimeter, the area of each class and subclass into which the reserves are categorized must be tediously measured by hand with the planimeter over the entire mapped area. Then all of the individual planimetered measurements are summarized to define the total area underlain by each of the subclasses of thickness or reliability. The area of each subclass must then be multiplied by the appropriate factors for the weight of coal per cubic foot, tabulated and checked to obtain the breakdown of reserves into the required number of thickness classes.

The time required to planimeter maps used for the calculation of coal reserves depends to a large extent upon the number of subclasses of thickness or reliability into which the estimate is to be divided. Reserves are commonly classed as proved, probable, or inferred by constructing arcs representing a radius influence at certain established distances from each drill hole or other point where the coal seam thickness is known. Many other constraints, such as maximum ash or sulphur limits, may be imposed in making the estimate. The measurement of coal reserves, using the polar planimeter, is very tedious and time consuming, and it becomes quite costly if one wishes to revise a previous estimate of reserves to incorporate new data acquired through additional drilling or other newly acquired information about the thickness or continuity of the coal seam. Often as mine feasibility studies progress, it is desired to make a separate estimate of reserves by sulphur content, ash content, etc., or to make several separate reserve estimates based on alternative approaches to the mining. Although the value of compiling the reserves estimate in a number of different ways may be recognized, it seldom can be accomplished because of the time and expense required to re-planimeter the maps to obtain revised estimates of the reserves. Using the computerized approach to coal reserves mapping, one can make a much more comprehensive study of the reserves, because the computer offers so much more versatility in formulating the data for the reserves computation. Because of the speed with which reserves can be measured and summarized by computers, it is practical to compile coal reserves maps based on a much wider range of parameters and to tabulate the reserves in a variety of ways that may be most useful for economic analysis and mining feasibility studies.

Mapping Techniques

While the establishment of a fully integrated coal data base that is capable of producing structure contour maps, coal isopach maps, and compiling reserves estimates from randomly distributed data points should be the ultimate objective of most users of computers for coal resources evaluation, it should be realized that conversion of existing records to a form suitable for computer processing will take some time. However, after the framework for a data base is established, it can be built up a module at a time so that the modules can later be combined to provide a fully integrated system. For example, the geological data from drill holes could be organized to provide maps for preliminary mine layout and for reserve studies. This could then be combined with topographic data, chemical analysis, property data, etc., that would permit many other maps to be made as the planning and development of a mine proceeds.

Many useful maps can be compiled using computer-graphics techniques long before a fully integrated coal data base has been established. As soon as a suitable mapping base has been selected, the land lines and other planametric elements for a base map can be digitized, these can then be entered into the data handling system to produce property maps and other base maps that will be useful for exploration planning. When additional data becomes available, it can be fed into the data base so that the capabilities of the system will increase as the data base is expanded.

A unique feature of a system devised for a comprehensive coal reserves study in the State of Illinois, which will be illustrated with a number of slides, was the use of interpreted coal thickness data as computer input in place of actual point data.[7] Such techniques provide the opportunity to take full advantage of the computerized routines for data manipulation such as volume integration and the drawing of maps while working toward a fully integrated coal data base capable of producing from point data the isopach maps required for estimating coal reserves.

In the recently completed computerized approach to coal reserves calculation at the Illinois State Geological Survey, nearly 100 billion tons of reserves in the two principal coal seams throughout the state was evaluated and mapped. Before the isopach lines were digitized, they were assigned a two digit decimal value representing coal thickness. As these lines were digitized, the coordinates of successive points along the isoline were recorded on magnetic tape together with the coal thickness represented by the isoline. Other lines representing features such as coal outcrop lines, boundaries of mined-out areas, and limiting lines for various classifications into which the reserves were divided also were digitized to form a separate file. Thus, one file contained the coordinate locations of coal thickness information, and another file contained information relating to features such as previously mined areas, coal outcrops, etc.

The computer program accesses these files of coordinate data relating to coal thickness and limiting features, and in several steps, the program builds a numerical surface that represents the geological information portrayed on the hand drawn coal reserves maps. From this numerical surface the volume integration routine in the program calculates the reserves and categorizes the reserves into classes. By interfacing the coal surface with ILLIMAP, a statewide computerized mapping base for Illinois,[8] the reserves can be calculated by county, township, and seam thickness.

An important feature of the computerized coal resources mapping system developed at the Illinois Geological Survey is its ability to draw maps from the coal data base showing coal reserves and related features at various scales. The estimates of coal reserves and the digitized data relating to coal-feature lines are then stored in magnetic tape files where they can be accessed for correction and modification as additional studies are undertaken.

REFERENCES

1.   Gomez, Manuel, and Donald J. Donovan, 1974, "Forecasts of Chemical, Physical, and Utilization Properties of Coal for Technical and Economic Evaluation of Coal Seams," U. S. Bureau of Mines Investigations 7842.

2.   Office of Coal Research, U. S. Department of Interior, 1975, "Computer Applications in Underground Mining Systems," Research and Development, Report 37.

3.   Cargill, S. M. and others, 1976, "PACER - Data Entry, Retrieval, and Update for the National Coal Resource Data System (Phase I)," U. S. Geological Survey Professional Paper 978.

4.   "Surface Techniques, Annotation and Mapping Programs for Exploration, Development and Engineering (STAMPEDE) Program Documentation," Program Number 360D-17.4.001, IBM Corporation.

5.   Miller, W. G and W. H. Smith, 1975, "Coal Data Interpretation and Mapping Using Computer-Graphics Techniques at the Illinois State Geological Survey," AIME Preprint 75-AY-376, SME fall meeting, Salt Lake City, Utah.

6.   Junemann, P. M. and L. M. Kaas, 1973, "Computer Uses for Coal Mine Engineering and Geology," AIME Preprint 73-AR-100, AIME Annual Meeting, Chicago, Illinois.

7.   Smith, W. H. and J. B. Stall, 1975, "Coal and Water Resources for Coal Conversion in Illinois," Illinois State Water Survey and Illinois State Geological Survey Cooperating Resources Report 4.

8.   Swann, D. H., P. B. Dummontelle, R. F. Mast, and L. H. Van Dyke, 1970, "ILLIMAP - A Computer-Based Mapping System for Illinois," Illinois State Geological Survey, Circular 451.

Section Seven Discussion

S. POLEGEG: (Mining University of Leoben, Austria): You mentioned the accuracy of the surface presentation of your data where a computer writes programmes. If you connect two sampling points you can do it in different ways. You can draw a straight line, you can draw a draft line. The only point on which I don't agree with you is that accuracy cannot be increased by enlarging the grid size within the two points. What I would like to ask you is what about the range of the accuracy of your surface presentation?

W.H. SMITH: You are absolutely correct. You have only a certain amount of data. That data is going to be the controlling factor. Your accuracy cannot exceed the accuracy of your data. What I was speaking of is the ability to subdivide the data, for example, if you are using 1 foot contour interval you might use a certain grid size but if you wanted to reduce your contour interval and look at the same information on say 6 inch contour intervals, your grid size would not be compatible with that. You could have half the grid size and get a more accurate evaluation within that category, but the overall accuracy of your figures would not be any greater than that of the raw data, the primary data that you have, that is correct.

A. RABITZ: (Geologisches Landesamt Nordrhein-Westfalen, F.R. Germany). My question concerns tectonics. Have you no faults and no other great tectonic structures in your coal fields? The question is important in applying the computer to coal reserves. When we try to do computer work on coal reserves in Europe, we are very handicapped by the tectonical structure of the deposits.

W.H. SMITH: Yes there are tectonic features which are not shown on that map. There is another map, the geologic map of the State of Illinois, which clearly shows the tectonics. There are a lot of faults particularly in the southeastern part of the coal field, but many of those are only 10 or 20 foot faults and those that are known have been mapped. They don't influence the quantity of coal. They will in the future influence the mineability, but in this overall estimate there was no consideration of taking away from the total reserve, an amount for the possibility that some of the coal

might not be mineable because of the tectonics. We did
however, as you can see, subtract for previously mined areas
or areas that were underlain by oil pools. We felt there is
no present technology for mining over those pools and for
the more known features like those sandstone channels and
geologic conditions, which we did know, but you are correct.
We did not deplete the reserves by some factor on the basis
of the tectonics because in this coal field although there
are tectonics, they are not a serious deterrent to mining.
Only on a local scale.

R.N. PRYOR: (Royal School of Mines, London) May I ask Mr.
Smith if he knows if packages are available for the economic
optimisation of the sequence of open pit mining using his
database information? If I could just enlarge on that question, I speak as one more experienced in copper mining and
we use the geological database to generate the ore reserves
into rectangular blocks and then that information can be
used for studying total pits and the sequence of mining, and
there is quite extensive literature in that field, but I
have recently been looking for literature in the coal field
and found it rather scarce.

W.H. SMITH: You are speaking of programmes that can integrat
the information, relative to surface mining?   Yes. I think
that there are some people working in that area. I have not
personally been involved in that type of work with programmes
that can integrate on the basis of slopes and compute the
volumes. I know some work that has gone on to do volume
integration. It mainly relates to people who are concerned
with things like highway cuts and things of that sort, that
can do it on a very small scale.

A.M. MACE: (Houilleres du Bassin du Centre et Midi, France)
The colliery I am concerned with has five million tons of
coal reserve the cost of which would be about £10, another 5
millions, the cost of which would be about £20, and another
250 million tons, the price of which would be under £100 per
ton. We sell it on average £15 per ton. So which figure
will you suggest I strike as our reserves.

K. WHITWORTH: Could you give us the details of that question
again?

A.M. MACE: In open cast mining the cost is only £10 per ton and we sell at £15. In our present mining, the cost is around £20 per ton and so we make a loss from it, but we know that there are 250 million tons of proved reserve. What would you estimate our reserves to be, this is the general problem of the relationship between reserves and cost of production.

W.H. SMITH: Well, I think that what you are probably referring to is to assess the strippable reserve on the basis of what would be your optimisation, you are looking at an optimisation technique of considering that reserve, the relationship between mining cost and recovery at different depths. If that is your question the obvious thing is that once you have the data that you are working with into a system of this sort, you have the capability of doing a much more in depth analysis and looking at it in many more ways to optimise your different stripping ratios, than you could do by hand unless you want to spend a great deal of time doing it.

F.W. PROKOP: (Otto Gold GmbH, F.R. Germany) Mr. Smith, I would be interested to find out whether these database programmes or respectively the data is available to the public and at which cost?

W.H. SMITH: In the bibliography of the paper I have given, some of the publications give information on the availability of programmes. Virginia Polytechnic Institute, has done some for the U.S. Bureau of Mines and the U.S. Office of Coal Research, and the U.S. Geological Survey. Most of those programmes are available at little or no cost. That is one of the things that the U.S. Bureau of Mines is endeavouring to do, to make the programmes for this type of work available at essentially no cost, to try to optimise the working capability of mining in the United States, because they recognise the vast amount of work that is needed if we expect to double the coal production perhaps within the next ten to twenty years, and the engineering effort is going to be enormous and this the Bureau of Mines recognises is one way to make the labour force stretch further.

F.W. PROKOP: A short question: Did you ever use the information of the survey's computer mapping for a recalculation of

samplings to reach these centres? (the distances from
borehole to borehole). Did you recalculate this by your
computerised programmes? So, if you can enlarge your sampling
to each size you have to make a more intense sampling for
each size than optimisation of sampling?

W.H. SMITH: I didn't quite understand the end part of your
question. It is obvious that if you can increase the number
of drill holes then you can re-run your programme and come
up with a secondary evaluation which is going to be much
enhanced over your preliminary, but was that your question?

F.W. PROKOP: What I mean is, if you get an interpolation
between your drill holes, this is the purpose of your computerised matter? You may have a feedback from this interpolation so that you can say I can enlarge my sampling
distances, for example, my drill hole distances. Has this
been done?

W.H. SMITH: Not a great deal of it and I think that I
personally who have been involved in siting drill holes and
working with that type of information most of my life, have
always been impressed with the fact that, to my knowledge,
no-one really has made an in depth evaluation of how many
drill holes in a certain situation are required to arrive at
a certain degree of probable reliability. You will find
that one group will make twice as many drill holes to arrive
at an optimisation as another group will and as to what is
the optimum number, that is a very good question and a very
good way to arrive at a sampling problem and people have
worked with it but, in the matter of drilling density there
is very very little in the literature that you can go to to
find information on that. It is more or less done by just
value judgements on the basis of the people involved in the
engineering.

SECTION 8

15.
**Coal Exploration Techniques and Tools
to Meet the Demands of the Coal Industry**
by Walter W. Svendsen

16.
**The Development and Adaptation of
Drilling Equipment to Coal Exploration**
by Keith Shaw

Section Discussion

15

COAL EXPLORATION TECHNIQUES AND TOOLS

TO MEET THE DEMANDS OF THE COAL INDUSTRY

Walter W. Svendsen
Technical Director, Longyear Company
Minneapolis, Minnesota, United States

In the post-war period of 1945 and 1946, there were reasons to forecast a heavy demand for equipment to be used for coal exploration, and suppliers of such equipment were "gearing-up" to meet the apparent need. However, in 1947, coal mining companies and labor unions became involved in lengthy negotiations, which, coupled with the growing avail-ability and greater convenience of oil and gas, led to cur-tailment of coal production and consumption, followed by an immediate and drastic reduction in coal exploration.

Some exploration for coal in the U.S.A. has been carried out on a continuous basis since 1947, but it was only the shocking realization that our reserves of oil and gas are limited (a fact, incidentally, that was well-known for many years by our geologists and petroleum engineers) that re-vitalized coal exploration and brought it into its current more active and vital role. The limited demand for coal ex-ploration caused us in the drilling industry to ignore the problems associated with coal drilling in order that we might concentrate our efforts on the development of equipment and techniques for the then more actively-conducted search for iron ore, lead, zinc, copper, gold, and other mineral re-sources. Fortunately, the results of this research have pro-ven to be of considerable value for all facets of exploration drilling, and many of the new tools and techniques used in

Table No. I (1)

Production of Bituminous Coal, by Type of Mine
(000 Tons)

Year	Strip Mining	Auger Mining	Underground Mining	Production
1945	109,987	.....	467,630	577,617
1946	112,964	.....	420,958	533,922
1947	139,395	.....	491,229	630,624
1948	139,506	.....	460,012	599,518
1949	106,045	.....	331,823	437,868
1950	123,467	.....	392,844	516,311
1951	117,618	205	415,842	533,665
1952	108,910	1,506	356,425	466,841
1953	105,448	2,291	349,551	457,290
1954	98,134	4,460	289,112	391,706
1955	115,093	6,075	343,665	464,633
1956	127,055	8,045	365,774	500,874
1957	124,109	7,946	360,649	492,704
1958	116,242	7,320	286,884	410,446
1959	120,953	7,641	283,434	412,028
1960	122,630	7,994	284,888	415,512
1961	121,979	8,232	272,766	402,977
1962	130,300	10,583	281,266	422,149
1963	144,141	12,531	302,256	458,928
1964	151,859	13,331	321,808	486,998
1965	165,241	14,186	332,661	512,088
1966	180,058	15,299	338,524	533,881
1967	187,134	16,360	349,133	552,626
1968	185,836	15,267	344,142	545,245
1969	197,023	16,350	347,132	560,505
1970	244,117	20,027	338,788	602,932

mineral exploration are now being used to improve drilling performance in the rapidly-expanding search for and delineation of minable coal beds.

Bituminous coal production in the United States, according to the U. S. Bureau of Mines, increased, generally, from 1940 to 1948, when a decline started that continued until the first signs of an upward trend appeared in 1970. However, it is important to note that production by strip mining doubled from 1950 to 1970, during which period production from underground mines decreased markedly. During this period, a new method of coal recovery, known as auger mining, was introduced, which grew from no recorded production in 1950 to 20 million tons of coal produced by this method in 1970. This equates to approximately 8 per cent of the coal mined in the United States by strip mining in 1970. Estimates for 1975 indicate that coal production in the United States will be approximately 5 per cent greater in 1970, with a 20 per cent drop in production by auger mining and a 10 per cent drop in underground mining production, accompanied by a 30 per cent increase in strip-mining production.

These trends in mining influence, to a marked degree, the types of tools and the techniques required to meet the demand of the coal-producing companies in their search for coal deposits. For example, the expansion of strip mining has called for more mobile drills, which can be quickly rigged up and down with minimal effort. Such drills must be capable not only of drilling relatively shallow holes at high rates of speed, but also must contain the necessary sophisticated instrumentation to ensure maximum recovery of core sample from the coal seam(s). Core samples of larger diameter are also becoming of greater importance. Today's drills must, therefore, be capable of achieving high penetration rates, using tool diameters in the 96-mm-and-larger ranges in the drilling of either inclined or vertical holes. A variety of relationships exist between these mining methods - auger mining, strip mining, shaft mining, degasification or underground exploration, offshore exploration - and the exploratory drilling equipment and techniques required to meet the needs of the various methods.

AUGER MINING

Exploration of coal seams to be extracted by auger mining techniques may be conducted by angle or vertical holes and, in certain instances, it may be desirable to check the coal seam by drilling a few hundred feet into the coal outcrop along the plane of the seam. The exposed face may be hand-sampled and the seam extending into the hillside is tested for quality and thickness by core drilling verti- cal, angle, and - in some cases - holes drilled along the plane of the seam.

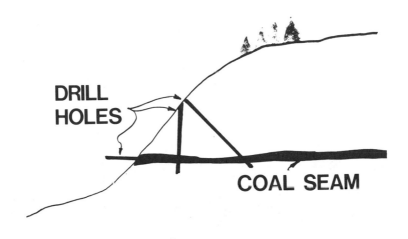

AUGER MINING

Cross-sectional view of a coal deposit and terrain where the initial mining may best be accomplished by the augering method. Figure No. 1.

Since auger mining is limited to a zone of about two hundred feet wide paralleling the outcropping face of the coal seam, the exploration drill holes are correspondingly shallow. In cases where the face of the coal seam is ex- posed on a steep hillside, angle holes may be required. Such a program could best be executed by a small, light- weight drill readily moved up or down steeply sloping ter- rain, possibly by men, animals, or helicopter. The drill would, of necessity, be provided with an angle-type head, and would have a light-weight, fabricated mast, or steel-or- aluminum tripod. Since the holes would be of relatively shallow depths, drilling could probably be efficiently done with conventional tools. However, since a single bit would normally suffice for one or more holes, it would probably be advantageous to use wire-line equipment. Small, light-weight drills of the type normally used for such a program are manufactured by a number of companies throughout the world.

In areas where overburden is thin enough and the geology is favorable, <u>open-cast</u> mining is common. Exploration of such deposits is performed with long-feed drills, normally using roller-rock bits and employing air as a circulating medium. Sampling of the coal seam may be achieved by collecting chip samples, or by coring with double-tube, swivel-type, air barrels. Several drills have been designed and built in the United Kingdom for this type of drilling. The drills are normally tractor-mounted and quickly and efficiently moved from hole to hole.

Samples of friable core. Figure No. 2.

STRIP MINING

Drills adapted to this method of exploration should be highly mobile and, preferably, should be totally self-contained, having a suitable mast and circulating pump as integral components, as well as all necessary controls and instrumentation to maximize and to ensure high core recovery.

Since strip mining entails complete overburden removal, it may be desirable, in some holes, to recover representative samples from strata overlying the coal seam. The drill rig and its array of tools should have the capability and flexibility of dealing with a wide variety of strata to be drilled and sampled. Bits may vary from roller-rock bits and diamond bits to augers, with final sampling done by diamond core drilling. Mobility can be achieved through a variety of means, including mounting the equipment on trucks, trailers, tractors, or scows. When severe weather is encountered, it may be necessary to enclose the drill. The type of mounting selected may, also, be influenced by environmental restrictions which, in some cases, may result in a completely new drill design. Where exploration is in roadless and environmentally-sensitive areas, it may be mandatory to move the rig by helicopter - a method of transportation requiring a drilling outfit that is very light in weight or one which is easily broken down into movable components.

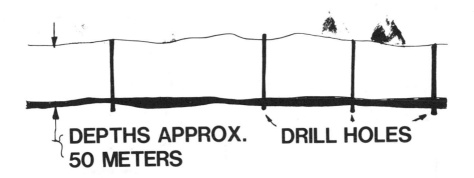

DEPTHS APPROX. DRILL HOLES 50 METERS

STRIP MINING

Cross-section of an area that would be a "candidate" for a strip-mining operation wherein the coal beds to be mined are normally within a depth range of 50 meters. Figure No.3.

Exploration and/or development drilling for data needed in strip-mining planning may require sampling of the overlying formations, as well as of the coal seam(s). The deeper drilling involved in securing information needed for

developing an underground coal mine, on the other hand, will require different equipment and methods, including the use of:
- drilling units and ancillary equipment of greater depth capacities.
- less emphasis on - or abandonment of - core sampling of beds overlying the coal seam(s) and the use of non-coring methods in the penetration of such beds.
- core samples of larger diameter.
- deflecting wedges or whipstocks to permit recovery of multiple core samples from each of the deep holes.

SHAFT MINING

With the universal and growing need for coal, deep deposits are being explored and mined today, some of which, ten years ago, would have been considered economically unattractive.

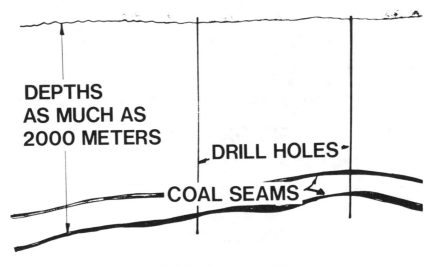

DEPTHS AS MUCH AS 2000 METERS

DRILL HOLES

COAL SEAMS

SHAFT MINING

A typical cross-section of an exploration program that would be carried out prior to opening up an underground coal mining operation. Figure No. 4.

Exploratory drill holes for coal deposits in the Donetsk Basin of the Ukraine are often of the order of 2,000 meters in depth, while those drilled in the United Kingdom and in the Saarbruken, Germany, coal fields, as well as in Eastern France, are in the depth range of 1,400 meters. Exploratory holes may be drilled to depths in excess of 2,000 meters. Due to the high cost of such exploration, the demand for reaping more information from each drill hole continues to increase.

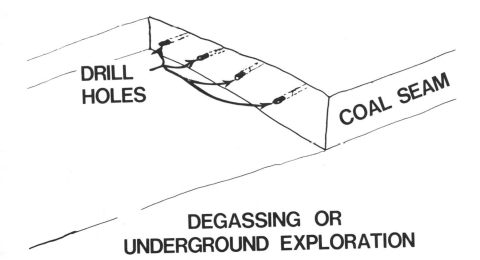

A series of long, flat holes drilled within coal seams are connected to a collection-and-exhaust system. Figure No. 5.

In addition to the demand for a high percentage of core recovery, there is, also, a desire to recover representative samples of gases contained within the coal seams. A considerable amount of research in the design of tools for this purpose has been carried out in the Soviet Union and articles on this subject have been published by the U.S.S.R. Ministry of Geology (2). The sampling of gases in the coal seam provides valuable information in determining the potential for tapping the methane gas prior to mining, as well as in assessing methane hazards prior to the mining operations.

A recent program for drilling in the United States is directed towards degasification of coal deposits prior to mining. Although gasification and degasification of coal deposits has been researched for many years in the United

Kingdom, the existence of methane gas in coal mines in the United States has been considered and treated as a nuisance. Today, with the emphasis on energy conservation, methane gas is being viewed as a potential asset. It now appears that methane gas may be extracted from coal seams for the dual purpose of marketing the gas at a profit and of reducing lost mining time due to excessive gas concentration.

Several methods of methane gas extraction are currently under study by the U. S. Bureau of Mines (3). A portion of this research includes drilling a series of long, flat holes within coal seams and connecting these holes to a collection-and-exhaust system. The Longyear Company, working as a sub-contractor, developed a hydrostatically-powered test drill with a 3-meter feed, which was used to determine the feasi-bility of drilling 600-meter horizontal holes, while remain-ing within the confines of a 1.5-meter thick coal seam. Since that time, Longyear has designed and built a more com-pact drill with a 2-meter feed, featuring semi-automatic break-out and hydraulic chucking, which is now in use in a special degassing program in an Eastern United States coal mine. Exploration of deep coal deposits is done by several methods. One method involves drilling the upper strata with roller-rock bits, followed by the installation of casing and completion of the holes through the coal seam(s) by conven-tional coring procedures. A second method includes contin-uous sampling from surface, using conventional coring pro-cedures. A third method, more recently introduced, entails core drilling from surface using wire-line drilling tech-niques.

EXPLORATORY DRILLING EQUIPMENT AND TECHNIQUES

Wire-line Systems

Wire-line drilling is performed using a string of thin-walled drill rods through which an inner tube can be lowered and latched into position within the outer core barrel to re-ceive the core as it is cut by the bit. After completion of a coring run, a device is lowered on a light-weight wire cable to unlatch the inner tube and its contents of core samples so that it may be hoisted to the collar of the hole, with the drill rods, outer barrel and bit remaining in the hole until they need to be withdrawn to replace a dull bit. The use of the wire-line system results in vastly-improved drilling progress over conventional coring systems in which approximately 70 per cent of the rig time is spent in pulling and lowering rods, compared to the approximately 30 per cent of rig time spent in pulling and lowering the inner tube, when using the wire-line system. Wire-line drilling is especially desirable whenever geological formations are resistant to penetration by roller-rock bits. A study of all factors, including drill rig size and cost, hole size, bit costs, anticipated progress per shift and core sample adequacy should be made before reaching a decision regarding the drilling system that is best-suited for any given drilling program.

In areas where the geology is less favorable, core drilling is performed with conventional, or with wire-line core barrels, using air, drill muds, or water, as may be required, for cooling the bit and removing the cuttings. The Longyear Company recently developed a drill with many features adapted to this type of exploration. Called the "HC-150", the drill is an all-hydraulic unit with a 1.8-meter feed, hydraulic self-centering chuck, and angle-hole mast, as well as with a main hoist and wire-line hoist. The HC-150 Drill is also provided with instrumentation displaying bit speed, bit weight, and torque. Additional gauges monitor oil volume, oil temperature, and filter condition. The HC-150 can be mounted on a variety of vehicles.

As formations become harder and denser, the life of rock bits is shortened, and, often, due to lack of adequate bearing capacity, the smaller sizes of rock bits are totally unacceptable. If larger bit sizes are used, the size of drill rig and in-hole tools will also need to be enlarged at correspondingly higher costs. In hard, dense formations, it is becoming common practice to employ diamond bits and wire-line drilling techniques, which has shown marked improvements in cost reduction, as well as in core recovery. A Saarbruken, Germany, drill job used "PQ" wire-line tools to drill holes as deep as 1,300 meters through hard, dense strata. These tools drill a hole of 122.6 mm in diameter and produce 85 mm diameter cores. Drilling on the Saarbruken program is done with a small oil-field type drill modified to limit the torque applied to the drill rod. The drill is also equipped with special make-up and break-out tools to provide for pre-torquing of the drill rod, which is an absolute necessity in the achievement of maximum performance of thin-wall, wire-line rods. The procedure on the Saarbruken contract embraces operating the drill continuously for one week, after which the rods are pulled, the bit changed, and the assembly re-lowered, preparatory to starting the next week's drilling program. Prior to the use of diamond bits and wire-line barrels, the Saarbruken area was drilled with roller-rock bits and oil-field drill pipe. The resulting bit costs, high attrition of drill pipe and low progress rates prompted the change to wire-line systems.

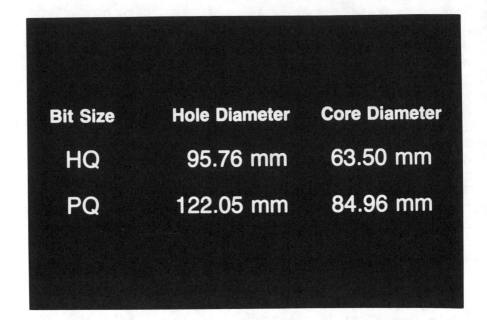

Bit Size	Hole Diameter	Core Diameter
HQ	95.76 mm	63.50 mm
PQ	122.05 mm	84.96 mm

In the last three years, there has been a very marked trend toward the demand for larger-diameter cores. Currently, the HQ and PQ larger-diameter, wire-line tools are being used in the United Kingdom, as well as in Canada, Poland, Australia, and the United States. Figure No. 6.

Offshore Coal Exploration near Newcastle - pioneered by a group from Wimpey Laboratories, Limited, of the United Kingdom - illustrates the successful use of wire-line tools in a drilling project which doubtless rates as one of the more difficult exploration undertakings, calling for considerable research and development, as well as close co-operation between suppliers, drill operators, ship's crew, geologists and the customer. Not only was it necessary to overcome the problems inherent in drilling through water depths in excess of 50 meters into a wide range of unconsolidated material, but it was also necessary to operate from the deck of a ship in the notoriously turbulent North Sea! In this North Sea site, the holes are now being drilled to depths of approximately 700 meters and in water depths of more than 100 meters. As reported in the Industrial Diamond Review of

Removal of soft core sample obtained with use of wire-line
technique aboard the Wimpey Laboratories, Ltd. Drill Ship
in the North Sea. Figure No. 7.

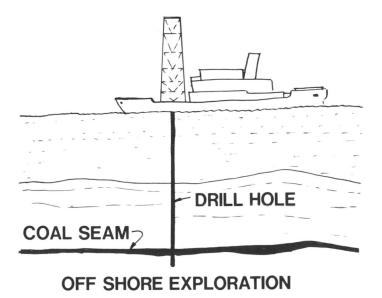

OFF SHORE EXPLORATION

Cross-section of North Sea Offshore Coal Exploration and the
use of the wire-line systems. Figure No. 8.

May 1975, the first six holes drilled resulted in a core recovery of 100 per cent in the coal seam - with an over-all core recovery of 98 per cent! Recovery of this high order of completeness aids considerably in assessing as accurately as feasible the permeability of the various strata before any underwater mining can be considered.

A most important and ingenious feature of the drilling set-up is the counter-weight system developed by Wimpey Laboratories, Limited, of the United Kingdom. This counter-weight system maintains a relatively constant weight on the bit in spite of tide or wave action. The close control of the weight allows the bit to cut cleanly through the various formations with a minimum of blocking and with maximum core recovery. This system has provided excellent bit-weight control in shallow holes and the ship is now equipped with a heave compensator for heavy-duty, or deep-hole drilling. Early drilling was carried out with the ship anchored in place, but more recent drilling has been done with the ship positioned and kept over the hole by dynamic positioning.

Taking cognizance of the many and varied types of drilling programs that must be carried out, it is obvious that no one drill or drilling system can be expected to produce maximum results under all conditions. There are, however, certain features that are desirable in any drill which are in harmony with the drilling objectives to locate and sample coal seams. These features, or operating capabilities, are generally agreed upon by experienced drillers throughout the world and may be described as instrumentation, feed control, long-feed-and-chucking, circulating pumps, and core-bits-and-barrels.

Instrumentation

In years past, the driller was taught to drill by "feel" and "sound" and it was truly amazing to watch an experienced driller identify in-hole conditions by the changes in engine sound or by the vibrations of the drill rod. Today's drill - equipped with hydraulic drive, fluid couplings, and large power units - calls for instrumentation to measure by gauge such drilling information as torque, bit weight and bit speed. The advantages of such instrumentation are such that the driller can quickly identify changing conditions and make necessary adjustments to achieve maximum results and to correct in-hole problems.

Feed Control

When drilling coal, it is important that the driller understand the best procedure and have the feed control features necessary to achieve maximum results. If the bit pressure is too great, it is possible to block the bit and grind up the coal sample. Conversely, drilling too slowly in a coal seam may permit the circulating fluid to wash the coal away before it enters the core barrel. It is, therefore, necessary to drill cleanly into and through the coal seam with proper bit pressure, a procedure that can best be done with a drill fitted with a suitable feed system.

Long-Feed-and-Chucking

When coring with smaller tools, such as the 60 mm and 76 mm size core barrels, the possibility of blocking and grinding in the coal seam is much greater than when coring with larger size barrels. This blocking is often experienced when rotation has been stopped and the driller goes through the re-chucking operation. The ideal condition would be to drill through the coal seam without stopping but, since this is often impossible, it is desirable to complete the re-chucking operation as quickly as possible and with a minimum change in feed rate or bit pressures and without exposing the core to excessive washing. This can best be done with fast-acting, semi-automatic chucks and the combination of long-feed and fast re-chucking is a desirable feature in any drill.

Circulating Pumps

Circulating pumps are an important part of the drilling program and, more especially, when drilling formations that are soft and easily washed away. A good drill pump must have adequate capacity for flushing of the hole during each phase of drilling and the pump should be fitted with a transmission or other means to allow adjusting the output flow without by-passing fluid. A closed-loop system from the pump to the bit will ensure that the fluid being pumped is reaching he bit. If the pump is not adjustable for output, the driller will be forced to by-pass fluid and this creates a potentially dangerous situation that may result in a burned bit and possibly a lost drill hole.

Core-Bits-and-Barrels

It is difficult, if not impossible, to discuss core bits

and barrels without generating differences of opinion as to
which is the best "bit" or "barrel". These are selected on
the basis of experience in an area, the type of formation
being sampled, the fluid being circulated, and the type of
drill being used, as well as on many other considerations.
It is a well-known, and proven, fact that double-tube swivel barrels have improved core recovery over the single tube
and double-tube rigid barrels and a number of developments
have been made to further the results obtained with these
barrels. Such features as inner tubes with chrome applied
to the inner surface, triple-tube barrels, shut-off systems
to indicate core blocks, special core lifters, and other
novel ideas have all contributed to improved coring results.
Obviously, these many features are not all available in any
one barrel, but it is important to pay close attention when
selecting a core barrel, since this is one of the most important parts of the drill system. Core bits are, also, a
matter of choice, but they should be selected with as much
consideration as the core barrel. Care should be taken to
ensure that the bit supplier is aware of the type of barrel,
the operating features of the barrel, the circulating fluids
being used, the formations being drilled, as well as the operating characteristics of the drill. A properly designed-
and-produced bit will give maximum results, while a poorly
designed-or-produced bit can be the downfall of an otherwise
successful drill program.

These are but a few of the highlights relating to the
rapidly-advancing exploration demands - and the equipment
being developed to meet these demands. The work of the supplier is only beginning! If the prediction - that the
United States will use two and one-half times as much coal
in the year 2000 A.D. as it consumed in 1974 - also applies
to other countries in the world, then it truly can be stated
that this gathering was timely. Hopefully, the exchange of
information during our meetings will create an attitude of
understanding and co-operation that will allow us to meet
these demands.

GLOSSARY OF DIAMOND-DRILLING TERMS

Auger. 1. A short spiral-shaped tool run on a torque bar
to drill soils and soft rocks, serving also as a platform to retain the cuttings for removal by raising the
auger to the surface.
2. A drill rod with continuous helical fluting, which
acts as a screw conveyor to remove cuttings produced by
an auger-drill head.
3. The process of drilling holes using auger equipment.

Auger drill. 1. A drill using an auger run on a torque bar.

Bit. Any device that may be attached to, or is, an integral
part of a drill string and is used as a cutting tool to
bore into or penetrate rock or other materials by utilizing power applied to the bit percussively or by rotation.

Bit blank. A steel bit in which diamonds or other cutting
media may be inset by hand peening or attached by a mechanical process such as casting, sintering or brazing.
Also called Bit shank, Blank, Blank bit, Shank.

Bit cost. Bit-use cost generally expressed in monetary
units per foot or per hundred feet of borehole drilled.
For a specific diamond bit the bit cost per foot drilled
is usually calculated in the manner shown as follows:

$$\frac{(R-S)Z+(CO+BL+ST-SC)}{Y} = X$$

where
R = Diamonds in original bit, in carats
S = Resettable diamond salvaged, in carats
Z = Diamond cost per carat, in dollars
CO = Cut-out charge, in dollars
BL = Cost of bit blank in dollars
ST = Setting charge in dollars
SC = Credit value of scrap diamonds in dollars
Y = Number of feet bit drilled
X = Bit costs in dollars, per foot drilled

Bit crown. (Crown. 1. As used by the drilling and bit-set-
ting industries in the United States, the portion of the
bit inset or impregnated with diamonds formed by casting
or pressure-molding and sintering processes; hence the
steel bit blank to which the crown is attached is not
considered part of the crown. Used in some countries
other than the United States as a synonym for Bit.)

Bit, coring. (Core Bit). An annular-shaped bit designed
   to cut a core sample of rock in boreholes. The cutting
   points may be serrations, diamonds, or other hard substances inset in the face of the bit.

Bit, noncoring. (Noncoring Bit). A general type of bit
   made in many shapes that does not produce a core and
   with which all the rock cut in a borehole is ejected as
   sludge. Used mostly for blasthole drilling and in the
   unmineralized zones in a borehole where a core sample is
   not wanted. Also called Blasthole bit, Plug bit.

Bit, roller. (Roller-cutter bit). A type of rock-cutting
   bit used on diamond and rotary drills. The bit consists
   of a shank with toothed, circular, or cone-shaped cutter
   parts affixed to the head of the bit in such a manner
   that the cutters roll as the bit is rotated. Generally
   used for drilling 3-7/8-inch-size or larger holes in
   soft to medium-hard rocks such as shale and limestone.
   Usually noncoring and not diamond set. Also called Cone
   bit, Rock bit, Roller cone bit, Roller rock bit, Roller
   cutter bit.

Casing. 1. Special steel tubing welded or screwed together and lowered into a borehole to prevent entry of
   loose rock, gas, or liquid into the borehole or to prevent loss of circulation liquid into porous, cavernous,
   or crevassed ground.
   2. Process of inserting casing in a borehole.

Chuck. 1. The part of a diamond or rotary drill that grips
   and holds the drill rods or kelly and by means of which
   longitudinal and/or rotational movements are transmitted
   to the drill rods or kelly.

Chuck, automatic. (Automatic chuck). A hydraulically actuated drill chuck. Also called Hydraulic chuck.

Core. 1. A cylindrical sample of rock and/or the process
   of cutting such a sample by use of an annular (hollow)
   drill bit. Sometimes incorrectly called Bit core.
   2. The central portion of a bit mold, that forms the
   inside diameter of the bit.

Core barrel. A length of tubing, usually 10 feet long, designed to form the coupling unit between the core bit and
   reaming shell and the drill-rod string. It carries or
   contains the core produced until the core can be raised

to the surface. The barrels are of single or double tubing and of swivel or rigid type.

Core lifter. DCDMA name for a split, fluted ring of spring steel used in a core-barrel assembly to hold and retain core while the core barrel is being hoisted from a borehole.

DCDMA. Abbr. Diamond Core Drill Manufacturers Association. A group of drilling-equipment manufacturers associated for the purpose of standardizing drill equipment and fittings in the United States.

Degasification of coal seams. A process of removing methane gas from coal deposits through a collection-and-exhaust system.

Diamond Drill. Any one of a number of different sizes and kinds of machines designed to impart a rotary and longitudinal movement to a string of hollow rods to which a bit having inset diamonds as the cutting points acts as a rock-cutting head capable of drilling either vertical or inclined boreholes, sometimes to great depths. Water pumped downward through the hollow, sectional rods acts as a bit coolant and washes away the rock fragments produced by the abrasive, rathern than percussive, action of the bit upward out of the hole. The machine may be driven by diesel or gasoline-combustion engines or by electric-, steam-, or airpowered motors and generally is equipped with a hoist capable of lifting and handling, on a single line, a specific-size string of drilling tools equal in length to the footage specified as the capacity of the drill. Such drills generally are used in mineral prospecting and development work but also are used to drill blastholes and to do various types of soil and foundation-testing work. Also called Adamantine drill, Core drill, Diamond core drill, Rotary drill.

Drill. 1. Any cutting tool or form of apparatus using energy in any one of several forms to produce a circular hole in rock, metal, wood, or other material.
2. To make a circular hole with a drill or cutting tool.

Drill rig, Drilling rig. A drill machine complete with all tools and accessory equipment needed to drill boreholes.

Drill rod; Drilling rod. Hollow, externally flush-coupled rods connecting the bit and core barrel in a borehole to

the swivel head of a rotary-drill rig on the surface.
Unit lengths of rod are usually 10 feet long and composed of two threaded parts, (a short pin-threaded
coupling and a box-threaded length of heavy-wall steel
tubing) connected together. The term "drill pipe" is
applied to rods used in a similar manner on rotary rigs
in petroleum-drilling operations. Also called Diamond-
drill pipe, Diamond-drill rod, Drill pipe.

Feed cylinder. A hydraulic cylinder and piston mechanism,
such as that on a diamond-drill swivel head to transmit
longitudinal movements to the drive rod and chuck to
which the drilling stem is attached. Also called Hydraulic cylinder.

Hydraulic feed. A method of imparting longitudinal movement to the drill rods on a diamond or other rotary-type
drill by a hydraulic mechanism instead of mechanically
by gearing. (See Feed cylinder.)

Overburden. 1. Clay, sand, boulder clay, and other unconsolidated materials overlying bedrock. Also called
Burden, Cover, Drift, Mantle, Surface.
2. The worthless material covering a body of useful
mineral.

Wire-line core barrel. Double-tube, swivel-type core barrels the outside diameters of which are of sizes made to
be used in various sizes of diamond- and rotary-drill
boreholes, and designed so that the inner-tube assembly
is retractable. At the end of the core run, the drill
string is broken at the top joint so that an overshot
latching device can be lowered on a cable through the
drill-rod string. When it reaches the core barrel, the
overshot latches onto the retractable inner-tube assembly
which is locked in the core barrel during the core run.
The upward pull of the overshot releases the inner tube
and permits it to be hoisted to the surface through the
drill rods; it is then emptied and serviced and dropped
or pumped back into the hole, where it relocks itself in
the core barrel at the bottom.

Wire-line drilling. The drilling of boreholes with wire-line
core-barrel drill-string equipment.

REFERENCES

1. 1972 Coal Mine Directory, United States and Canada, Published by Keystone Coal Industry Manual, Mining Informational Services of the McGraw-Hill Mining Publications, "Production of Bituminous Coal, by Type of Mine," SOURCE: U. S. Bureau of Mines, Tables prepared by the National Coal Association, McGraw-Hill, Inc., 1221 Avenue of the Americas, New York, New York, 10020, p. 155.

2. Лачинян, Л. А. и Угаров, С. А., Конструование, Расчет и Зксплуатация Бурилъных Геологоразведочных Труб и их соединений, Нздательство Недра, 1975.

3. Deul, Maurice and Kim, Ann G., "Methane in Coal: From Liability to Asset," Mining Congress Journal, November 1975, pp. 28-32.

4. Long, Albert E., "A Glossary of the Diamond-Drilling Industry," Bulletin 583, Bureau of Mines, United States Government Printing Office, Washington, D. C , 1960, passim.

16

THE DEVELOPMENT AND ADAPTATION OF

DRILLING EQUIPMENT TO COAL EXPLORATION

KEITH SHAW C.ENG. F.I.MIN.E.
CHIEF EXPLORATION ENGINEER
NATIONAL COAL BOARD
GREAT BRITAIN

INTRODUCTION

With recent exciting discoveries of oil in Britain's offshore areas in what for this country are previously unheard of scales, it is easy to overlook the fact that at the end of the day, British coal reserves are still likely to be at least thirty times as great as its oil reserves.

Although the quantity of coal in the ground is finite and reduces in quantity only by extraction or damage, reserves of coal are assessed on the economics of exploitation and marketing and are therefore directly affected at any point in time, along with other factors, by the competitive influence of other fuels.

The economic pressure from oil was responsible for the trauma that the world's coal mining industries experienced starting in the early 1960s, one effect of which has been a substantial reduction in the economically workable reserves available to existing capacity. The situation in Britain is illustrated by comparing the assessment of economically workable reserves made in 1962 when they stood at 15,000 million tons with the 1974 figure of 3,500 million tons, a depletion of 1000 million tons per annum.

FIG.1
Relationship between coal reserves and rate of exploration.

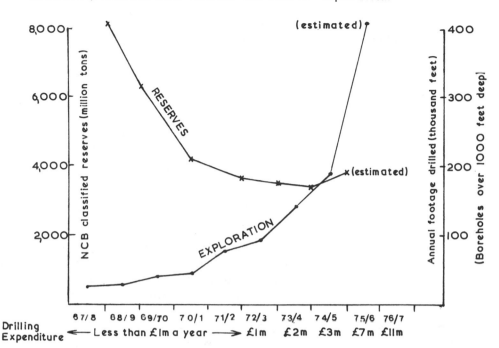

(estimated)

RESERVES

x(estimated)

EXPLORATION

NCB classified reserves (million tons)

Annual footage drilled (thousand feet)

(Boreholes over 1000 feet deep)

8,000

6,000

4,000

2,000

400

300

200

100

67/8 68/9 69/70 70/1 71/2 72/3 73/4 74/5 75/6 76/7

Drilling Expenditure ⟵ Less than £1m a year ⟶ £1m £2m £3m £7m £11m

Fortunately for the coal mining industry the pendulum in recent years has swung the other way and coal's competitive position has now started to reflect itself in an increase in the economic reserves. Changed fortunes, and brighter horizons will however not restore the vast reserves lost during the last fifteen years as a result of colliery closures on economic grounds.

In the early 1970s it was realised that there would have to be a rapid increase in exploration if the reserves depletion rate was to be curtailed, and this to be followed by major investment to establish new primary capacity.

In 1974 the then Parliamentary Under Secretary of State for Energy, the Right Honorable Eric Varley, announced the Government's approval of the National Coal Board's "Plan for Coal" as the broad strategy for the Industry's future development.

The Plan postulated investment to provide 42 million tons per year of new capacity by 1985 from:-

9 million tons per year from extensions to the lives
of collieries which would otherwise exhaust.
13 million tons per year from major projects at
existing collieries.
20 million tons per year from new collieries.

Each of these three categories of new capacity is dependent;
along with other resources; on the availability of adequate,
well proven, and economically exploitable coal reserves.
To ensure that this criteria is achieved, exploration of
reserves is carried out by proving drivages, blocking out
layouts, development faces, exploratory boreholes from
underground workings - both upwards, downwards and within
the seam, seismic surveys and surface exploratory boreholes.

These exploration methods are not necessarily alternatives
each to the other but are the various techniques which it may
be possible to adopt dependent upon the particular circumstan
associated with the possible reserves area under examination.

The drilling of boreholes through strata to record
sequence and to locate deposits for future exploitation has
been, and as far as can be foreseen will continue to be the
most used form of exploration. In coal mining they are the
first stage of discovering new areas of coal, and then for
proving the extent of the reserves both horizontally and
vertically. This information when used in conjunction with
seismic surveys can now provide an advanced picture of the
degree and pattern of faulting.

Some idea of the usage to which boreholes are put can be
obtained from the fact that a $1\frac{1}{2}$ million tons per annum
new mine requires some 20 to 30 boreholes with the coal
measure section fully cored to "discover" the field and to
ensure that it extends sufficiently to justify the costs of
providing access - which on to-day's costs could well be of
the order of £35 million. In general terms a minimum
proving of over 50 million tons of extractable coal is
required to justify such sinkings. (1)

This paper is intended to review the development and
adaptation of drilling equipment to the particular and
sometimes peculiar exigencies of coal exploration.

Exploration by means of boreholes in the coal mining
industry falls into two distinct but complementary categories
each with its own techniques and problems. For the purpose
of this paper and being primarily a mining engineer, I have

chosen to discuss firstly underground drilling then surface
drilling.

UNDERGROUND DRILLING

Although little or no exploratory drilling for new colliery
prospects can be accomplished from underground sites,
underground drilling to generate reserves at existing mines
is in general cheaper and can be accomplished quicker than
surface drilling.

Underground exploratory drilling has increased from about
43,000 feet in 1971/72 to some 80,000 feet in the financial
year just ended.

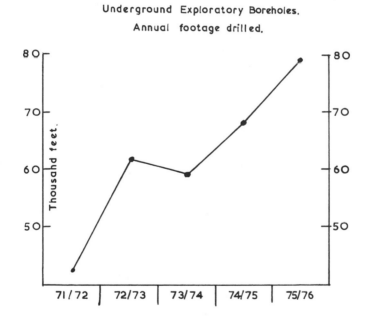

Underground Exploratory Boreholes.
Annual footage drilled.

FIG.2

The choice of sites underground is limited and is
dictated by the availability or otherwise of access
roadways in the vicinity of the area of overlying or
underlying reserves to be explored. The need as in surface
exploration to persuade reluctant land owners and farmers
to permit rigs to operate on their land is avoided as is the
need to obtain Town & Country Planning permission and to
satisfy the requirements of local water authorities and
other bodies.

Because of the very high regard placed on safety in the British coal mining industry any activities that take place underground; and indeed many on the surface within the colliery precinct; are controlled by legislation contained within the Coal Mines Act and Regulations and by Board issued directives and instructions.

Underground drilling practices, techniques, drill rigs and associated down hole equipment have to conform to and operate within constraints imposed both by legislation, physical environment, production priorities and trade union agreement.

Legislation not only prohibits the use of naked lights underground but also stipulates that all electric motors and switchgear be flameproof, prohibits the use of aluminium, legislates in favour of fire resistant fluids in hydraulic equipment, under certain circumstances requires drilling to be through a stand pipe, and stipulates minimum standards in respect of certain fittings and hoses etc.

By the very nature of underground conditions, dimension and weight limitations restrict the size of rig components and of rod and casing it is possible to wind in the shafts and to transport underground.

Dependant on the location of the borehole site underground travelling times can reduce the effective length of a working shift to possibly less than $6\frac{1}{2}$ hours.

Although colliery management readily accept the need for exploratory boreholes, site preparation and servicing do tend to take second place to production needs.

Wage rates agreed with the National Union of Mineworkers do not allow differentiation between the experienced driller in charge of an underground exploratory rig and his assistant, nor can any incentive payment be made.

Swivel Head Drilling Machines

A great change has taken place in the design of underground drilling machines since the 1940s. At that time it was usual for exploratory boring in excess of 250 metres to be carried out by drilling contractors. Since the rigs used were of conventional surface type belt driven rotary table design it was necessary to construct a chimney above the mine roadway to provide a total height of up to 12 metres to accommodate the derrick type mast. Shorter

boreholes were usually drilled by colliery labour with a
simple screw feed, hollow spindle, column mounted type rigs,
often compressed air driven.

BOYLES BBS 25
UNDERGROUND SWIVEL HEAD DRILL

BOYLES BROS.

FIGURE No.3

Nationalisation of the British mining industry in 1947
resulting in an impetus in production and a greater need for
exploration coincided with the introduction of the swivel
head drill from across the Atlantic and the Board acquired
a number of Joy-Sullivan diamond core drills as part of
lease lend assistance.

These rigs were basically surface rigs adapted for
underground use by the replacement of the diesel engine with
an air motor or as air became less popular with a flameproof
electric motor.

Although the rig required a site 27 feet in height and
12 feet wide to operate it was readily adopted and used to
drill down holes to 350 metres and upholes to 170 metres.

The swivel head concept was adopted by other manufacturers
including Boyles Bros., Atlas Copco Craelius, Longyear,
E.D.E.C.O.

The earlier machines were equipped with gearfeed swivel head but following the development of the hydraulic pull down swivelhead for surface operating core drills these started to be incorporated into later designs. Drilling machines of the swivel head design perform well on down hole exploratory drilling within their design capacities, but performance is restricted on vertical or steeply inclined upward holes due to slow rod and core barrel handling by virtue of restricted space and to drilling fluid cascading around the drilling area in spite of the use of standpipes, casing, and grouting. This creates unpleasant working conditions and does not improve brake linings and other vulnerable parts of the rig. Natural mine waters can be encountered with comparatively high acidic or alkaline contents which will result in rapid corrosion problems as well as being injurious to health.

Slide Carriage Drilling Machines

About 600 of the 800 or so underground drilling machines in use by the Board are employed drilling   boreholes for the extraction of methane.

Until the mid 1950s a large proportion of the drills in use were of German manufacture either Turmag or Hausherr and were air powered long stroke machines with usually two speeds and a rack and pinion feed motion. Although satisfactory for methane drainage work they were not ideal for exploratory boreholes this usually being carried out with swivelhead diamond drills. At about this point in time a dual purpose drill was designed and produced by the English Drilling Equipment Co.Ltd. working in collaboration with Board ventilation and drilling engineers.

This type of drill is of the general description category of Slide Carriage Rig and comprises four main sub-assemblies viz. the drill head and carriage slide, the erection frame, the power pack and the hose assembly.

The Hydrack Drill was an electro-hydraulic machine with a 5 foot hydraulic feed, having high thrust and a six-speed gearbox providing speeds up to 700 revolutions per minute. Besides being a powerful and efficient methane drainage drill it has been widely used underground for exploration core drilling up to distances of 150 metres.

In 1963 the Hydrack was superseded by the Mini-Hydrack, a lighter, more compact version which rapidly became

E.D.E.C.O
MINI-HYDRACK

Figure No. 4

established as the standard. Basically the machine comprises
a skid mounted power pack, a separate or combined control
table and the drill frame unit itself which is usually
mounted in a goalpost type erection frame but has also been
fitted on 'A' frame mountings and, on a fully mobile crawler
unit. The hydraulic system operates on fire resistant
emulsion with further protection provided by relief valves
and temperature and level control units.

The Mini-Hydrack has a stroke of 2 feet 6 inches, can
exert a thrust of more than 4 tons and is fitted with a
three-speed gearbox giving speeds of 62, 98 and 186
revolutions per minute. It was designed first and foremost
as a methane drainage drill but the top speeds are quite
suitable for diamond core barrels working in friable
formations. It is usual for 5 feet long core barrels
75 millimetre diameter to be employed but the standard NXM
or larger series can also be used to advantage.

The drill rods used are specially designed taper threaded
rods with clamp flats 41·3 millimetre by 762 millimetre long
although half length rods of 380 millimetre are also
available. Rods incorporating ball non return valves are
used for upward drilling. The drill incorporates a
telescopic nose piece which enables drill rods to be

connected and disconnected without accurate location of the drill head in relation to the pitch of the rod thread, thus virtually eliminating thread wear since no axial load is applied when making up or breaking joints.

Rigs of this type are readily transported underground in their component parts and assembly takes about 90 minutes. Since a 2ft. 6in. rod is inserted within the length of the drill carriage the need for major roof excavation is reduced to the minimum and the 4 foot square floor area can be accommodated in most underground roadways and still allow working room around the rig.

It is probably true to say that proving can be carried out more rapidly by conventional diamond core drills but where the distance of penetration is within about 150 metres, the Mini-Hydrack is a most useful machine capable of operating within its own erection height. Two features of the Mini which facilitate core drilling are the sliding rod clamp and the hinged gearbox assembly. The rod clamp may be moved aside when running core barrels and with the gearbox assembly swung to one side, clear access is provided to the drill hole.

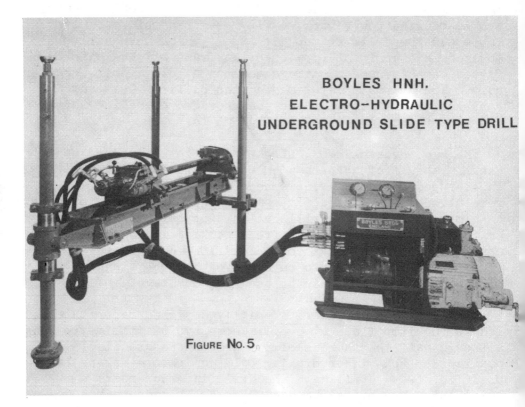

BOYLES HNH.
ELECTRO–HYDRAULIC
UNDERGROUND SLIDE TYPE DRILL

FIGURE No. 5

Another example of design co-operation between customer and manufacturer is exampled in the Boyles "H.N.H." electro-hydraulic slide carriage drill rig which was originally developed for drilling long straight in-seam boreholes for experimental work in underground gasification. Continuous design improvement over some fifteen years has resulted in an easily transportable and readily erected rig suitable for both operational and exploratory drilling.

The drill head is powered by two hydraulic motors which can be engaged individually or jointly. Rotational speed range and output torque depend on the selection of motor combination, but can be up to 300 revolutions per minute and 1200 foot lbs torque. The drive sleeve is mounted on the drill head assembly in taper roller bearings and incorporates a built-in water swivel and reduction gearing. It is hinged so that it can be swung to one side to permit the insertion of large diameter casing and core barrels in the borehole. A thrust of 9,500 lbs can be imparted to the drill head by means of a 4 inch bore hydraulic feed cylinder and roller chains. The drill carriage assembly allows a stroke of 5 foot $9\frac{1}{2}$ inches which is adequate for the use of 5 feet rods. The roller box assembly incorporates rollers to act as rod steady and is also fitted with hydraulically actuated rod clamps which are used to hold the rods when breaking joints under power or when drilling vertically.

The machine is rates at 270 metres with N.W.L. rods.

In 1972 the Board expressed interest in an underground exploration rig more powerful than those then available and capable of operating within the constraints of the coal mining industry and restricted headroom of 16 feet maximum.

Hydraulic Drilling Equipment Ltd. demonstrated their "Minor" rig, a 5 ton rated machine for surface work and agreed to produce a prototype of a similar capacity underground rig capable of recovering good cores from both up and down holes drilled to 500 metres. After field testing and subsequent design improvements resulting therefrom the drill rig is now being successfully marketed under the name of the "Sherwood".

The rig and power pack are assembled from a number of transportable units comprising:-

    Base unit.

    Three section mast unit for drilling with 10 feet

rods or 5 feet rods by removal of the middle mast section.

Rotary head, rod clamp, power breakout, power pack, control panel and hose assembly.

The single pole mast is $15\frac{1}{2}$ feet high for drilling with 10 feet rods and $10\frac{1}{2}$ feet high when adapted to use 5 feet rods. A hydraulically operated pulldown and hoist system governs the rotary head movement over an 11 feet or 6 feet power stroke with an infinitely variable capacity of 0 to 6 tons and speed of 0 to 60 feet per minute. The mast mounted rotary head has a stall torque of 960 lbs per foot clockwise rotation and a maximum of 170 revolutions per minute. The head is guided in the mast by large area adjustable guides and when drilling down is capable of pivotting to facilitate the make up and break of tool joints. There is clear hole centre of 5 inches diameter fitted with a $1\frac{1}{2}$ inch diameter water swivel. To provide increased headroom above the stand pipe or to run a wire line sheave system the rotary head can be swung to one side to give clear access on the centre line of the hole. For drilling in an upward direction the rotary head is reversed within its carriage. A hydraulically

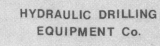

HYDRAULIC DRILLING
EQUIPMENT Co.

HYDREQ SHERWOOD
UNDERGROUND DRILLING RIG

Figure No. 6.

operated rod clamp and guide bush system attached to the mast can be swung to either side to facilitate the setting of casing or standpipe, or can be transferred to the top of the mast when drilling upwards. A tandem fixed displacement hydraulic pump with 65 horse power electric drive caters for all operations on the rig. The hydraulic oil reservoir has a capacity for 100 gallons of fire resistant fluid and to assist in cooling when operating in ambient temperatures of up to 80 degrees fahrenheit cooling tubes are fitted. The control unit can be either incorporated in the skid mounted power pack or supplied as a separate unit.

The rig is rated at 410 metres with HW and HY rod, 460 metres with HQ rod, and 670 metres with NW rod.

The Board are currently operating five Sherwood rigs and the following features have been proved in practice.

Fast conversion from down hole drilling to uphole drilling on the same site without moving the base and mast.

Fast recovery of good core by conventional barrels and by wire line barrels from both up and down boreholes.

No personnel required stacking at mast head when changing rods since these are broken out in the horizontal plane.

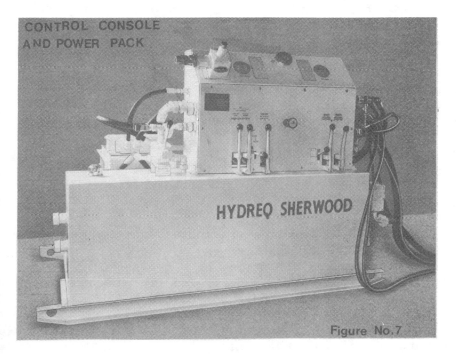

CONTROL CONSOLE AND POWER PACK

HYDREQ SHERWOOD

Figure No. 7

Wire Line Drilling Underground

Following the successful application of the wire line
system to drills operating from surface, it was only natural
to expect that there would be a demand for similar equipment
that could function with underground rigs. This created no
problem when underground holes were drilled vertically down
or within an angle range of 30 to 40 degrees from vertical
in a downward direction. However some boreholes drilled
underground are close to the horizontal and many are drilled
at an upward angle, often vertically.

Because the wire line technique was originally developed
around a system that allowed the inner tube and the overshot
to reach the hole bottom by gravity, it was obvious that a
means of forcing these components into place was required if
horizontal or up holes were to be drilled by this technique.
The most logical means of achieving this was by using the
circulating fluid to pump the tools into place and although
this might seem to be a simple solution there were a number
of very important operating features to be considered which
made the design problems far more complex.

A special overshot with pump in packing was needed, and
in order to allow this to be pumped into place it was
necessary to design a special quick connecting packing gland
to provide a seal around the overshot cable. This gland is
also equipped with a fitting to connect to the circulating
pump in order to pump the overshot into place.

The core barrel also required to be sealed against fluid
bypass while pumping into place, but also capable of opening
to provide free fluid flows during drilling operations and
subsequent retrieval of the inner barrel.

To achieve maximum efficiency from the system, it was
also necessary to develop a system of plumbing that would
allow quick changes of the hose lines as well as valves to
direct fluid flows.

The advantages of wireline are most evident on deep holes
when the round time for the recovery of the core barrel is
much less than that of the string of a conventional drill
rod and core barrel. On an underground rig of the swivel
head type the HQ wire line core barrel can be withdrawn
from a 350 metre borehole, a new barrel inserted and drilling
recommenced in 40 minutes or less. This compares with
about 5 hours work to remove, reinsert, and recommence

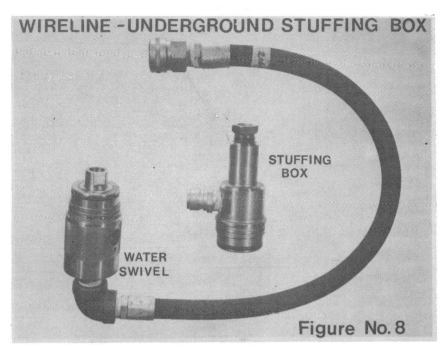

WIRELINE -UNDERGROUND STUFFING BOX

STUFFING
BOX

WATER
SWIVEL

Figure No. 8

drilling using conventional N.W. rod in 10 foot lengths.

The other advantages applicable to the use of wireline on underground boreholes are:-

Less caving in the hole due to the avoidance of need to pull rod except to change bits.

Less of a hazard to operators especially when tripping rods in an up hole.

Less fatigue on operators and less wear on equipment.

Longer bit life since with bit and rods remaining in the hole the chance of damage from debris is virtually eliminated.

Less chance of premature core blocking by debris.

Direct indication by changed pump pressure that the core barrel is full resulting in longer core runs and avoidance of damage to core.

The obtaining of an acceptable coal core from a vertical hole drilled in an upward direction has for a long time

Table No. I   Core Barrels

	Outside Diameter (millimeter)	Inside Diameter (millimeter)	Weight (Killograms per 10 feet)
AQ	44·5	34·9	14·1
BQ	55·6	46·0	18·2
NQ	69·9	60·3	23·1
HQ	88·9	77·8	34·9
PQ	117·5*	103·2	46·5**

* coupling OD
** with coupling

Table No. II   Diamond Coring Bits

	Core Diameter (millimeters)	Hole Diameter (millimeters)
AQ	27·0	48·0
BQ	36·5	60·0
NQ	47·6	75·7
HQ	63·5	96·0
PQ	85·0	. 122·6

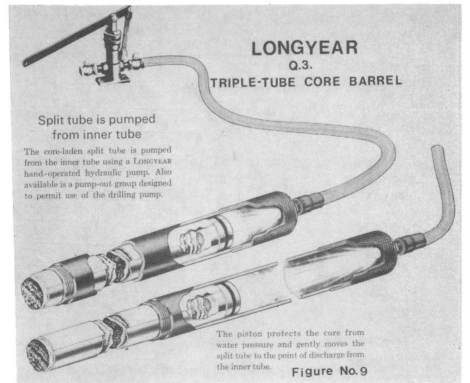

LONGYEAR
Q.3.
TRIPLE-TUBE CORE BARREL

Split tube is pumped from inner tube

The core-laden split tube is pumped from the inner tube using a LONGYEAR hand-operated hydraulic pump. Also available is a pump-out group designed to permit use of the drilling pump.

The piston protects the core from water pressure and gently moves the split tube to the point of discharge from the inner tube.

Figure No. 9

presented a problem which until recently had not been successfully solved by any equipment manufacturer. To quote two firms "There is no known proven method of obtaining a coal core in good cylinders from vertical up over holes".

Various ideas have been tried and modest success was achieved with a Hallprene oil seal placed at the top of the inner core tube and pushed down by the entering core, however the introduction of the wireline triple tube core barrel has resulted in a distinct improvement in core recovery.

In this system a third tube is added to the two tubes of the normal wireline barrel. This tube with its inside surface plated with hard low friction chrome for smooth core-entry is split lengthwise and fits snugly inside the inner tube. This arrangement permits recovery of the core from the inner tube in one piece by allowing this third split-tube to be hydraulically pumped from the inner tube. Once clear the upper half can be removed leaving the core in the lower half in a virtually undisturbed condition.

In Seam Exploration

In practice it is extremely difficult to drill boreholes in a coal seam maintaining direction and keeping within the confines of the seam. Apart from operational holes of this type drilled for such purposes as pulsed infusion shotfiring, water infusion, methane drainage, gasification etc. it is often necessary to drill such holes to locate and prove the safety or otherwise of old workings. This type of drilling, by virtue of generation of reserves, is considered to be as much of an exploratory nature as is other in-seam forward probe drilling designed to prove the presence or otherwise of face stopping faults and other geological hazards.

The surveying of underground boreholes is presently restricted to single-shot non electrical instruments such as the Tro-Pari, but surface testing of three models of multi-shot instruments is currently in hand prior to requesting permission to start underground trials. The important information needed of an underground "in-seam" borehole however, is not its position relative to ordnance datum or magnetic north, but its progress in the vertical plane during drilling in relation to the lie of the seam. Although there is at present no instrument available that can determine this relationship, it would seem a practical proposition that one could be developed from the nucleonic sensing device which is used in conjunction with an anology control system to automatically steer ranging-drum shearers.

At present only close examination of drilling cuttings and careful comparison with the visible roof, seam and floor stratum: if necessary by simple on the spot float and sink analysis comparisons: coupled with the possible use of a manometer can give guidance to the driller.

Over the years there have been many attempts to design rig and down hole equipment specifically for in-seam drilling, including as previously mentioned the original concept of the Boyles H.N.H. in the 1950s when the contract drilling side of this firm was employed by the Board on long hole drilling for experimental underground coal gasification at Newman Spinney near to Chesterfield.

The Victor "Precision" long hole drilling machine and its special lead rod were also specifically designed for operational drilling. The requirement in this case was for a rig to drill a straight borehole between the two advanced headings of a 50 metre long coal face for use with pulsed infusion shotfiring. The borehole was required to keep within the confines of the seam and to be parallel to the line of the coal face.

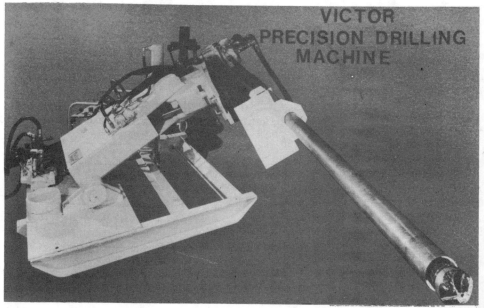

Figure No. 10

An interesting feature of design is the lead drill rod
which in essence comprises two concentric tubes with the
inner carrying the flushing fluid while the annulus between
the inner and outer conveys the return liquid and the drilling
debris. The 'E' type rod is 1·8 metre long and has an outer
diameter of 50 millimetre precision machined over the full
length with hard faced areas ground concentric with the
drilling axis. The drill bit design was evolved after long
experimentation and produces a hole which accommodates the
lead rod with the minimum of clearance. The theory behind
this design of rod and bit is that by achieving the minimum
possible clearance between lead rod and hole size, deviation
from intended direction and inclination will not exceed
1 in 150.

The drilling machine imparts a thrust of up to 6,000 lbs
and can operate at a torque of 300 foot lbs at 320 revolutions
per minute. It is a feature of the machine that it can be
operated at half torque and twice the revolutions per minute
and that it will automatically switch over to the higher
torque and lower speed if it encounters high resistance.
Although in many ways conventional the control of drilling
includes a facility for controlling the rate of feed
separately from the control of thrust.

The machine was designed to incorporate devices which
would facilitate very accurate setting in both the horizontal
and vertical planes.

Setting to the desired angle in a vertical plane is
automatically achieved under hydraulic control of a pendulum
actuated servo valve, whilst setting in the horizontal
plane is achieved manually by sighting through an optical
device.

Because coal seams rarely have either consistent gradient,
composition, thickness or parallel roof and floor over even
reasonably short distances, the object of long hole
exploratory in-seam drilling is usually not to drill a
borehole "as straight as a gun barrel" to a pre-determined
inclination and direction, but to achieve a hole which
follows the contours of the seam in the vertical plane
whilst maintaining as accurate a direction as possible in
the horizontal plane.

Effective control could almost be said to begin with the
ability to contain a drill bit to a straight line advance,
and valuable data was recorded on the experimental long hole

drilling at Newmans Spinney (2).

It has long been known that a fulcrum immediately behind a drill bit alters the direction of advance of the borehole if thrust on a rod is varied. However, increased thrust does not always induce rise nor reduced thrust fall. For example when a borehole has started to rise in a slow arc increased thrust increases the rate of rise and similarly on a falling hole increases the rate of fall.

FULCRUM AND BIT ASSEMBLY

FOR RISING BOREHOLE

FIGURE No.11

A "rose" bit will always drill a rising hole, but by virtue of its design it is susceptible to sharp deflection by hard bands, ironstone or pyritic nodules.

A "flat-nosed" bit used in conjunction with a 9 inch long, ribbed and fluted fulcrum, fitted immediately behind it and having a diameter 1/16 inch less than the bit, has been proved by many trials to have the least inclination to wander, and control can be largely exercised by the experienced driller varying thrust on the rods as he acquires "feel" of the particular borehole.

With the present trend of concentration of production to fewer high production faces any means of reducing the risk element and obtaining assurance of continuity of advance or early knowledge of any impediment to production by impending geological or other physical hazard is welcome. Although early investigatory work is being done on possible seismic approaches to this problem, the only means currently available of obtaining forward detailed information of the immediate seam economics are by in-seam drivages such as advance headings or blocking out type layouts or by in-seam forward probe boreholes. The Board welcome any advances in drilling machine, bit, down hole equipment design, or in improved techniques which will assist in this problem, and are currently engaged in conducting major equipment trials to this end.

One of the new drilling machines being tested is the Arrow manufactured by Hydraulic Drilling Equipment Ltd. and designed specifically for in-seam exploration.

The machine is of the hydraulic head slide carriage type with 30 horse power electro hydraulic drive providing infinitely variable and reversible torque capacity between 0 to 300 foot lbs. and rotational speeds between 20 and 370 revolutions per minute. The maximum pull-out force under non-stick conditions is 3500 lbs.

Figure No. 12

The control console and power pack can be positioned up to 30 feet apart.

In initial tests the equipment demonstrated its ability to bore $4\frac{1}{2}$ inches diameter holes to a depth of 30 feet for the purpose of setting standpipe and then to drill reasonably straight holes of up to 330 feet depth with 5 foot long AWT rods and 60 millimetre bit. Average drilling rates of 5 feet per minute were achieved, and 45 minutes were taken to pull and break a 330 feet string of rods.

SURFACE DRILLING

Surface drilling can be sub-divided conveniently into shallow/medium depth holes and deep hole drilling and for the purpose of this section of the paper holes over 2000 feet and down to 4500 feet are considered as deep holes.

Shallow/Medium Holes

The Board own and operate some 30 small surface rigs of widely differing size, design, vintage and lineage, ranging from the hand held Atlas Copco "Cobra" to truck mounted Boyles BBS 37. These rigs are used for such varied duties as locating and proving old shafts, shallow and medium depth core drilling, location of fault planes, spoil heap investigation, ground stabilisation etc. Board operated rigs handle the base load of such drilling activities with the peak load; generally concerned with coal exploration; being undertaken by outside drilling contractors using mainly truck mounted swivel head type rigs with conventional H.W.F.type corebarrels. A considerable amount of shallow hole exploratory drilling is carried out by outside contractors in association with opencast coal extraction and although opencast production represents an important and valuable part of the Board's activities; contributing about 10 million tons of output last year; since it is controlled by a different executive from deep mining I shall not expound further on this aspect of exploratory drilling.

Deep Hole Drilling

As stated earlier and illustrated in Figure No.1 since June of last year there has been an unprecedented increase in the volume of deep hole drilling. In the current year over 150 holes are planned with drilling continuing to be required at around this level into the 1980s.

Figure No.13 lists the contractors and Board operated rigs currently drilling as part of this programme.

The majority of holes are open-holed to just above the coal measures and then continuously cored throughout the productive sequence, however there are circumstances where the borehole is drilled either entirely openhole or entirely cored and there are other cases where spot coring of selective strata is carried out. In all cases prior to abandonment cement infilling, the hole is electrically logged using special equipment and programmes developed and applied

N.C.B. AND CONTRACTORS OPERATED DEEP HOLE (+3,000') DRILLING RIGS
ENGAGED ON EXPLORATORY DRILLING

Firm	Rig No.	Make	Mast Mounting	Capacity ft.	Notes
Foraky	2	Ideco H 40	Lee C Moore Mast	8,000	Oilfield core barrel
"	4	National T 20	Lee C Moore Mast	4,500	"
"	10	Failing 1500	Lorry Mounted	2,000	"
"	12	Failing 2000	Trailer Mounted	2,500	Oilfield core barrel
"	14	Failing 2,500	Trailer Mounted	3,000	"
"	15	Failing 2,500	Trailer Mounted	3,000	Oilfield core barrel
"	16	Failing 2,500	Trailer Mounted	3,000	Oilfield core barrel
"	17	National T 50	Fixed Shell Derrick	6,000	Oilfield core barrel
"	18	National T 12	Fixed Shell Derrick	6,000	Oilfield core barrel
"	19	Foraky XXII	Fixed Sands Derrick	3,500	Oilfield core barrel
"	47	Failing 2,500	Trailer Mounted	3,000	PQ Wire line string
Thomson	10	Walker Neer 2000	Trailer 60' Mast	3,000	Oilfield core barrel
"	23	Cardwell TR 69	Trailer 110' Mast	5,500	Oilfield core barrel
"	29	Walker Neer 2000	Trailer 60' Mast	3,000	Oilfield core barrel
Kenting	17	Failing Strata 90	Trailer 80' Mast	4,500	Oilfield core barrel
"	12	Brewster N 45	Trailer 127' Mast	6,000	Oilfield core barrel
Thyssen		Salzgitter SA314	Trailer	5,000	PQ Wire line string
National Coal Board		Gardner Denver 2500	Trailer	4,500	PQ Wire line string
"		Gardner Denver 2000	Trailer	2,500	PQ Wire line string
"		Longyear 44	Skid mounted	2,300	HQ Wire line string

by the British Plaster Board Co. Since this technique is to be the subject of a paper by Mr. Reeves, Managing Director, B.P.B. Industries (Instruments) Ltd. at this Symposium I shall not further enlarge.

Prior to 1975: with only one exception: all rigs then employed in drilling deep holes were operating with conventional double tube swivel type corebarrels of 5,6 or 7 inches outside diameter. A.P.I. drill pipe of 2.7/8 inches or 3.1/2 inches was mainly used in 20 foot lengths, the height of masts generally allowing 40 foot pull.

At this time it was taking an average of about three months to complete a typical part cored borehole to a depth in excess of 1000 metres.

To be fair to the British contractors it would be true to say that at that time there was no expressed degree of urgency in the limited boring programme, rigs tended to be operated on five or five and a half day week basis rather than seven, a greater percentage of the boreholes were "difficult" being related to existing mines and often passing through old goafs. Because of the sparcity of boring programmed the competitive element of contracting was absent, there had been little capital investment, most rigs were at least twenty years old, British drilling had tended to become insular.

In an attempt to improve productivity and based on continental experience, one rig: a Failing 2500: had been equipped in 1974 with a Longyear PQ wire line string and standard 10 foot long PQ core barrel. To avoid excessive use of casing and to increase the life of couplings the contractor decided to use special oversize couplings: 4.7/8 inches in diameter: which in turn required the use of correspondingly oversized diamond crowns. The contractor manufactures the oversize couplings himself and they give an average life of three to four months with constant usage.

A great deal of experimentation has been applied to bit design and encouraging results with cube stone settings have been achieved with penetration rates in typical coal measure strata at times in excess of 30 metres in 8 hours. The contractor confidently expects to achieve penetration rates of over 45 metres in the shift.

HRISTENSON — SERIES 250P
IILFIELD CORE BARREL

- Safety Joint Pin
- Spring
- Friction Ring
- 'O' Ring
- Safety Joint Box
- 'O' Ring
- Cartridge Cap
- Shims
- Bearing Retainer
- Thrust Bearing
- Outer Tube Sub
- Cartridge Plug
- Inner Tube Plug
- Steel Ball
- Outer Tube
- Pressure Relief Plug
- Inner Tube
- Inner Tube Shoe (upper half)
- Inner Tube Shoe (lower half)
- Core Catcher

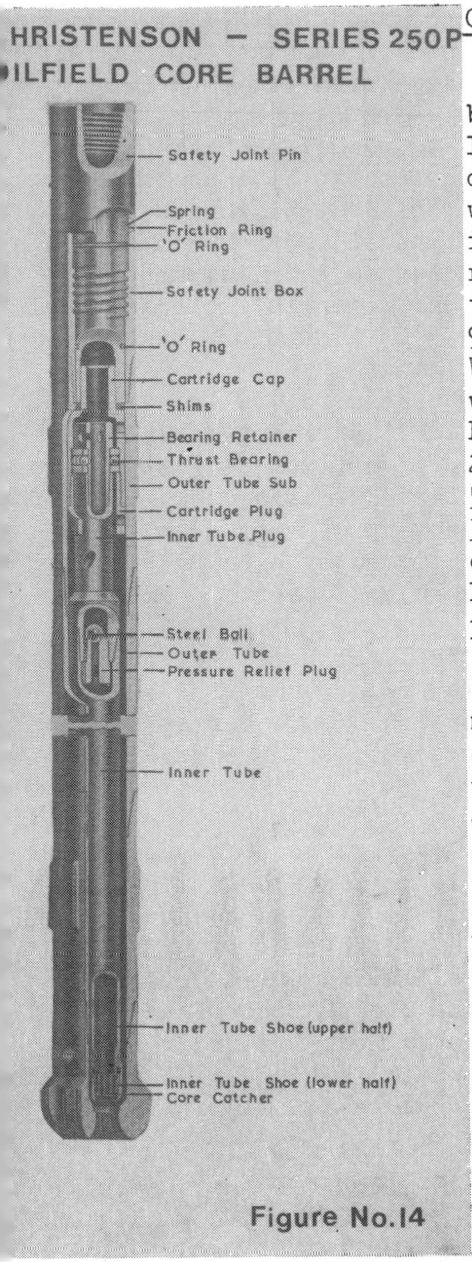

Figure No. 14

Limited contracts on a trial basis were entered into in June 1975 with two Canadian firms of drilling contractors, both of whom had backgrounds and expertise in oil drilling operations. Three rigs were initially imported, these being the large Failing Strata 90 and the two smaller capacity Walker Neer 2000s. These rigs were equipped with Christensen Diamond Products oilfield type 250.P series core barrels; $5\frac{3}{4}$ inches outside diameter, $3\frac{1}{2}$ inches internal diameter; operating with $3\frac{1}{2}$ inches I.F. flash welded oilfield drill pipe of 13·3 lbs per foot.

Coal is a notoriously difficult material to core, often being extremely friable, prone to washing away, scrubbing and to all other factors leading to core loss.

Reservations were expressed by senior Board officials: including myself: on the ability of oilfield type equipment and indeed oilfield orientated operators to achieve acceptable core recovery from coal seams. However I was given renewed confidence on observing the following notice prominently displayed in the dog house of the first rig to arrive:- "It must be borne in mind from start to finish that the production of good core is the only reason we are working here"

The Canadian contractors were optimistic of their ability to produce good core and in this

contention were supported and encouraged by the core barrel
and bit manufacturer. The firm's optimism was based on a
recent major break-through which had been achieved in the
Athabasca tar-sands in north east Alberta where initial
results from oilfield equipment and techniques were so
successful that almost all subsequent exploration for these
oilmines was with oilfield type rigs and equipment. The
oilsands exploration success resulted in many Canadian mining
companies turning to similar procedures and equipment in other
areas and for exploration for other minerals including coal
seams.

In the event the initial boreholes were drilled at rates
which were unprecedented in coalfield exploration in this
country, rates which were four to five times as fast as those
which had for many years been accepted as the norm. The other
important achievement was the recovery of very acceptable
coal seam cores, subsequently confirmed by electric logging
to represent an average recovery rate of over 90 per cent.

The successful results from the first boreholes encouraged
the mobilisation of other Canadian rigs and of the equipping
of several British operated rigs with C.D.P. series 250.P
core barrels with impressive improvements in penetration rate
without loss of core quality.

The series 250.P barrels have outer tubes and outer tube
subs manufactured from 4142-H cold rolled steel tube with
internal/external flush joints at each thread connection, the
threads being cadmium plated. Inner tubes are manufactured
from smooth bore alloy tube. The fluid annules between the
inner and outer tubes and the flow through the bearing
assembly have been designed to carry normal fluid volumes
without increased pump pressure which is a useful feature in
coal coring where minimum pump pressure and fluid flow are
a pre-requisite of good core recovery.

Because parts are standardised the 250.P barrels can be
put together in multiples of 30 feet up to a maximum of
180 feet. In some of the Board's exploration areas 60 foot
core barrels have been successfully used, but overall the
30 foot barrel is the most widely used because of the possible
risk of damaging a coal seam by having perhaps 40 or 50 feet
of strata in the barrel when a seam is penetrated. Because
the seat earth below the coal seams is often very
unconsolidated and friable a condition can exist whereby the
weight of core in the barrel could exert sufficient pressure
to cause this material to cause a core block or jamming in

the barrel. With penetration through a coal seam at the rate
of 1 metre in 2 minutes, blockage may not be immediately
apparent on the surface.

All 250.P barrels are provided with a safety joint enabling
the operator to withdraw the inner tube and core if the bit
and outer tube become stuck in the hole.

In general the rigs are operating at 180-200 revolutions
per minute, and are applying weight of up to 15,000 lbs with
$4\frac{3}{4}$ inches to $5\frac{1}{2}$ inches drill collars. Drilling fluid at
150-170 gallons per minute is circulated at 550-700 lbs per
square inch pump pressures.

Diamond Crowns used with C.D.P. series 250 core barrel

The joint achievement of the best penetration rate coupled
with longest possible bit life has presented many problems
in the coal measures where there is little comparison between
drillability of the strata in the Board's several areas of
exploration.

An example of the difference that the contractor may require
to face on successive holes is Nottinghamshire sandstone

CHRISTENSEN C12 FACE DISCHARGE BIT

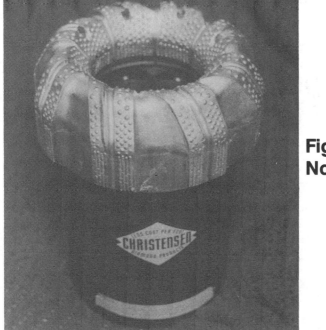

**Figure
No. 15**

which cores at about 6 metres per hour and South Wales
pennant sandstone where penetration is less than 1 metre
per hour.

Diamond crowns 6.7/32 inches outside diameter set with
cube stones have been reasonably successful but not absolutel
reliable. Penetration rates in excess of over 100 per cent
improvement on round stones have been achieved in mudstones,
but the presence of any ironstone considerably shortens the
life. Presently experiments are in hand to arrive at a happy
compromise combination of cube and round stones.

There are three basic bit designs in general use, the 'B'
crown, the double cone crown and the step bit. The contracto
is provided with strip logs of the anticipated geological
column by the geologist from which bit selection is decided.

The 'B' crown is generally used at the commencement of
coring in the sandstones and has proved to give good
penetration rates through conglomerates.

The step type bit is probably the most popular choice for
softer mudstones and silt stones. It has a flow area of
about 45 per cent, a spiralled pattern with large cube
diamonds and operates best at faster rotation and higher
weights.

The double cone crown tends to be used in formations too
hard for the step bit. The double cone design has a self
stabilising effect making it suitable for faster rotation.
Six stones per carat are used for these crowns for silt stone
and sandy mud stones. It is interesting to note that igneous
rocks in the Belvoir exploration area were restricting
penetration to about one metre per hour until laboratory
testing of strata samples resulted in a new design of double
cone crown with twelve stones per carat which resulted in
improved penetration rates in excess of 4 metres per hour.

Diversion Drilling

When electric logging confirms that it is necessary to
carry out diversion drilling the Canadian contractors apply
oilfield practices by using Dyna-Drill equipment with a bent
sub rather than a whip stock wedge.

The Dyna-Drill is essentially a multi-stage moyno-pump
used in a reverse application and comprising about half the
total length of the tool. The motor consists of a spiral

passage containing a steel rotor which moves eccentrically while rotating. Shaped in a regular recurring wave form the rotor is free and unconnected at the upper end, while the lower end is attached by a connecting rod to a hollow drive shaft and thence to a conventional bit sub.

Drilling fluid is pumped down the drill string, entering the motor under pressure and so turning the rotor and in turn the bit at high speed.

In application the borehole is cemented up to about 100 feet above the section of strata required to recore. When the cement is set at least as hard as the immediate strata the top is polished with a conventional bit prior to introducing the turbine drill followed by a 2 degree bent sub then standard collars and drill pipe. After the turbine has kicked away and penetrated some 60 feet, it is replaced with the oilfield core barrel and penetration continues with "bad drilling" practices of using over weight collars and slow rotation in order to maintain the hole diversion. Dependant on problems encountered a typical divergance for a target seam can usually be executed in between two and four days.

Offshore Drilling

An interesting, although comparatively small part of the Board's exploration activities is offshore prospecting to determine the extent of variable reserves to collieries developing in a seaward direction, in Scotland and the counties of Northumberland and Durham.

The Board's offshore prospecting for coal commenced as early as 1955 from a fixed platform rig built specially for the Board and operated by Messrs Foraky Ltd. of Nottingham.

It could be claimed that the success of our activities in this sphere of drilling paved the way for subsequent operation of larger and more advanced offshore rigs in the North Sea search for gas and oil reserves.

The original rig was later supplemented by the purchase of a second rig and over seven years a total of 24 holes were drilled to an average depth of 700 metres. The deepest hole was to 1500 metres and holes were put down up to four miles from the shore. The rigs were designed to be capable of operating in water to a total depth of 40 metres, to withstand 80 mile per hour gales and 10 metre high waves. The actual boring equipment was identical to that then in use by Foraky Ltd. for land based drilling and comprised an 18 metre high derrick mast, a rotary table working at 80 revolutions a minute driven by a 30 horse power Ruston diesel engine via a friction clutch and V-belt drive. (3)

To protect the drill rod casing down to the sea-bed 24 inc diameter tube was driven through the unconsolidated sea-bed t the rock head and was slung beneath the drilling table.

By 1964 sufficient information had been acquired within th sea depth limitations of these type of rigs. One rig was sol for scrap and the other gained honourable retirement by being purchased by Trinity House and refurbished to serve as an offshore lighthouse in the Humber estuary.

The next step in the search for offshore coal reserves was the use of a prospecting vessel, and here the Board used the services of Wimpey Laboratories Ltd. and in particular their drill ship "Wimpey Sealab".

The vessel was previously the merchant vessel "Elizabeth Bowater" a 6000 ton displacement pulp and pulpwood carrier with three holds forward and one aft, owned by the Bowater Steamship Company. It was originally designed and used to transport raw materials for the paper industry.

The vessels length is 100 metres, beam $15\frac{1}{4}$ metres and operating draught 6 metres. The main engine is a Sulzer six cylinder SD 60 diesel - 2700 BHP operating a four blade single screw heliston type propeller to give a speed of 10 knots. Wimpey Laboratories Ltd. purchased the vessel in 1972 with the intention of operating it as a site investigation and geological exploration vessel. Major modification were necessary to convert the vessel for this purpose in the light of experience gained by the company over a number of years with their small exploration vessels "White Thorn" and "Briar Thorn".

The main modifications and installations undertaken were the provision of four retractable 1000 horse power thrusters, a 9 metre square moonpool, mooring winches, power generators 33 metre high drill derrick, draw works, active/passive heave compensator, mud tanks etc. Extra crew accommodation to a total of 61 berths has been provided to house scientist: engineers, drillers and navigation location control specialists, and a helicopter platform has been erected on the aft deck.

The drilling equipment comprises a 33 metre high derrick with Gardner Denver GD 700E drawworks with 700 horse power variable speed electric drive. The main drum has a capacity of 915 metres of 28 millimetres wire and the sandreel holds 610 metres of 14 millimetre wire. Duplicated 200 horse power Keelavite electro-hydraulic power pack operate a variable speed Dando power swivel. The mud-pumps consist of a 500 horse power Gardner Denver PY-7 Triplex, an 80 horse power Ideal Oweco Duplex, and duplicate 15 horse power Boyles BBP 40 Triplex pumps.

The rig is equipped with 1000 metres of $4\frac{1}{2}$ inches drill pipe and PCQ, HQ and NQ wire line strings.

The advanced dynamic position keeping system utilises duplicated short baseline acoustic and taut wire position referencing systems. The exact position of the ship relative to a point on the sea floor is fed into a digital computer, together with information on wind speed and direction and the ship's heading. The computer automatically controls the operation of four 1000 horse power retractable thrusters, each fitted with variable pitch propellers and capable of 360 degrees rotation, so as to keep the vessel on station. The computer can be instructed to hold the ship stationary at a predetermined heading or can be requested to cause the ship to rotate about the fixed position to a heading requiring minimum thrust from the thrusters. The value of dynamic position keeping in deep water areas, where conventional mooring systems become impossible, is obvious. In shallower areas, however, dynamic positioning enables the vessel to rapidly set up on new locations and to move off with relative ease when the job is finished or in times of emergency. The percentage of actual working time on station is thus appreciably increased and the higher daily charges will still be economic for many activities when compared with a conventionally moored vessel.

The Satellite Navigation System enables the vessel to position itself accurately at predetermined locations and has an integrated Doppler Sonor System for traversing predetermined paths to very high orders of accuracy. Other features include acoustic guidance for lowering tools and caters for re-entry of a borehole.

To date a total of 36 offshore holes have been drilled by Wimpey Laboratories Ltd. for the Board of which "Wimpey Sealab" has been responsible for 22. The programme is scheduled to continue in the coming year.

CLOSING OBERVATIONS

British coalfields have provided about 25,000 million tons of coal since Roman times, and the nation still has about 100,000 million tons of reserves in the ground.

I opened this paper by making reference to the National Coal Board's Plan for Coal, and consider it therefore appropriate to close on the same note by quoting from Sir Derek Ezra's introduction to that Plan:- "The size that we can make Britain's mining industry rests ultimately on our ability to remain competitive, particularly with oil. That means improving our performance in tons and productivity. We have the capacity to do that now as we enter a new era for coal".

REFERENCES

1. Rees, P.B., Exploration of Deposits: Exploration The Mining Engineer, April 75. Vol. 134. No. 171

2. Baxter, S.J., Drilling of long borehole in coal. The Colliery Engineer, Dec. 1959.

3. National Coal Board. Information Bulletin No. 56/172

Section Eight Discussion

DR. U. KLINGE: (Ruhrkohle A.G.) My question is addressed to Mr. Svendsen as well as to Mr. Shaw and it concerns horizontal drilling in-seam. If I understood you well, you succeeded in reaching about 2000 feet roughly 700 metres in seam. Is this correct?

W.V. SVENDSEN: I probably went to roughly 300 metres, however the machine was designed to go to 600 metres.

DR. KLINGE: We are very much interested in this technique and we succeeded to 300 metres, not on account of using the correct type of bit, on the contrary, the type of bit we are using is primarily designed for staying in the seam. My question is now: is your equipment suitable for underground exploration, for instance in advance of the face, considering weight, dimensions and other requirements underground, and could you please give just a rough summary of the difficulties you have had in going further than 300 metres?

W.V. SVENDSEN: To answer the first part of the question, I would say that the work that we have done has been strictly experimental. The first machine that I showed you was built for the Bureau of Mines under contract. The tests were conducted by the Bureau of Mines and a paper has been published and is available from the U.S. government printing office, describing this work in detail. It tells the procedure that was followed and the results obtained. It has photographs, and I think you would find the article very interesting. The second part of your question is: do we have a machine that is available or do I know of a machine available for doing advance exploration work in a seam like this? No, I do not. At the present time as far as I know, work is strictly experimental. We have only built one machine which is in use at the present time.

K. SHAW: The Coal Board are under remit to find equipment which will do advance probe drilling within the seam in order to avoid the Lofthouse type situation, or try to avoid it. To that end, pretty well every manufacturer who expressed interest, has been invited to participate in trials which will be taking place in the very near future, starting at the end of this month at one of the collieries in the northeastern area, where the object of the exercise is to endeavou to drill a borehole some 450 feet (150 metres) in length, keeping within the coal seam, in a reasonably straight direction within the working time of a coal shift - round about 6/6½ hours. The machine is also required to fit within the confines of the gate end of a normal production unit, to be erected at the close of the Friday night production shift, to drill a planned series of boreholes over the weekend and then to be out of the way for the normal start of the Monday morning shift. These boreholes will require drilling through stand pipes because obviously, if there is the slightest possibility they are going to meet some old workings which could contain water or gas, you must of necessity drill through a stand pipe. Here again, we have another problem that on the Monday morning we shall be faced with a number of stand pipes at each end of the face which must not impede normal coal production and present thoughts are to produce some sort of a stand pipe that the coal face machine will cut through, i.e. plastic of some description or other. Our other thoughts are that this work could be carried out possibly from advance headings on the face but these tests are shortly to take place and perhaps Dr. Klinge would like to know some results of these afterwards. The longest that we have drilled on some tests that were carried out about 12 months ago, was 320 feet in I think, 3½ hours.

PROF. PETER O'DELL: Mr. Shaw in his answer has indicated that the National Coal Board and Ruhrkohle could probably share the experience of this kind of thing in line with the remarks of the Chairman of the Coal Board this morning. After all, much of the British industry and the industry of the Ruhr have so much in common in terms of the development of the industry, its structure, mining conditions, that this seems to be one of those fields where cooperation would appear to be a self-evident requirement.

D.C. YATES: (National Coal Board) I was pleased that Mr. Svendsen and Mr. Shaw have spoken about the actual drilling operation, particularly in view of the vast amounts of money being spent in the U.K. to obtain good core recovery. My question is of a very practical nature: What plans do the National Coal Board have for training drillers in view of the rapid expansion in both surface and underground drilling operations? Perhaps Mr. Svendsen would also like to comment on the method of training drillers in the United States of America.

K. SHAW: The answer to your question is very complicated and concerns politics perhaps. As I mentioned in the paper, we have a ridiculous situation as far as I am concerned and as far as Mr. Yates himself is concerned, that we pay the man in charge of the rig exactly the same as his helper. You can have a man with twenty years experience as a driller underground and somebody starts tomorrow as his labourer, and is paid the same money. This is because of the NUM agreement on wages. We do have schools, certainly when I was in North Notts, we started these schools and trained drillers, and fitters and maintenance fitters on rigs and they were very successful and indeed they are still being carried on. But, I think we have got to first have some differentiation between the skilled man and the labourer. From the surface aspect, this is again slightly different and the Coal Board now having taken a more active interest and participation in actual drilling itself, in the North Notts area, we do at least have the chance of giving the people in charge of the rig sensible salaries, because the only way to overcome the situation was to make them members of the management structure and so the man in charge of the rig is on the management grade. So is the toolpusher on each shift and, I'll use the term 'his assistant' as opposed to say, derrick man or hoist man. I am certain that if we go that way, we shall certainly set up and carry out an efficient training process for drillers, toolpushers, derrick men and hoist men.

E. GRAY: (National Coal Board) Mr. Svendsen, you referred to long hole drilling in coal. Could you tell me if the holes that you refer to in your paper - are they being drilled on production faces, or has a section of a seam underground been made available for this purpose?

W.V. SVENDSEN: Well the entire programme was experimental in nature, and started out by drilling exposed faces in a pit area. The holes were drilled as I said, in excess of 1,000 feet in depth. There are two programmes being carried out right now, on horizontal holes underground. There are also additional vertical holes being drilled, large diameter holes for methane drainage, but with the horizontal holes that I referred to there are two programmes, one: a continuation of the experimental programme under the direction of the Bureau of Mines and working with one of the private mining companies where they are actually taking out the methane and marketing it. The second programme is in an operation where they are using the tunnel machine to make a drift, and in this particular case the drill is being used to remove the methane gases before the tunnel bore is advance into the coal seam.

E. GRAY: I would like to point out that the methane drainage in the North Notts coal field, requires one shift operation for the erection of the machine, drilling the hole, putting in a stand pipe, connecting up the vacuum pump and dismantlin - all in one shift, and we must not interfere with coal produ tion.   I think that is important to know. My second query really was to assist our European colleague here. Mr. Shaw didn't have time to read his paper right through, but if you look at Fig. 11, you will see drilling referred to at Newman Spinney. In fact I had the privilege of working on that job in the 1950's and we drilled 600 foot holes and kept in seam. I am sure Mr. Shaw could provide copies of the paper by J.S. Baxter at some early date if necessary, which might help you.

K. SHAW: I would very much like some cooperation with the Ruhrkole on this particular problem because if they have some experience of bits that will keep you in seam, then we are looking along these lines at the moment.

W.V. SVENDSEN: If there is anyone interested in obtaining a copy of the report prepared by the U.S. Bureau of Mines, if you will give me your card or your address, I will forward the address of the U.S. government publishing office which has this, along with the number of the document, so that you can write for it. It is a rather interesting paper and I think those of you who are involved in methane drainage would certainly get something out of it.

K.SHAW: It is all right to talk about machines that can push forward a borehole a certain distance in a certain time, the important thing however in in-seam mining, is that that borehole is in the seam. As long as the chippings show coal, everybody's happy. Immediately stone or dirt chippings come out, then you have a problem. You say to yourself, 'have I gone in the roof, have I gone in the floor, have I gone through a fault, or have I hit some old working'. One of two things is needed: either a bit which will guarantee when it hits the roof or the floor it will come back, or alternatively, some sort of measuring device to tell you where you are.

17.

Main Principles of Exploration of Coal Deposits in the USSR and New Problems of Exploration Methods at the Present Stage
by E.V. Terentyev

Section Discussion

17

MAIN PRINCIPLES OF EXPLORATION OF COAL DEPOSITS
IN THE USSR AND NEW PROBLEMS OF EXPLORATION
METHODS AT THE PRESENT STAGE

E.V. TERENTYEV

Candidate of Geological and Mineralogical Sciences,
Ministry for the Coal Industry of the USSR

The coal industry of the USSR is systematically developing
and steadily increasing the coal production. Although in
the post-war year of 1946 the coal output amounted to 164
million tonnes, in 1960 it was boosted to 510 million
tonnes and in 1975 reached 701 million tonnes. According
to "The Main Directions in the Economic Development of the
USSR for 1976-1980" adopted at the 25th Congress of the
CPSU, it is envisaged to increase the annual level of coal
output up to 790-810 million tonnes.

The main task of geological exploration being conducted in
the coal basins of the Soviet Union has been always the
large scale increase of the explored coal reserves, which
takes into account the planned and steady growth of the
coal output.

At first sight there is nothing unexpected in this process -
growth of the coal production naturally requires increase
in the scale of the exploration of the reserves. But
nevertheless, there is some unexpectedness at the present
stage, since the need of explored reserves increases by
substantially higher rates than coal production itself.
Comparatively recently, i.e. 20 to 25 years ago the adequate
level of the increase of the reserves was several hundreds
of millions of tons per year. At present, in spite of the

fact that the recovery ratio has increased on an average to 75-80 percent, annual rate of the depletion of the explored reserves reached a level of about one billion tonnes. In order to simultaneously compensate the depletion of the reserves at the existing enterprises and explore an adequate quantity of mining fields for the construction of new mines, at present it is necessary to prepare annually several billions of tonnes of explored reserves.

The explanation of this fact lies in the peculiarities of the current development of the coal industry (1.4). The principal feature is the increase of production capacity of underground and open-cast coal mines and concentration of production at a minimum number of coal faces. For instance, in 1950 an average colliery annual production in the country was about 300 thousand tonnes of coal, and an average annual open-cast mine production was approximately 900 thousand tonnes. At present collieries with an annual capacity of 4.5 to 6.0 million tonnes of coal have been commissioned, for example, Raspadskaya Colliery in the Kuzbass, Vorgashorskaya Colliery in the Pechora basin, Krasnoarmey-skaya Colliery in the Donbass and other. Annual production capacity of open-cast mines amounts to 8-12 million tonnes, and in Kazakhstan there exists Bogatyr' open-cast mine, production capacity of which has reached 35 million tonnes per year. Naturally, such large mines require the availability of vast explored coal reserves within the boundaries of their mining fields in order to justify the investments. It is considered, for instance, that for the above mentioned open-cast mines coal reserves occurring within one field must be from 500-600 million tonnes to 2-3 billion tonnes.

Concentration of production and increase of the production capacity of the mines and capacity of mechanisms in the coal industry of our country is not an accidental temporary phenomenon. Such a process is the most peculiar and important feature of the development of social production in the era of the scientific and technical revolution. It is enough to mention practical examples of the construction of power plants and iron and steel works where production capacities not only of the whole enterprise but also of turbines, generators, blast and open-hearth furnaces, converters, etc. are steadily growing. In such conditions the coal industry, similar to the above mentioned industries, is faced as never before with acute problems of ensuring reliability and stability of coal production and efficiency of utilisation of higher-capacity machines.

It is evidently clear that under the conditions of a colliery or an open-cast mine stability of production is influenced not only by the reliability of mechanical design but also by mining and geological conditions in which they operate. The experience gained indicates that in most cases the latter is a decisive factor. The reason for this lies in the fact that modern coal-winning and transport machines are specialised according to mining and geological conditions of application, i.e. seam thickness and dip, hardness of coal and adjoining rocks, stability of roof and floor strata. Each type of machine (power-loader, hydraulic roof support or conveyor) works only in a narrow range of specific parameters.

For example, power-loaders are specialised according to seam thickness ranges 0.7 to 1.3 m, 1.5 to 2.2 m, 2.5 to 3.2 m. They can work seams of such thickness ranges, working thickness being changed insignificantly.

Coal-winning complexes equipped with hydraulic powered roof supports cannot be successfully applied in seams dipping over 20 to 35 degrees, even if such angles of dip occur at the coal face temporarily. Seam hypsometry must not be complicated by undulations with curvature radius less than 35 meters. Seam floor must bear loads from 7.5-8 to 15-27 kg/cm^2 and the roof must withstand loads up to 5-7.5 kg/cm^2. Roof strata must not be friable, and there must not be prolonged delays by roof fall.

Changing mining and geological conditions necessitate replacement of mechanism types, change of mining systems, etc., i.e. expensive measures which, as a rule, cannot be implemented without output losses. It is unexpected changes in mining and geological conditions that particularly affect mine operation, which is often the case when tectonic ruptures occur. Coal face output may be limited by other facts also, for instance, gas content, water inflow, etc.

Thus, there arises the problem of ensuring reliability of forecasting mining and geological conditions of coalfields. Since a modern higher-capacity and highly mechanised enterprise requires large investments up to scores of millions of roubles, convincing proofs of the efficiency of such high cost enterprises are necessary prior to the beginning of construction. For this purpose the appropriate Institutes carry out thorough calculations and comparisons of a great number of variants of technical solutions for a construction project. In order to find out an optimum variant methods of mathematical analysis are used. Extensive practice indicates that priority of an optimum variant accounts, as a rule, for only several percent, rarely over 5 to 6 percent of the construction costs. But the above figures account for hundreds of thousands and even millions of roubles and significantly exceed all the exploration costs, so "the game is worth the candle". However, high degree of accuracy of technical and economic calculations, naturally, requires the same accuracy of exploration data.

The above mentioned modern strict conditions of a coal-mining enterprise necessitate thorough programming of its operation and, first of all, coal winning operations, drivage of development workings, maintaining the total length of coal faces. Programming must be operative as well as precise and with minimum reserves, which is stipulated by the very nature of modern large-scale, high-capacity and high-speed production in the coal industry. Taking into account the present state of exploration technique and the level of development of scientific and methodological problems of exploration, it has been possible, in the main, to ensure such a level of programming, however certain efforts were needed, particularly for solving methodological and economic problems.

Necessity to increase the reliability of geological fore-casts required not only quantitative but also qualitative growth of geological information, improvement of the methods of its obtaining, processing and summing up.

The above mentioned peculiar features of the operation of coal-mining enterprises and the industry as a whole raised new tasks for geological exploration which can be generalised as follows: at the present stage it is necessary not only to explore a certain quantity of coal reserves but also to work out qualitative forecast of mining and geological conditions of the occurrence and working of these reserves. Certainly, at present, as well as in the past, in order to plan and construct a colliery or an open-cast mine it is necessary to find sufficiently large coal deposit and determine the quantity and quality of coal it contains. But experience gained indicates that, at least, under the conditions prevailing in the Soviet Union, where there is a sufficiently large number of coal-fields, in order to solve this first problem a comparatively small amount of exploration work and costs in needed. The predominant part of exploration work is required for forecasting mining and geological conditions of developing and working coal seams, including the following:

- establishing the morphology of seams (finding out changes in seam thickness and structure, substitutions, layering and wedging out);

- study of structural conditions of seams and constructing maps of tectonic ruptures;

- study of physical and mechanical properties of floor and roof strata and forecasting their stability in mine workings;

- prospecting of hydrogeological conditions and determination of the amount of possible water inflows;

- qualitative assessment of gas content of coal seams and adjoining rocks for use as initial data for ventilation calculations;

- determination of the degree of silicosis danger of rocks, liability to spontaneous combustion of coal, possibility of rock bumps or sudden outbursts, temperature conditions and safety and health working conditions for coal-miners.

The above list is steadily changing and, as a rule, from time to time is extended depending upon new methods of coal mining being adopted (open-cast method, hydraulic mining or underground gasification), increase in mining depth and new social factors. For example, much attention being attached to the problems of environmental control necessitates a fuller and more detailed chemical analyses of gases, underground waters and surrounding rocks, etc.

It should be mentioned that the above conditions must be not only described but also determined qualitatively and, what is most important, geometrized in space and represented in accurate detailed drawings. In the Soviet Union scale of such drawings, or graphical materials as we call them, is accepted to be predominantly 1:5000 or 1:2000, sometimes it is 1:1000 for very complicated deposits, or 1:10000 for some deposits very simple and large in area.

To ensure the coal industry development in the Soviet Union a large amount of exploration work is regularly done. Many-year experience has shown that such work is mostly efficient only on the condition that a sufficiently high level of research has been achieved in the field of coal geology, organisation of planning and methods of exploration.

In this report we shall deal in detail only with the problems of exploration methods. As far as research in the field of coal geology is concerned, we would like to mention only that at present the Soviet Union possesses detailed monographic descriptions of all the coal basins and deposits. These monographies were published for general use. A forecasting map of coal-bearing areas located throughout the USSR territory has been drawn up and published several times; the scale of the map is 1:2500000. State estimates of the reserves are carried out systematically. The last estimate was made in 1968; hundreds of specialists working directly in coal basins took part in this work. As is well known, according to this estimate geological coal reserves occurring on the territory of the USSR are 6.8×10^{18} tonnes. The above mentioned material permits the planning of geological exploration work properly and in due time. The State system for early development of spare and fully explored fields has been worked out and implemented.

Particular attention is paid to the working out and scientific substantiation of exploration methods which imply principles and complete set of means for carrying out geological surveys on the geological evaluation of coal deposits and obtaining geological materials for the construction and operation of a coal-mining enterprise. Such attention can be explained by the fact that methods of exploration determine not only the completeness and validity of geological data, but also, and mainly, exploration costs. The latter is dictated by the fact that exploration methods outline a system of prospecting (number, depth and spacing of exploration boreholes and prospecting mining workings), types and amount of samples of coal and adjoining rocks, amount and types of analysis, etc.

In the Soviet Union methods of exploration of coal deposits have been developed on the basis of four general principles of prospecting, formulated by V.M. Craiter as early as in the thirties (5):

1) Successive approach; 2) completeness of investigation; 3) equal proving (uniformity) and, 4) minimum costs and time requirements.

The principle of successive approach is implemented by planning and carrying out prospecting work in stages.

Four main stages can be distinguished: search, preliminary exploration, detailed exploration, operating exploration. The system of stages gives at minimum costs the most promising fields for industrial development and excludes from exploration unpromising ones. The efficiency of this system is ensured by the fact that after completion of each stage geological record is drawn up in which estimates of coal reserves are included, and on the basis of such record technical and economic calculations of the feasibility of industrial development are made. The next stage is fulfilled only if the results of geological surveys and feasibility studies are successful. One of the numerous examples of this principle is the case history of the discovery of L'vovsko-Volynsky coal basin and the development of coal production there.

Work aimed at the discovery of Carbonic coal deposits in the Western regions of the Ukrainian SSR commenced on a large scale in 1948 and completed in 1950 when L'vovsko-Volynsky bituminous coal basin was discovered. In 1951 the first areas prospected in detail construction of collieries commenced, and in 1955 the first 200 thousand tonnes of coal were produced. By 1965 coal output increased up to 9.8 million tonnes. At present the basin has become a large fuel and power centre in the south-west of the country and the basis of the international power system. Such high rates of basin development and coal output growth have become possible due to the fact that exploration started from wide-scale searching. Two years were spent for search during which the main characteristics of geological structure of the 20 thousand square kilometer closed territory were outlined; this permitted the establishment of the industrial coal-bearing capacity of Namurian deposits (instead of presupposed principal coal-bearing capacity of Visean deposits). On the above mentioned territory, a comparatively small industrial coal-bearing area, approximately one thousand square kilometers located in the extreme west, was contoured, prospected and recommended for development (fig. 1), this area accounts for only 5 percent of the total territory explored. Later on our Polish colleagues taking into account these results discovered the new Lyublinsky basin containing coal reserves of several scores of billions of tonnes.

An opposite example is the case history of the Dombarovsky region in the South Urals where in 1932-1933 construction of collieries commenced only on the basis of the availability of separate intersections of working seams, however, the

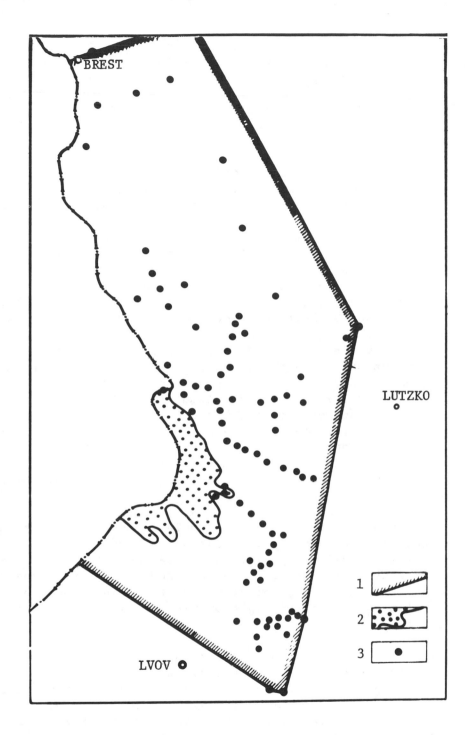

<u>Fig. 1</u> Comparison of areas of search and detailed explora-
tion during the finding of the L'vovsko-Volynsky bituminous
coal basin: 1, contour of the searched for area; 2, explored
in detail workable area of the basin; 3, prospecting boreholes.

general geological structure was not revealed. In fact, stages of preliminary and detailed exploration were omitted and searching was not completed. In the course of construction it was found that in spite of high coal-bearing capacity of Carbonic measures and the occurrence of the anthracite it was not possible to mine it because of tectonic faulting of the seams.

The principle of survey completeness has been also of great importance. However, this principle is of particular significance at the present stage in view of the huge growth of the geological information required, as mentioned previously. The implementation of this principle is controlled by the special State classification of the reserves according to the degree of their exploration (proving), by establishing standards of the minimum of reserves of high category exploration which are required for various types of deposits at each stage of exploration, and other measures. Requirements for completeness and accurateness of surveys in case of detailed exploration of coal deposits have been established on the basis of the experience gained in the construction and operation of underground and open-cast mines. Results of detailed exploration are considered and approved by the State Commission on Reserves. Investments for the construction of a new coal-mining enterprise or the reconstruction of an existing one are allocated only if there is the minimum of the reserves required by the above mentioned standards and a favourable decision has been issued by the Commission.

In order to provide the required completeness of the surveys and to prevent unnecessary things that may occur in the course of exploration, special methodical guides for exploration projects have been drawn up and used. These measures guarantee the implementation of the fourth main principle of exploration, i.e. the principle of minimum time requirements and costs.

In working out exploration projects particular importance is attached to the questions of feasibility studies of the system of exploration openings at each of the explored units (region, deposit, mining field, area). Under the term "system of exploration openings" we imply their spacing in area and length, the total number and frequency required for each of the geological parameters studied separately. In the past, when particular attention was paid only to the evaluation of the quality and quantity of coal, the system of exploration openings, as a rule, was a rectangular or square uniform network. In case of well-distinguished folding of seams this network consisted of uniformly located lines of exploration openings. At the present stage such exploration networks or lines are still used as the basis of exploration systems, but they are not sufficient because obtaining the data to solve the above mentioned new problems of exploration is possible in some cases only by a closer system of exploration openings than the main network. For this purpose, new methods of investigation of a proportion of boreholes are becoming more widespread, for instance, application of thermometry in deep holes, extending the types of analyses of core samples, etc. However, a high degree of alteration of mining and geological parameters in the area being explored and in depth, their failure to coincide from seam to seam require, as a rule, drilling of special systems of additional holes. It is also necessary to take into consideration the fact that some of the investigated parameters lay down mutually exlusive requirements for exploration boreholes as far as the methods of drilling or location are concerned (e.g. structural and hydrogeological holes). In order to find out other parameters special constructions of boreholes and special methods of investigation are required (e.g. for obtaining engineering and geological data, technological samples of coal, etc.).

This circumstance necessitates special exploration systems. A typical example in this connection may be specialised systems for surveying structural conditions of seam occurrence, which will be described in more detail later on. Particularly specialised systems, which are not connected with the evaluation of the reserves, are holes drilled for prospecting physical and mechanical properties of overlying strata and for determining stability parameters of open-pit banks. For instance, during the completion of the more detailed exploration of the Bachatsky open pit the system of the appropriate boreholes substantially exceeded the boundaries of the area of coal reserve evaluation (fig. 2).

In a number of cases sound and proper integration of exploration methods provide the require data in the main system of exploration openings and use the minimum number of additional workings. For example, very often hydro-geological boreholes are located in the main network and solve this particular task, as well as the general task of exploration. In this respect, a typical example is the exploration of Svobodny coalfield in Priamurye (fig. 3), where extensive hydrogeological studies (100 boreholes with a total length of 7.7 thousand meters) have been carried out, mainly, over a

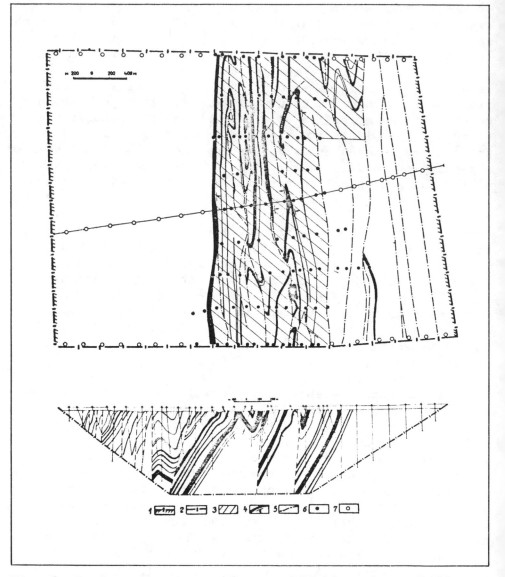

Fig. 2 Exploration system for establishing the stability
parameters of the pit banks at deep Bachatsky open-cast mine:
top, plan; bottom, vertical geological section along line 9-1
1, boundaries of the open-cast mine on the surface;
2, boundaries of the block along the strike and in depth;
3, coal-bearing area of the open-cast mine; 4, outcrops and
dip directions of the seams; 5, main tectonic ruptures;
6, exploration boreholes of the main grid; 7, additional
boreholes.

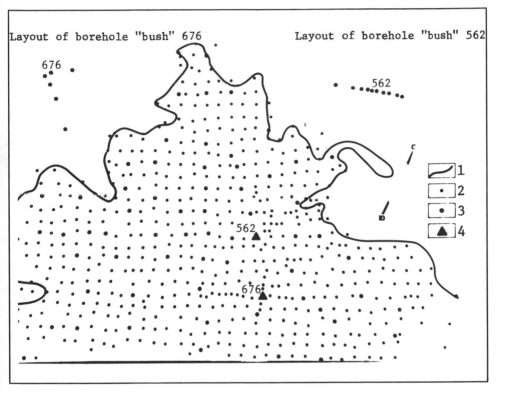

Layout of borehole "bush" 676   Layout of borehole "bush" 562

Fig. 3 Exploration system for investigating hydrogology of the Svobodny coalfield. At the top--location of hydrogeological boreholes in groups. 1, contour of coal seam occurrence; 2, exploration boreholes; 3, single hydrogeological holes; 4, groups of hydrogeological holes.

single system of hydrogeological and prospecting boreholes.

Systems used for more detailed exploration for gas content also pertain to the coincident exploration systems. In such a case rather comprehensive exploration work is fulfilled. For example, in order to investigate gas content of the mine field of Capital'naya N 40 Colliery in the Vorcutsky coal-field 53 holes were drilled; boreholes are 1100 meters deep and have a total length of 36.7 thousand meters.

The coincident systems allow the saving of exploration costs and increase exploration efficiency, however, detailed studies of mining and geological conditions on the whole increase exploration costs.

Thus, under present conditions exploration and prospecting grids are formed as complexes of differentiated exploration systems. In this brief report it is not possible to describe methods of constructing all differentiated systems. We shall deal in more detail only with those systems which are used for structural and tectonic mapping due to the fact that tectonic investigations are particularly significant at the present stage of exploration and also because there are still unsolved problems.

The significance of detailed investigation into structural conditions of mineral occurrence was proved long ago by mining theory and practice. For example, B.V. Bokiy, a well-known mining specialist, pointed out (2, page 411): "The choice of a mining system is influenced by many factors and, first of all, by depth and shape of the deposit". The author considers angle of dip to be the second significant factor after seam thickness and intensive tectonic faulting takes the first place among other conditions which determine the choice of a mining system.

The tectonics of coal deposits is a determining factor of the structural conditions of coal seam working. Tectonic pattern influences seam depth; angles of dip, character and degree of their alteration; size and shape of folded structures; degree and nature of tectonic ruptures in these structures and, consequently, size of production blocks.

Seam depth determines the method of opening, development and mining of the deposit, i.e. underground or open-cast. When underground mining is used a number of other mining and geological characteristics depend upon seam depth, i.e. gas content, geothermic characteristics, stability of strata in mining workings.

Geometric shape and size of folded structures and location of tectonic ruptures, space orientation of all these elements determine the size of mining fields at collieries and open pits, location of their technical boundaries, size of production blocks and, thus, the system of opening and development and the mining system.

Absolute values of seam dip are of paramount importance because they determine the choice of mining systems and methods of working out mining fields, systems and methods of coal and waste transport, types of mining equipment and machines used.

In recent time particular importance is being attached to the degree of low-amplitude tectonic faulting of coal seams (amplitudes which may be compared with seam thickness or exceed it insignificantly). Small tectonic ruptures which are not "dealt with" by winning complexes sharply reduce coalface output. For example, investigation carried out in the Kuzbass have indicated that due to cutting of the un-faulted length of a mining field from 600 down to 300 meters monthly coalface output reduces from 22.5 to 15 thousand tonnes, or by 33 percent (9). Strong tectonic faulting has an adverse effect on the recovery of the explored reserves. Naturally, that without the appropriate surveys of tectonic structures it is impossible to estimate correctly the quality of workable reserves which is the most important initial index for con-struction projects of collieries and open pits.

Zones of tectonic ruptures frequently become the sources of sudden inrushes of large quantities of water and gases, and may cause rock falls and mass-caving of rock into mine workings. Recently the genetic relationship between karst (3), sudden outbursts of coal and gas, gas seepage and such zones has been established. Thus, the tectonics of coal deposits directly influence safety of mining operations.

The principle method of investigating the hypogene tectonic structure of a coalfield is the drilling of detailed structural profiles and the forming of vertical geological sections, and on the basis of the latter the construction of detailed hypso-metric plans of coal seams. The above mentioned drawings are a component part of all geological records, and the reserves are estimated on hypsometric plans.

To ensure that vertical geological sections precisely

represent the real tectonic structure, it is necessary to keep to two main conditions - firstly, to provide their appropriate resolving power and, secondly, to avoid errors in the interpretation of drilling results. Sources of probable errors can be presented theoretically on the drawing, which shows various possible shapes of seam occurrence between two very remotely located holes and the principle scheme to find out the real structure. Particularly flagrant errors in structural constructions are possible if angles of strata dip are unknown or if these angles are well-known but are not taken into account. Improvement of the resolving power is promoted by reducing the distance between boreholes on the profile and by drilling such holes to obtain a core. Profiles on which boreholes are particularly dense are called "supporting". At present they are widely used not only in the final phase, but also in the initial phases of exploration. In order to reduce costs for creating "supporting" profiles it is unnecessary to drill holes down to the coal seam. Sufficiently reliable data can be obtained if some boreholes are drilled only down to some overlying marker level, for instance, to a limestone layer, as happened in the mining field at Obukhovskay Colliery in the Donbass (fig. 4).

The practical efficiency of the described systems for structure exploration can be illustrated by the example of the mining field at Vorgashorskaya Colliery in the Pechora basin. In the mining field at this colliery exploration grid oriented only to the estimation of the reserves, contained in the non-undulating seam about 3 meters thick, could not delineate the real tectonic structure. The latter was delineated only after drilling a special system of holes, the total number of which was 600 and the total length 102 thousand meters, that exceeded the amount of previous work by more than three times (see fig. 5). After this more precise work was done the construction of the colliery with an annual capacity 4.5 million tonnes of coking coal was successfully completed.

In spite of the certain successful results obtained in the field of tectonic investigations, it is necessary to point out that at present exploration is not able to fully meet the requirements of the industry. Boreholes can establish tectonic ruptures with amplitudes no less than 10 to 15 meters under very favourable conditions. Even in boreholes it is difficult to establish tectonic rupture with amplitude less than 5 meters. Improved resolution in exploration can be ensured by developing geophysical methods of studying

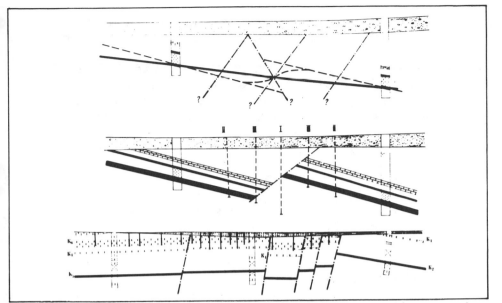

Fig. 4 Layout of operations to establish the real tectonic
structure on exploration line along the seam dip. At the
top—possible section variants when spacing of holes is un-
acceptably large. At the bottom—more detailed exploration
of faults in the mine field at Obukhovskaya N 1 Colliery with
mapping the structure according to the supporting stratigraphic
level of contiguous limestones K_4-K_5,-K_5^1,-K_6 overlying coal
seam K_2.

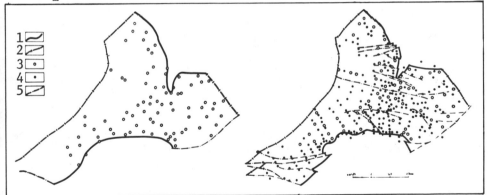

Fig. 5 Exploration system for establishing the tectonic
structure of the mine field at Vorgashorskaya N 1 Colliery.
At the bottom—the structure of the coal deposit more precisely designated due to the results of additional exploration. At the top—data of the detailed exploration.
1, outcrop of the Moshchny seam under Quarternary measures;
2, mine field boundaries; 3, boreholes for detailed exploration; 4, additional structural holes; 5, tectonic ruptures
of the seam.

between the holes and directly from development and production workings. Such investigations commenced 10 years ago.

It is necessary to improve the methods of investigating and forecasting other mining and geological factors - water inrushes, stability of adjoining rocks, etc. We believe that the present Symposium will promote the successful solving of these problems.

REFERENCES

1. Bratchenko B.F. Coal of the Eighth Five-Year Plan. Publishing House "Nedra," Moscow, 1971.

2. Bokiy B.V. Mining Engineering. Publishing House "Gosgortechizdat," Moscow, 1959.

3. Gazizov M.S. Karst and its influence on mining operation Publishing House "Nauka," Moscow, 1971.

4. Grafov L.E. Technical Reequipment of the Coal Industry of the USSR. "TSNIEIugol," Moscow, 1971.

5. Craiter V.M. Search and Exploration for Minerals, Publishing House "Nedra," Moscow, 1969.

6. Mironov K.V. Search and Exploration for Coal Deposits, Publishing House "Nedra," Moscow, 1966.

7. The Main Directions in the Economic Development of the USSR for 1976-1980. Moscow, Publishing House "Politizdat," 1976, p. 26.

8. Terentyev E.V. Tasks for the Exploration in connection with the New Requirements of the Coal Industry. Coll. "Geology of Coal Deposits," v. 1, Publishing House "Nauka," 1969.

9. Krylov V.F., Kulakov I.S. On the Improvement of the Coal-Winning Complexes Equipped with Powered Supports due to the Optimization of the Length of Production Fields, "Ugol," N 6, 1971.

Section Nine Discussion

PROF. PETER O'DELL: Perhaps the way in which the coal industry in the Soviet Union has continued to enjoy development provides the essential background to the story of continued development also in the exploration and the exploration methodology, theoretical as well as practical, on which you touched during your presentation. One point caught my attention particularly, and this was your description at one point in your paper of the estimates you have made recently of the coal resources of your country. The Soviet Union appears to have something like 100,000, times the coal resources of the United Kingdom which obviously constitute an energy resource of the greatest magnitude and of interest surely not only to the Soviet Union itself, but also to other parts of the world as well, where as we heard this morning, there does appear to be some concern as to the possible development of an energy gap before the end of this century.

That forms one aspect of the work of this conference and the kind of information you have brought to us this morning is of immense significance from that point of view, as well as it being of immense significance to the renewed and redoubled efforts that are now being made in this country, in western Europe, in North America and elsewhere, to develop exploration methodologies for evaluating the coal resource base.

A. ORHEIM: (Store Norske Spitzbergen Kulkompani A/S) Sir, in the northern part of your country (I am not referring to Spitzbergen where you also have a coal company but on the mainland), you have severe climatic conditions. I believe that you have underground coal mines with permafrost. The permafrost offers some advantages, such as better roof compared with normal conditions, there is no water to be pumped out and the working temperature is better. On the other hand, you can't use water to keep down the dust which is getting more and more a problem because of mechanisation, and if you are wanting to have a cleaning plant of some sort, it must be based on a dry process, unless you put in a drying plant before storage. I put two questions to you: The first one, in equal conditions, would you prefer a mine to be situated in permafrost or not? and the second, do these severe climatic conditions influence your mining so much that when you pick your areas for exploration, you take the climatic conditions into consideration or do they influence the priority you give to different areas?

E.V. TERENTYEV: We have mines working in permafrost and this is no problem.

D.C. ION: The Chairman in saying that USSR reserves were 100,000 times those of the UK. is comparing the Soviet second category of reserves used by Mr. Terentyev with a different category of reserves used by Mr. Shaw for UK. reserves. The important point is that in the year 2000 the USSR production may or may not be 10 times that of the UK., it will not be 100,000 times as great.

PROF. PETER O'DELL: I wonder what it will be in the year 3000? Because, hopefully our society in one form or another will still be around. If we listen to some of the Jeremiahs, then some time after 2000 on the basis of some of the reserve figures that are given, including the coal reserves of the United Kingdom as well as the oil reserves of the world, we are going to be cold. It was nice to hear from our Russian colleague that at least there is one part of the world where this energy gap doesn't appear to be a problem in terms of the ultimate resource base as opposed to the reserve production ratio.

SECTION 10

18.
Exploring Coal Deposits
for Surface Mining
by Charles E. Wier

Section Discussion

18

EXPLORING COAL DEPOSITS

FOR SURFACE MINING

Charles E. Wier

Exploration Manager, International Coal

AMAX Coal Company

INTRODUCTION

Exploration for coal poses many of the same problems and employs similar techniques as does exploration for other fossil fuels, metals, or industrial minerals. It is quite similar to looking for one of the industrial minerals, such as limestone, clay, or sandstone. Large areas or basins are well defined and known to contain megatons or gigatons of the desired commodity. The geologist must find that small part of the basin where the quantity is large enough, the quality is good enough, and mining and preparation costs are low enough, that the prepared product can be sold for a cent or two per pound. Thus, exploration tends to be mostly involved in target areas and puts great emphasis on small variations in depth, thickness, structural features, mineral matter, and other selected chemical or physical characteristic

Exploring for coal that can be economically mined by surface mining methods is only slightly different from exploring for deeper coal. Because the coal required for surface mining must be near the surface, it is likely that people who were in the area in earlier years have dug out small amounts of coal and that something is already known about the coal seams. Except for areas that have a thick cover of younger sediments, geologic mapping is likely to be an excellent tool to provide general information on the distribution of shallow coal seams. Coal seams are commonly noted in water well records by the driller, because of their relationship to water supply. In some areas water flows easily through joints in the seam or through thick fusain-rich bands. In seams that are rich in pyrite, the water will have a sulfur taste and oxidation of the iron in pyrite causes the water to have a reddish color. On the other hand, records from deep drill holes such as oil and gas tests do not usually furnish good information; both drillers logs and geophysical logs tend to be absent or poor in that interval of about 60 meters below the surface.

Exploration consists of four stages: 1) Preliminary investigation 2) Reconnaissance of favorable area, 3) Detailed exploration of project area, and 4) Evaluation. After each stage a decision is required to determine if the program will continue. There are many procedures and decisions within each stage. See Table No. I. If the results look favorable, after the evaluation stage, a complete feasibility study is made and, if still favorable, the project goes to development.

Each prospect or project is different from the others and the exploration program must be designed specifically for each. Depending on the location of the deposit, availability of information and the time frame necessary to complete the study, any of the four stages could require a major part of the effort. As a general rule, however, more than half of the effort and costs are involved in detailed exploration – especially in drilling. If sufficient data were available at the beginning of the program, conceivably but unlikely, the work could consist of only the evaluation stage.

Table No. I

Exploration Procedure

- Decide on Area of Interest
  - Location of Market
  - Required Specifications of Coal
- Make Preliminary Investigation
  - Library Research
  - Regional Reconnaissance
- Do Reconnaissance of Favorable Areas
  - Reconnaissance Geologic Mapping
  - Reconnaissance Drilling
- Restrict to Project Areas
- Investigate Project Area in Detail
  - Detailed Geologic Mapping
  - Drilling Program
    - Core Study
    - Geophysical Logs
  - Sampling
  - Analyses
- Evaluate
  - Maps
    - Thickness of Overburden
    - Thickness of Coal
    - Overburden-Coal Ratio
    - Structure
    - Overburden Material
    - Assay Isopachs
  - Cross Sections
  - Reserve Calculation
  - Report

PRELIMINARY INVESTIGATION

Selection of Area

If the location of the expected market and the specifications of the required coal is known, the job of selecting an area of interest is simplified. Assume that the market for the coal is not now designated. If the object is to find coal anywhere in the world that is suitable for surface mining, one may get a copy of the most recent "Survey of Energy Resources" prepared by the World Energy Conference[1] or the map of "World Coal Resources and Major Trade Routes"[2] and look for the areas of large coal resources. It is obvious that the lion's share of world coal resources are in the northern hemisphere. North America has 28.1; USSR, 53; Peoples Republic of China, 9.4; and Europe, 5.9 percent of total world resources. Many countries both in the northern and southern hemispheres contain about 0.1 percent.

However 0.1 percent of total world coal resources is 10,000 million tons. If we assume that only 10 percent of that will be economical to mine, then 0.1 percent is 1,000 million tons. If we assume that only 10 percent of that could be mined by surface mining, then we are looking at 100 million tons. We might be satisfied with only half of that - assuming that there are no expensive problems in producing and marketing such coal. Thus, most countries that are listed as containing coal in these world summary reports do have some potential for at least one surface mine. Note that the term, resources, includes not only those areas of coal reserves where some drilling and mining have been done, but also those interpreted areas that are believed to contain coal, based on adjacent areas and a **knowledge of sedimentation, stratigraphy, and structure.** It is in this latter area that the exploration geologist can do the most good for his company. Areas of known reserves are usually already controlled by competing companies or else have a serious problem in mining or marketing.

Library Research

Assume that we have chosen a country, province, or state for which we have no significant information in our files. There are many sources of information. Commonly there is a department of mines, a geological survey, or similar agency that keeps records of past geologic mapping, drilling, analytical work, etc. If it is a country where coal mining has not been well developed, probably reports have been written by geologists of earlier years from England, France, Spain, Portugal, Germany, The Netherlands, or other countries Many of these reports are available in geology or geoscience libraries of large universities. If a desired publication is not available locally, probably it can be borrowed from another library through the interlibrary loan system. Bibliographies are prepared annually by some agencies that index by topic and country in such a manner that it is simple to find references to those publications that, say, cover coal in Afghanistan or coal in Zambia. Special compilations also are made such as, for example, "Selected Sources of Information on United States and World Energy Resources"(3) and (from the U.S. viewpoint) "Foreign Literature on Coal."(4). These compilations tend to list serials and individual publications that are generalized or cover larger areas. In a general reference such as "Coal Resources of the United States"(5) there are listed in References Cited many papers covering individual small areas in detail.

In addition to published books and articles, some consulting companies will furnish, for a fee, a study of the coal resources in a coal basin or a country. At any rate, work by consultants and library research will usually restrict the area of interest to less than 10 percent of the original area.

Regional Appraisal

After the initial library research, or simultaneously, a geologist should visit the area of interest, contact local officials and/or owners and obtain local information. And, during this preliminary investigations, the same kind of information must be collected that is needed for the later final feasibility study which is made to enable an executive decision whether or not to proceed through development. However, the information in the preliminary investigation must, by necessity, use approximate numbers and often vague information. Despite such limitations, information should be obtained on quantity, quality, depth, thickness, and structure of coal seams and infrastructure (especially transportation facilities), exploration and development laws and regulations, local political philosophy and practices, available manpower, wage, scales, access to area, etc.

At this stage a decision is based on vague information but a decision must be made whether to forget about the area, to do geologic mapping, or start a drilling program. Often, decision making is helped by specific requirements of marketing. Reserves may not be large enough or demand not large enough to pay for expensive infrastructure; market is too far away and transportation cost too high to sell an inferior coal at a competitive price; ash or sulfur content may not be reduceable by available preparation methods to fit in the required market, etc. The attitude of local civic groups or government officials may not have crystal-lized at this point but they certainly are an important consideration.

RECONNAISSANCE OF FAVORABLE AREA

If the decision is to proceed, then the exploration program is designed to obtain more information, not only to learn how much coal is present but also to furnish information that will be required in a feasibility study. Commonly available information is not good enough nor reliable enough for a proper decision, but the decision must be made whether to stop or proceed. In some cases, the decision may be to proceed in a favorable area with a reconnaissance mapping or drilling program that will allow a reduction in size of the area or furnish addi-tional information. In order to do this, accurate base maps are required to properly plot the information obtained.

Available geologic maps, topographic maps, and aerial photos of the area may be available for purchase. If only generalized maps are available, the best information may be the LANDSAT (formerly ERTS) imagery. With proper interpretation LANDSAT imagery can help determine the general structural features and, perhaps, allow the map-ping of large rock units. Either at this stage of explora-tion, or the next, a contractor should be hired to fly the area and furnish appropriate photography. During recon-naissance mapping, scales of 1:250,000 to 1:62,500 are satisfactory, but a scale of 1:20,000 or 1:24,000 is more useful.

In addition to reconnaissance mapping, it may be necessary to drill some test holes. This drilling program will include a large number of holes on a widespread grid pattern covering the entire area in order that the geo-logist can find the best looking part of the favorable area. Some questions that may need to be answered include: Are earlier drill records reliable? What is the quality of the coal? Do slumped outcrops show representative thickness? etc. In order to answer those questions, perhaps only a few drill holes at selected locations are required. Based on the additional information obtained, the decision can be made whether to proceed with a detailed and expensive exploration program in the project area.

DETAILED EXPLORATION OF PROJECT AREA

If geologic mapping were done during the reconnaissance stage, perhaps more detailed work could be done in the restricted area. However, in most coal basins, the coal-bearing rocks have a limited number of exposures and detailed mapping at a scale larger than 1:20,000 is not useful.

In conjunction with geologic mapping, special geophysical surveys may be useful. If, for instance, earlier valleys in the bedrock are completely filled and concealed by unconsolidated material, refraction seismic work can allow the delineation of such valleys. If the coal seam is cut out in these valleys, such work is required. Under certain conditions, gravity surveys can do as well. If igneous rocks are intruded into the coal-bearing rocks or form a concealed boundary, these areas can be mapped by the use of gravity or magnetic techniques. Geologic mapping is a tool that will help understand the three-dimensional distribution of coal, but, even with the aid of geophysical methods, the required detailed three-dimensional grid can be obtained only through a well-planned drilling program (Fig. 1).

Drilling Program

Although the early stage of a drilling program may have a goal of outlining the best part of the project area, the main goal is to delineate measured coal reserves. The closeness of the drilling pattern varies, depending on the complexity of the geology. For instance, the greater variation in thickness of the seam, the more abundant the contemporaneous or post depositional erosional features, and the more intense the folding and faulting patterns, the closer the drill hole spacing. Normally, drill holes would have a spacing of 800 meters or less and reserves should be established to that accuracy where future drilling would not change the tonnage numbers by as much as 20 percent. The spacing of drill holes may be increased in parts of the area because of sharp variations in quality, if those variations are critical to limits dictated by market requirements.

In planning the sequences of drill holes, unless there is a good reason to do otherwise, the initial effort should be expended in the area of best information or presumed thickest coal and data is added as drilling progresses in all directions. This initial set of holes should be deeper than one would reasonably expect to mine in order to provide the additional information that the geologist needs to correlate the coal seams and establish the sedimentational and tectonic framework that will help to anticipate and understand variations in thickness and quality of the coal.

When doing deep drilling it may not be necessary to core the complete hole. Common practice is to use the less expensive rock-bit technique, which produces only cuttings, for the upper part of the hole above the coal seam. Then only the coal seam and about 6 meters of roof rock are cored. When drilling the shallow holes for surface mining, the holes should be cored all the way - especially in the initial grid spacing. If a large volume of core is not required, wireline coring is less expensive and just as satisfactory as regular coring. Cores are not required for a large number of holes that are closely spaced, and are needed only to determine position and thickness of coal.

A standard system of core description and record-keeping is necessary in order to allow integration of work done by different geologists and by the same geologist, done at different times. A photograph of the core is an additional record that is useful when questions arise later. Even more useful is the X-ray photograph that shows excellent contrast between coal and partings.

At the same time that the geologist is studying the cores obtained from drilling, he can start the preliminary evaluation of slope stability for the highwall in the surface mine. Steep dip of individual beds, well developed sets of joints, or thick sections of clay, or soapstone add to possible stability problems. In most cases, the more serious part of the stability problem are those unconsolidated or poorly consolidated materials that overlie

Rotary drilling rig coring coal. Courtesy of Indiana
Geological Survey. Figure No. 1

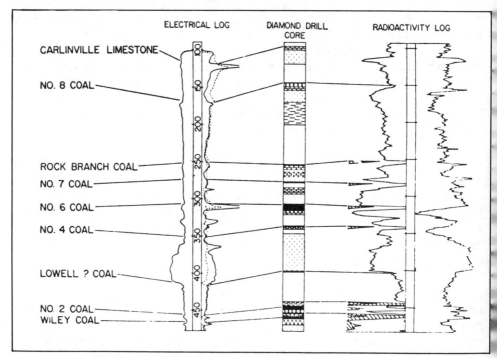

Correlation of log from a core with an electric log and
a radioactivity log (6). Figure No. 2

the coal-bearing rocks. This upper part is not normally cored because special equipment is required. But the driller and/or the geologist should list thickness of different kinds of materials such as loose silt, sand, or gravel and unconsolidated clay till, lacustrine clay, marine clay, residual soil, etc. Any observations as to water content or hydraulic head may be helpful later. It is not up to the exploration geologist to be proficient in rock mechanics and soil mechanics or to design maximum slopes that could safely be maintained. It is the geologist's responsibility to provide information that will alert the mine design engineer to potential problems in maintaining the highwall.

Overburden rock has other ramifications in the mining operation. Different kinds of rock, depending on grain size, thickness of bedding and kind of cementation require different plans for blasting the highwall. It may also be helpful in planning to know where, in the coal field, the right kind of rock is available for road construction, dam construction, etc.

Geophysical Logs

Soon after the drill hole is completed, geophysical logging of all holes is essential, but not all available kinds of logs are required. Basically, the geologist needs the logs to determine the exact depth and thickness of each coal seam. Even if a core is furnished, a check is needed to determine if any of the coal is included in a core loss. If only cuttings and a driller's log are available, then the geophysical log is the only precise way to determine coal depth and thickness. Most kinds of logs that can be interpreted to show coal thickness can also be used by the geologist to help in correlating the individual coal seams. Quite often in a restricted area, one seam will have a characteristic signature that will serve to identify it.

Different kinds of geophysical logs work best in different areas. Radioactivity logs can be used to identify and determine the thickness of mineable coals (Fig. 2). In some places, a simple, and much less expensive, self potential log does just as well. In areas where there has been a great deal of petroleum activity, a large file of logs containing self potential, two different spacing of resistivity, and a lateral resistivity log are available (Fig.2). These are useful correlation tools but depth to and thickness of coal cannot usually be interpreted closer than plus or minus 0.15 meter. In order to be able to interpret geophysical logs, the geologist must be familiar with the rock sequence and understand the range of possible lateral variations in each rock type.

Combinations of logs, including bulk density, are sometimes run to evaluate rank of coal, percent of ash, sulfur and volatile matter, and kind of overburden material. Normally, these can be used only in conjunction with analytical information from adjacent cores.

Sampling

One of the reasons that cores of coal are taken in the initial stage of drilling is to provide representative samples of coal seams for analysis. Sampling of coal is a critical exercise that must keep in mind the end purpose. It is well known that coal must be placed in air-tight containers to preserve the natural moisture and to minimize oxidation, but samples are not always divided in units that can be fitted into mining plans. Although, to be sure, it requires some background of mining practices in the area, the geologist should sample the coal seam into the sample units that might go to the preparation plant. That is, solid units of nearly pure coal plys are separated from rock plys that are too thick to be included, and from impure coal or mixture of coal and rock plys that may or may not be profitable to mine. After analyses are made, units can be kept separate or lumped depending on methods of mining and preparation.

If coal cores are approximately 50 millimeters (or about 2 inches) in diameter, they contain a large enough volume of coal to provide meaningful analytical results. However, larger samples may be required for washability tests that are needed for final design of the preparation plant. Larger samples can be obtained by drilling a few holes of 150 millimeters (6 inches) or 200 millimeters (8 inches) in diameter. Where coal seams are deep, bulk samples can be obtained from a drill hole by hydrofracturing and pumping out the fragments. When bulk samples are required for shallow coals, the best method is trenching to the coal seam.

If sampling is done where the coal seam is exposed in active or abandoned mines, the channeling method is used. That is, a channel of uniform width and depth is chopped along the vertical face. Because coal is heterogenous and different bands may have vastly different characteristics, collecting the same volume of material from each unit of

vertical distance is essential in order that the sample be
representative of the mine product. The vertical sequence
may be divided into more than one sample depending on amount
and thickness of impurities present.

Analysis

Analysis required depends to a great extent on the proposed use of the coal. Obviously, different properties are
critical depending on whether the coal is intended for a
steam plant, a coking oven, or a gasification or lique-
faction plant. If the geologist does not have a market restriction, a large variety of analyses may be run on the
early coal cores in order to adequately characterize the
coal. When it is obvious that the coal will have a specific kind of use, then analytical work will be directed
toward that use.

The coal samples collected early in the exploration
program should have, as a minimum, the long proximate
analysis made, that is, percent of moisture, volatile
matter, fixed carbon, ash and sulfur and the heating
value. These analytical results will allow the geologist
to characterize the coal as to rank and probable use.

If the coal is burned to produce steam, then the most
important characteristic is calorific value. Analyses are
listed in British thermal units per pound, calories per
gram or kilojoules per kilogram and a simple conversion
factor can change units from one system to the other. However, coal may be sold on the basis of a fixed price per
million B.t.u. with penalties for excess ash or sulfur.
Or, in fact, if either ash or sulfur is too high, the coal
may not be acceptable to the buyer. Because of the above,
not only must analyses for ash and sulfur of raw coal be
made in some detail, but washability tests should be made
to see what percent of each will be in the marketable product
at various specific gravities and at various percents of
recovery of the coal.

In some market areas the percent of sulfur is critical.
Local pollution laws may place a maximum in terms of percent,
or in weight of sulfur per unit of heat. In effect, this
can restrict the shipment and sale of a coal in an area if
the coal contains more than 0.6 percent sulfur; in other
areas, 1.0 percent. Other consumers may be concerned about
sulfur only from the standpoint of fouling tubes in boilers.

In order to evaluate sulfur problems early in the exploration program, some analysis should be made for sulfur in its three common forms: sulfate, sulfide, and organic. In most coals sulfate is quite small - less than 0.1 percent. If it is large, the coal probably has undergone oxidation. The sulfide type of sulfur is in the minerals pyrite and marcasite. Depending on the size of these minerals and the size to which the coal is crushed, half or more of this sulfur may be removed. The organic sulfur, however, is part of the hydrocarbon structure and can be separated from the coal only by burning or by solvent extraction. Thus, as a first look, the geologist can anticipate that the sulfur content of cleaned coal will be the percent of organic sulfur plus half of sulfur from the sulfides. Later washability tests will be more precise.

Other tests that are less important in the combustion of coal (unless the coal gives especially high or low numbers in one of the parameters) are percent of moisture, volatile matter, and potash, temperature of ash fusion, and grindability index.

Moisture and ash content affect the heating value of the coal; that is, the higher the percent of these impurities, the less heat can be produced per unit weight. Moisture and grindability indicate the energy required and mill capacity to produce the fine sized coal particles that are required for some uses, such as, in a pulverized coal-fired furnace. Ash fusion (the temperatures at which the ash will begin to deform, soften, or become fluid) determines whether the ash will fuse into clinkers in various types of furnaces. Low-fusing ash is desirable in cyclone furnaces and others where the ash is removed from the bottom in a liquid state, but is undesirable in static fuel beds because of the difficulty in removing clinkers. Composition of ash, especially silica, is useful in calculating slag viscosity. Other elements may indicate fouling or corrosion effects. Phosphorous is the most common deleterious element in ash.

If the coal is to be used to produce metallurgical coke, other analytical procedures are important. In earlier days, it was much simpler to determine if the coal could be sold as metallurgical coking coal because there were strict requirements that eliminate most coals. Volatile matter should be between 17.5 and 25 percent and sulfur less than 1.0 percent. Now that two or more coal seams are blended to balance strong and weak characteristics in each, a much wider range of coal seams are of potential coking use. This fact could be critical in planning later stages of exploration. High quality coking coals may command a price that is more than double the price received for a good steaming coal and even "soft coking" coal may sell for 50 percent more than if it were sold as steaming coal. Thus, the increase in price over steaming coal may allow the exploration program to consider coal that is deeper, thinner, or less accessible to the market.

The same proximate analyses that are made from samples of coal intended for steaming use serve to give a ball park evaluation of the coking use. For instance, percent of volatile matter indicates, in general terms, the rank of coal and medium and low volatile bituminous coals are most likely to make high quality coking coal. Using the U.S. ASTM system of coal classification, coals that contain less than 14 or more than 43 percent volatile matter, dry and mineral matter-free basis, are not likely to have good enough coking properties. If the volatile matter is between 15 and 31 percent, the odds of producing a coking coal are good. The effects of moisture and ash are mostly from the stand-point of dilutment. The higher percent of moisture the less useable coal per unit. The higher the ash in the coal, the higher it is in the coke, and the greater amount of slag that must be removed from the blast furnace. Coals sold for coking blends generally contain less than 9 percent moisture and 13 percent ash.

Percent of sulfur in the coal is also a critical characteristic. The larger the amount of sulfur in the coke, the greater amount of coke is required. An increase in percent of sulfur in the coke from 1.0 to 1.5 percent might increase the required coke charge by 250 pounds per ton - a 15 percent increase. If cleaning the coal will not reduce the sulfur to 1.5 percent or less, it is not likely that the coal will be useful as a coking coal blend.

If coal fits within the ball park of criterion mentioned above, then tests that characterize the plasticity are required. Free swelling index, or coke button index, is a simple and common test that gives some indication of amount of swelling when the coal is heated to the plastic state. Results are listed in numbers from 0 to 9. The higher the number above one, the greater the swelling. Some users require coal with numbers greater than 3, others greater than 5, etc. Tests that are better indicators of coking quality include Audibert-Arnu dilatometer, Gieseler plastometer, and Gray-King coke type. One or more of these

tests must be run depending on the criteria used by the
potential coking coal customer.

A new technique that has come into its own in the past
15 years is analysis through coal petrography. This technique has been developed and used by laboratories not only
to characterize the potential coking coal but to calculate
the optimum percent of two or more coals in a blend that
will make the best coke. In essence, analysis is done on
a representative small sample of crushed coal with a research
microscope. The rank, or amount of coalification, is determined by reflectance measurements of vitrinite. Percent of
each maceral is determined and these are lumped into reactive
or inerts - based on the temperature at which that part of
the coal will fuse when it is heated to make coke. This
information is used in formulas and results are compared
to and interpreted by means of graphs, charts, and tables.
Satisfactory results can be obtained only if the coal seam
itself has very little lateral variation and where standard
charts and tables that relate petrography to coking strength
and stability have been accumulated by empirical tests.

Despite all of the accumulated analytical results, the
final decision on coking coal use is made by the customer,
based on the empirical testing of different ratios of blends
with other coals available to him.

EVALUATION

As drilling information is received, it is plotted on
various interpretive maps. This may be done by hand by a
junior geologist or engineer or, when large amounts of data
are available, these data may be fed into a computer program that prints the appropriate information on a map at
the proper location and then constructs structure, contour,
or other iso lines.

Correlation of Coal Seams

The most basic task, and certainly a most important one,
is the correlation of coal seams. In some areas where there
is only one thick seam, this is simple. However, the geologist must be ever watchful that it is, in fact, a single
seam and not a situation where more than one seam is present
in the area but where only one thick seam is present in each
drill hole. Obviously, if the coal seams are not properly
identified, data collected from each drill hole cannot be

assembled to produce a meaningful picture of the coal field. Mis-correlations could cause serious mistakes in determining the amount of coal present, the structural position of individual seams, and the average physical and chemical characteristics of each seam. This, in turn, may cause wrong evaluation of the property, and problems in mining, preparation and marketing might develop.

There are numerous techniques used in correlating seams. They all start with a standard or known outcrop or drill record. Adjacent drill records or other points of information are used to match the coal seams with the type section. Geologists use a combination of physical characteristics to identify seams. Some are: thickness of interval between coal seams, thickness of seam, sets of persistent partings in the seams, unusual rock types above or below the seam, and unusual dull or bright bands within the seam. As drilling progresses and more information is available, the geologist gains more confidence in his correlations.

Geophysical logs are useful correlation tools. In many areas, self potential, resistivity, or radioactivity logs have a distinctive shape to the "kick" that represents the coal and the rocks immediately above and below so that after it is once recognized it is an extremely useful signature.

If there are real problems in correlation, perhaps a palynologic study would be needed. This technique concentrates the remains of plant reproductive bodies - spores and pollen - by dissolving a coal sample. With the use of a microscope, the palynologist identifies and calculates percentage of each type. In theory, each coal seam has its own peculiar assemblage. However, this technique is not commonly available to the exploration geologist.

Mis-correlations made in the field may be obvious and easily corrected when cross-sections and various kinds of maps are constructed, but, more commonly, when anomalous information is plotted more than one structural or depositional picture looks possible. A thorough understanding of the depositional environment is necessary in order to make the correct decision. This judgement, often based on insufficient data, has caused some people to say that the process of correlating coals is an art and not a science.

Basic Maps

The three most basic maps for surface mining are those

showing thickness of coal, thickness of overburden, and overburden to coal ratio (Fig. 3). If only one seam of coal is
being considered, the maps are straightforward. If several
benches of seams are present, they may be lumped or considered separately, depending on expected use of such maps.

If the geologist constructs the isopach lines showing
coal thickness, he may use a bias of thickening or thinning
as related to structural or depositional features. Thus, a
structural map based on one or more coal seams is usually
included early in the evaluation. Topographic maps must
be utilized in the construction of overburden isopach lines.
If a suitable base map is not already available, it must be
made. Using the same overburden and coal thickness data, a
ratio of overburden to coal can be constructed in which coal
is given as unity, that is, ratio of 5:1, 20:1, etc. First
the geologist calculated these ratios as feet of overburden
to feet of coal (or meter to meter). This is not useful to
the engineer who, when planning the mine, relates cubic yards
of overburden removed to tons of coal produced. The over-
burden-coal ratio is now expressed in cubic yards of overburden to short ton of coal or cubic meters to metric tons.
Using the metric system, the calculation is simple, but in
the foot-short ton terminology a constant is required.

$$\text{Ratio} = \frac{\text{OB (meters)}}{\text{C (meters) X SG}}$$

$$\text{Ratio} = \frac{\text{OB (feet)}}{\text{C (feet) X SG X 0.843}}$$

Where OB is thickness of overburden, C is thickness of coal,
and SG is specific gravity of coal. Because square meters
and square yards are slightly different and metric tons and
long tons are also, the ratios do not have the exact same
value. Ratios in the metric system are about 0.8 that of
the cubic yard-ton system.

In the U.S., where specific gravity of the coal is not
known but rank is, the following numbers are used by many
geologists[5]:

Rank	Specific Gravity
Anthracite	1.47
Bituminous	1.32
Sub-bituminous	1.30
Lignite	1.29

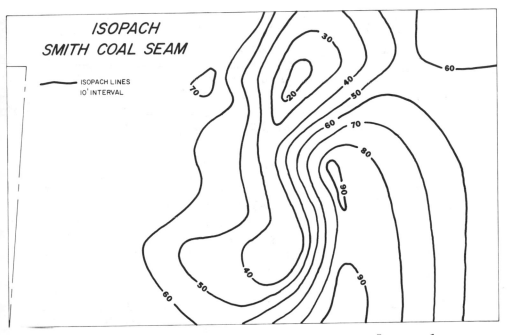

Isopach map showing variation in thickness of a coal
seam.   Figure No. 3.

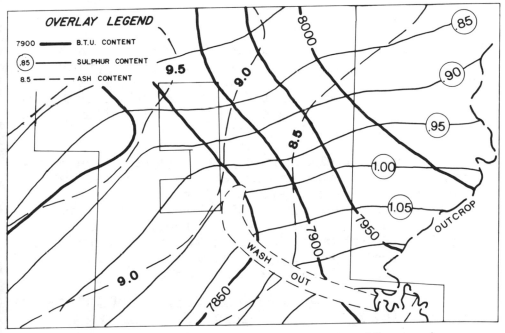

Isopach map showing variation in B.T.U., sulfur, and
ash content.   Figure No. 4.

However, the amount of mineral matter in the coal has a great effect. For instance, a "clean" coal that has a specific gravity of 1.3 and that contains 20 percent clay and silt at SG 2.2 and 2 percent pyrite at SG 5 would, in fact, have an SG of 1.54 - a decrease of nearly 20 percent in the ratio and an increase of 240,000 tons of coal for each square kilometer per meter of thickness. Thus, if possible, the specific gravity as determined from the coal samples should be used.

Reserve Calculations

Using the maps showing thickness of overburden, thickness of coal and overburden to coal ratio, the geologist can calculate coal reserves. He can divide the reserves into areas on the map that fit expected categories of mining. He may use alternate combinations of the following kinds of categories (thickness in meters):

Overburden Thickness	Coal Thickness	Overburden/Coal Ratio (maximum)
1-20	1-2	4:1
20-30	2-3	6:1
30-50	3-5	10:1
50-100	5-10	20:1

For example, one set of caluclations might use a maximum depth of 50 meters, maximum ratio of 20:1, and a minimum thickness of coal of 1 meter. Of course, in the area where the ratio of 20:1 and 50 meter depth were the controlling factors, the coal would have to be 1.6 to 1.9 meters thick, depending on its specific gravity.

In discussing coal reserves at this point we are considering those that are commonly referred to as measured or proven reserves. The categories of measured, indicated, and inferred reserves and additional resources (which when added together are called total resources) are defined in slightly different ways, depending on unit of measurement, emphasis placed on drilling versus outcrop, and mining information. Listed below are common usage:

Reserve Category	Distance between Points of Information	
	Miles	Kilometer
Measured	½	1
Indicated	1-1½	2-3
Inferred	> 1½	> 3

Note that the exploration program would already have done enough drilling that points of information were close enough to call the reserves measured. Either in the later stage of exploration or in the follow-up development stage, additional drilling must be done to outline the crop line edge of the proposed surface mine and, depending on the variability of the distribution of the coal, an additional set of holes will be required to bring the drill spacing to 400 meters (1320 feet) apart or even to 200 meters. Although at this point in exploration we are concerned with measured reserves in a potential mine area, we may wish to list indicated and inferred coal in adjacent areas of possible future exploration, that is, possible reserves for a future mine or for expansion of the planned mine.

Reserves are first calculated for coal in the ground. Using specific gravity determined in the laboratory, the formula in the metric system is:

$$\text{Tonnes} = C \text{ (meters)} \times A \text{ (meters}^2) \times SG$$
$$\text{M Tonnes} = C \text{ (meters)} \times A \text{ (kilometers}^2) \times SG$$

Where SG is specific gravity of coal, C is thickness of coal. A is area, and M is one million. The same kind of formula works for the foot-short ton system:

$$\text{Tons} = C \text{ (feet)} \times A \text{ (acres)} \times 1360.6 \times SG$$
$$\text{M Tons} = C \text{ (feet)} \times A \text{ (miles}^2) \times 0.8707 \times SG$$

When dealing with a coal field where the coal has a limited range in rank, an average specific gravity may be assumed and rounded numbers such as those listed below commonly are used[5]:

Rank	Specific Gravity	Tons per Acre per Foot of Thickness	Tons per Square Miles per Foot of Thickness
Anthracite	1.47	2,000	1,280,000
Bituminous	1.32	1,800	1,150,000
Sub-bituminous	1.30	1,770	1,132,000
Lignite	1.29	1,750	1,120,000

It must be borne in mind that even measured reserves are not precise, but are considered to be within 20 percent of true tonnage. As a rule of thumb, the drilling program should continue until additional holes do not change the amount by more than 10 percent.

There are many kinds of terms used to identify the amount of coal available. The geologist initially calculates reserves in the ground, or in situ. Because of irregularities or unpredictable absences of coal or unusual overburden problems, there is a loss of coal that we might call geologic loss. There is also a loss in mineable areas due to unpredictable mining problems and losses due to the normal mass-production-type of mining. This is mining loss. After the coal is mined, it is sent through a preparation plant where part of the impurities are removed. But some of the coal fragments that are fastened to large pieces of rock are also removed and discarded. This is a preparation loss. Thus, coal reserves in situ may be reduced by nearly 50 percent from in situ reserves to marketable coal. For example:

Coal Reserves	Tons	Loss	Recovery
In situ	1,000,000	Geologic 10%	90%
Mineable	900,000	Mining 15%	85%
Run of mine	765,000	Preparation 20%	80%
Marketable coal	612,000	Total 38.8%	61.2%

Generally, geologic loss and mining loss are based on mining experience in the area. Preparation loss is based on washability studies.

Additional Maps

Additional interpretative maps that relate various parameters of overburden or coal seams may be important if such variations need to be shown graphically as maps. For instance, thickness of rock between two seams, thickness of unconsolidated material, ratio of sand to lacustrine clay, sandstone to shale ratio, etc. It may also be desirable to show variations in analytical characteristics such as percent of ash and sulfur or heating value. If these variations are simple, they might be displayed more obviously by cross-sections. (Fig. 4).

After the exploration data are collected, tabulated, interpreted, and summarized, a decision can be made whether to continue into final feasibility study and development planning. In some cases, further exploration information may be required to fill in blanks in the acquired data. But, probably, further data acquisition will be part of the development program and our exploration program is finished.

REFERENCES

1.   "Survey of Energy Resources 1974," World Energy Conference, 1974.

2.   "Map of World Coal Resources and Major Trade Routes," World Coal, March 1975.

3.   Averitt, P., and Carter, M. D., "Selected Sources of Information on United States and World Energy Resources: An Anotated Bibliography," U. S. Geological Survey Circ. 641, 1970.

4.   Esfandiary, M. S., "Foreign Literature on Coal: A Guide to Abstracts and Translations," U. S. Bureau of Mines Info. Circ. 8063, 1961.

5.   Averitt, P., "Coal Resources of the United States, January 1, 1974," U. S. Geological Survey Bulletin 1412, 1975.

6.   Simon, J.A. and Smith, W.H., "Increasing Effectiveness of Diamond Drill Core Exploration for Coal," Proc. Illinois Mining Inst., 1962.

Section Ten Discussion

A.N. LANE: (University of Zambia):   Dr. Wier, you briefly
mentioned some effects of weathering on coal samples and I
am particularly interested in your experiences in Africa.
At this moment in Zambia we are starting a coal exploration
programme in the mid-Zambesi Valley and one of the problems
is sampling, particularly at the outcrops, and I would be
very much obliged if you could briefly summarise the effects
of weathering on coal analyses and coal properties and make
some comments on the depths of weathering in this particular
environment, bearing in mind the climate and topography. Is
there in fact a safe depth below which you know you get un-
weathered fresh coal?

C.E. WIER: That question is a quite difficult question. If
I can start with the latter part first, the depth of oxidation
in countries such as you are talking about where there is 20
inches or more of rain per year: in that case there is
enough moisture so that the oxidation goes fairly rapidly,
so if there is a natural outcrop and if the coal is primarily
in shale, the depth may be just a few feet. If it has sandstone adjacent to it, then it may go back for many feet. In
Botswana it was somewhat different in that the whole of the
coal area is overlain by Kalahari sand, some very slight
amounts, some several hundred feet, and therefore there is
the old erosional surface and in some areas I suppose there
are many tens of feet vertically where one does have oxidation
But I think from the standpoint of sampling the outcrop, one
can tell pretty well simply by digging in although it may
take a lot of digging, but you simply dig into it until the
coal looks like normal fresh coal. I recognise your concern
from the standpoint of the analysis if the coal does not
weather or oxidise at the same rate, and is not equally
affected. If it is oxidised it deteriorates all of the
important parts of the analysis that you are concerned with,
that is it certainly ruins the calorific value as much too
low, if it is possible coking coal it ruins all of those
nice tests that you are interested in and of course it does
ruin the moisture content. If you can't get fresh coal
samples it is really worthless as far as the consumer is
concerned.

V. DEQUECH (Geosol - Geologia e Soudagens Ltda. Brazil): I
saw in your paper that you used to start coring six metres
above the coal seam. As the distance between drill holes is

800 metres, I ask you if there is not the risk of drilling through the coal seam without coring. If the dip of the seam is only 1 degree different from what you expect, you have a difference of level of 17 metres each 1 km. and 14 metres in 800 metres. Then if the dip is 5 degrees different you have 70 metres difference. I think it better to have a gap of say, 10% of the distance between drill holes.

DR. WIER: Your question applies to the amount of rock above the coal when one starts coring?

V. DEQUECH: Yes, the point where you stop non-coring and begin coring for the seams.

DR. WIER: Yes. Well, you are exactly right in what you imply. One may attempt to save a great deal of money in the drilling programme by not coring the drill holes. I think the way to do it initially is to core the hole away until you get a good feel for it. Now, if then later one has the feel for the structure and the variation in sedimentation, the driller can start coring near the coal. Of course he has to have some kind of leeway, but a few leaders would be sufficient. Every once in a while he makes a mistake and he goes through the whole coal or he gets into the coal before he realises he is there and so has defeated the purpose somewhat of saving money by not coring. This is where the good geologist earns his money. You are exactly right in that if you don't have good control, if you don't know pretty well where the coal seam is, then you really haven't much of a chance of rock drilling down to some set of metres above the coal and then coring coal.

DR. DEQUECH: I think that even when we know well a deposit six metres is a very short distance because the other thing is that generally, the drill is made by contract. I am from a contractor, and the driller has instructions to stop when anything unexpected arises. Then if we reach the coal before the depth that is expected, we must stop and look for the geologist and perhaps lose time. We don't accept contracts which make conditions like that, without requiring that a standstill is paid by the firm. Another question I would like to present to you Dr. Wier is: can you explain to me hydro-fracturing?

DR. WIER: Well this is a technique of obtaining larger samples, a kind of bulk sample of the coal, when you have drilled about a 2 inch core. It doesn't normally apply to surface mining exploration but it may apply in deeper mining. For instance, if you had to obtain a core from a hole let us say 800 feet to the coal, then it becomes very expensive to drill large size cores in order to obtain large samples of coal and so, following some petroleum drilling techniques, one can then use a 2-3 inch hole and pump water under quite large pressures down into the coal fracturing it, and almost all coals have rather well developed cleat. The coal breaks up and washes back up the hole. Now this has its good point from a standpoint that is not nearly as expensive and you can get some pretty large samples that way, but it has the bad point that if you have lots of fine material for instance it washes out and is lost and you may not get quite the exact sample of coal that you would like.

DR. DEQUECH: I don't know if with common drills you can have the size of sample with only water pressure. We have under-reaming tools that can do that.

DR. WIER: Well we used it for bituminous coals in mid-western United States.

DR. DEQUECH: Well, that must be a particular kind of coal, soft.

SECTION 11

19.
Exploration and Geological Structure of Coal Measures in Western Canada
by Clive W. Ball

20.
Financial Considerations in Evaluating Newly Discovered Coal Deposits and Ventures
by John K. Hammes

Section Discussion

19

EXPLORATION AND GEOLOGICAL STRUCTURE

OF COAL MEASURES IN WESTERN CANADA

Clive W. Ball

Chief Geologist,

Canex Placer Limited, Vancouver, British Columbia

I. INTRODUCTION

In Western Canada coal measures are widely distributed and the most extensive fields occur in three geographical regions defined as the Plains, Foothills and Rocky Mountains In addition, coal is found in a number of isolated basins, principally in British Columbia Interior Region and the Yukon Territory. The bulk of the production including cokin and steam coals has come from the Cretaceous beds, but with increase demand for thermal power some large deposits of Tertiary age will be developed in the near future. In general distribution of the high potential coal deposits of Western Canada is shown in Figure 1.

II General Geology

The coal deposits of Western Canada were formed in Jurassic, Cretaceous and Tertiary times and occur in fairly well defined sub-basins. The stratigraphic sequence has been well established by field work and correlation by the Geological Survey of Canada and private companies. Table I is a typical correlation chart covering the main regions of Western Canada. The geological structure becomes progressively more intense from Saskatchewan through the Plains Region of Alberta to the Foothills and Rocky Mountains regio This is directly due to the Laramide orogeny during late

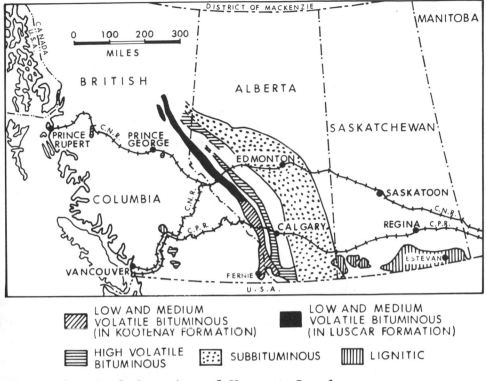

Figure 1. Coal deposits of Western Canada.

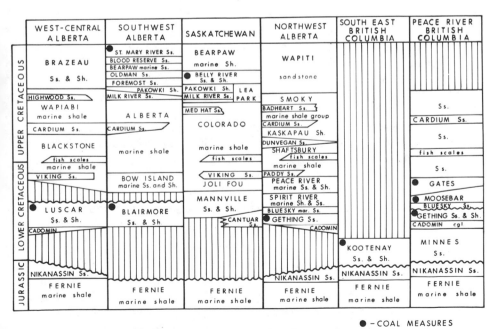

Table 1. Correlation chart. Mesozoic strata of Western Canada.

Cretaceous and early Tertiary times.

The intense dynamic forces which operated in the Foothills and Rocky Mountains region have resulted in isoclinal
and recumbent folds in the coal measures with dips of the
order of 50 degrees to 60 degrees being quite common. A
large number of thrust faults have been mapped and they tend
to slice and overthrust the folds. The structure in local
areas is still more complicated by normal faulting.

The increase in structural intensity from East to West
is accompanied by an increase in Rank although the important
coking coal deposits in the Foothills and Rocky Mountains
regions probably result more from the weight and compression
of superincumbent strata than to the effect of the forces
which caused the folding and faulting of the coal beds. For
the most part the coal has been formed in isolated basins
about 30 miles long by 15 miles wide. Some of the coal
measures in Western Canada were formed under estuarine
conditions where the coal was deposited in a deltaic sequence
as for example at Canmore, Alberta.

III Exploration

The methods applied depend largely upon the structure
of the coal measures and the quality of the coal. For
example, in the Plains region of Alberta and Saskatchewan
the coal was initially located by a few outcrops and subsurface intersections in test wells exploring for oil, gas
and water. In the mountains, however, most major coal
measures were first detected through outcrops and areas of
coal float.

In the Foothills and Rocky Mountains regions extensive
exploratory testing is required. This is primarily because
of the higher rank and quality of the various coal seams.
For example, the exacting specifications for coking coal
necessitate the application of very selective methods and
the intense structural complexity of the beds requires
testing on a closely spaced grid.

The first stage of exploration involves detailed surface
geological mapping on scales varying from 400 feet to 1,000
feet to one inch. Full use is generally made of air photos
and photo-grammetric maps. Field exploration utilizes small
crews of experienced geologists and prospectors concentrated
for work during the relatively short field season.

The second stage requires the use of bulldozers to strip the most favourable sections and carry out local trenching of the coal seams in order to expose sections of bedrock. Usually this is difficult in Western Canada on account of the steep terrain and widespread cover of glacial drift which in places ranges up to 200 feet in depth.

The third stage follows when encouragement is obtained in the previous stages and requires rotary, reverse-circulation and diamond drilling in order ot obtain sub-surface cross sections and samples of coal for preliminary testing. Almost coincident with close grid drilling on about 3,000 foot spacing, it is essential to make adit entries into the coal seams for bulk samples for analysis and testing. Adits are completed by driving cross-cuts and drifting or raising to expose the entire seam for sampling. The latter approach is particularly essential when testing the coal for its coking qualities and to obtain samples beyond the zone of oxidation.

Geophysical methods do not have a wide application for use in Western Canada coal deposits, but upon completion of each rotary test well or large diameter diamond drill hole, it is essential to run gamma-ray neutron logs, resistivity and density logs for correlation purposes. Attempts to apply various geophysical methods for surface exploration in Western Canada have not been successful to date. Induced Polarization surveys have been applied in a few areas and serve to delineate the structural pattern on a large scale. However, Induced Polarization has not been useful for complex structure and multiple seam definition. Very low frequency (V.L.F.) electromagnetic survey offers hope provided that the coal beds have a strike direction or orientation parallel or subparallel to the bearing from the transmitting station. Gravity and magnetometer surveys have been used to outline the coal basins.

IV   Reserves

Coal reserves of Western Canada are estimated at 118.7 billion tons of coal in place as shown in Table II and the reserves are divided as follows: Saskatchewan 10.1%; Alberta 39.8% and British Columbia 50.1%.

Estimated reserves by Rank and Province are tabulated in Table III.   The Rank classification is based on the American Society for Testing and Materials (A.S.T.M., 1966).

TABLE II

Coal Reserves of Western Canada by Province
(000's of short tons)

Province	Proven & Probable	Possible	Total
Saskatchewan	7,315,500	4,698,400	12,013,900
Alberta	34,300,000	12,940,200	47,240,200
British Columbia	18,504,000	40,953,000	59,457,000
Western Canada Total	60,119,500	58,591,600	118,711,100

TABLE III

Coal Reserves of Western Canada by Rank and Province
Low and Medium Volatile Bituminous
(000's of short tons)

Province	Proven & Probable	Possible	Total
Alberta	20,602,300	7,366,500	27,968,800
British Columbia	17,718,000	40,480,100	58,198,100
Rank Total	38,320,300	47,846,600	86,166,900

TABLE III (cont'd)

High Volatile Bituminous
(000's of short tons)

Province	Proven & Probable	Possible	Total
Alberta	6,278,600	3,043,700	9,322,300
British Columbia	146,000	172,900	318,900
Rank Total	6,424,600	3,216,600	9,641,200

Sub-Bituminous
(000's of short tons)

Province	Proven & Probable	Possible	Total
Alberta	7,419,100	2,530,000	9,949,100

Lignitic
(000's of short tons)

Province	Proven & Probable	Possible	Total
Saskatchewan	7,315,500	4,698,400	12,013,900
British Columbia	640,000	300,000	940,000
Rank Total	7,955,500	4,998,400	12,953,900
GRAND TOTAL	60,119,500	58,591,600	118,711,100

V Distribution of Coal Measures

The coal deposits of Western Canada lie in natural geo-graphical regions extending from Saskatchewan through Alberta westwards to the Pacific Coast of British Columbia and northwards through the Yukon Territory to the Arctic coast. The coal measures are vast in areal extent. However, exploration is costly on account of logistic features such as difficult terrain and lack of net works of roads and railroads. As a result many of the deposits are at present inaccessible and pre-production and development costs are therefore high.

Saskatchewan

Plains Region

Coal, of lignitic rank, is widely distributed across the extreme southern part of Saskatchewan and the deposits are on the northern fringe of a large basin centered in North Dakato, U.S.A. The coal is contained in the Ravenscrag Formation of Early Tertiary (Paleocene) age.

In the Estevan area, eight seams are known to occur and extensive stripping operations are continuing by working mainly on the upper four seams which normally range from 5 to 10 feet in thickness and lie within a depth of 200 feet below ground surface. The coal beds are relatively flat lying, the strata having a regional dip to the south-east at 25 feet to the mile. Although glacial erosion has completely removed some of the upper coal seams in certain areas they are readily outlined by exploratory drilling in advance of mine development.

Alberta

In the Province of Alberta, extensive drilling combined with geological mapping has enabled regional trends to be established. The progressive increase in rank of coal westwards from the Plains Region through the Foothills to the Mountain Region is shown in Figure 2. The increase in rank is accompanied by age of the rock formations. Likewise, the deformation becomes more intense in the Foothills and Mountain Regions. Coal bearing strata underlie a great area in the Province and each of Alberta's three physio - graphic domains (Plains, Foothills and Mountain Regions) contain different types of coal.

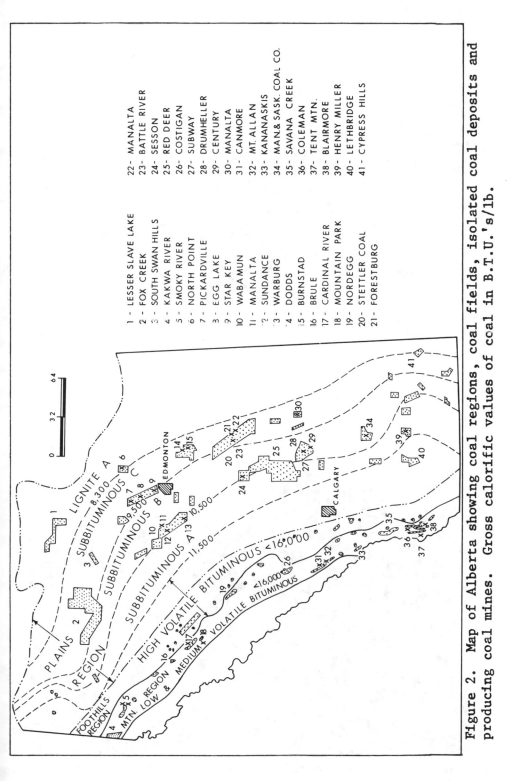

Figure 2. Map of Alberta showing coal regions, coal fields, isolated coal deposits and producing coal mines. Gross calorific values of coal in B.T.U.'s/lb.

1 - LESSER SLAVE LAKE
2 - FOX CREEK
3 - SOUTH SWAN HILLS
4 - KAKWA RIVER
5 - SMOKY RIVER
6 - NORTH POINT
7 - PICKARDVILLE
8 - EGG LAKE
9 - STAR KEY
10 - WABAMUN
11 - MANALTA
12 - SUNDANCE
13 - WARBURG
14 - DODDS
15 - BURNSTAD
16 - BRULE
17 - CARDINAL RIVER
18 - MOUNTAIN PARK
19 - NORDEGG
20 - STETTLER COAL
21 - FORESTBURG

22 - MANALTA
23 - BATTLE RIVER
24 - SESSON
25 - RED DEER
26 - COSTIGAN
27 - SUBWAY
28 - DRUMHELLER
29 - CENTURY
30 - MANALTA
31 - CANMORE
32 - MT. ALLAN
33 - KANANASKIS
34 - MAN.& SASK. COAL CO.
35 - SAVANA CREEK
36 - COLEMAN
37 - TENT MTN.
38 - BLAIRMORE
39 - HENRY MILLER
40 - LETHBRIDGE
41 - CYPRESS HILLS

Looking north towards No. 5 Pit, Grassy Mountain, Crowsnest
Pass, Alberta.

Coal seams, Grassy Mountain, Crowsnest Pass, Alberta.

Seam No. 2 - Pit No. 5, Grassy Mountain, Crowsnest Pass, Alberta.

Showing folded coal seam and sediments, Grassy Mountain, Crowsnest Pass, Alberta.

1. Plains Region

Sub-bituminous coal occurs in laterally persistent coal zones associated with Upper Cretaceous Belly River Group and Horseshoe Canyon Formation, and with the Upper Cretaceous-Lower Tertiary Paskapoo Formation.

True seam thicknesses vary up to 12 feet and except for a slight southwesterly dip into the Alberta syncline, the seams lie substantially flat and tend to be lensy. Regional dips range up to 20 feet to the mile. All of the formations are partially eroded resulting in reduced thickness and occasional removal of one or more seams. The bulk of current production is supplied from strip mines in the Halkirk-Forestburq and Wabamun Lake Districts.

2. Foothills Region

High volatile bituminous coal is associated with the Cretaceous Wapiti Group as well as with the Belly River and St. Mary River Formations. The reserves are considerable and plans are being developed to satisfy the needs of Ontario for generation of thermal power by utilizing the Foothills coal. A large number of isolated fields including the Coal-spur, Turner Valley and Cowley deposits occur within the Foothills region.

3. Mountain Region

Low and medium volatile bituminous coals occur in the Kootenay, Luscar and Commotion formations which are of Upper Jurassic to Lower Cretaceous age. In the Foothills and Mountain Regions the coalbearing strata are highly folded and generally faulted, so that repetition of rock sequences is common. As a result individual seams which normally average 8 to 10 feet have been locally thickened to as much as 40 to 50 feet. This is well illustrated in the cross-section of Luscar coal seam (Figure 3) and the Kootenay coal seams at Grassy Mountain (Figure 4). Typical folds and fault are present at Smoky River (Figure 5). The major folds and thrust-faults are aligned northwesterly following the trend of the Rocky Mountains.

Producing mines in this region are located at Coleman in the Crowsnest area, Canmore in the Cascade area, Luscar (Cardinal River Colleries) in the Mountain Park area and Grand Cache in the Smoky River area, the greater part having good coking qualities. The Mountain Region comprises a large area of unexplored land which is considered to contain the largest resource of coking coal in Western Canada.

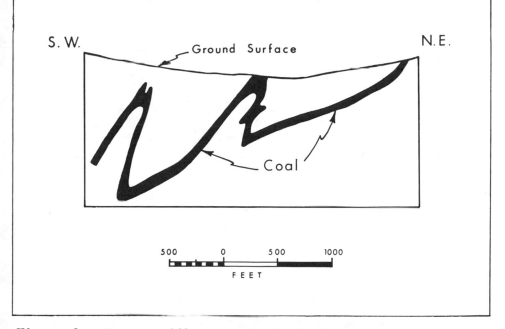

Figure 3. Luscar, Alberta. Geological cross-section showing accordion type folds in coal seam.

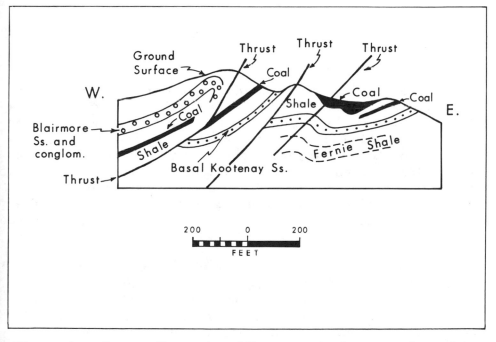

Figure 4. Grassy Mountain, Alberta. Coal seams shown in
relation to shales and sandstone marker beds.

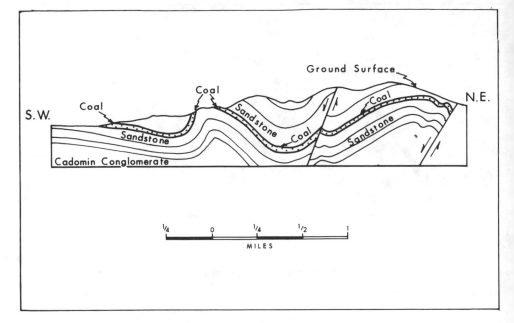

Figure 5. Smoky River Area, Alberta. Geological cross-
section showing coal-bearing Luscar Formation resting on
Cadomin conglomerate.

British Columbia

1. Mountain Region

The Kootenay Formation is well represented in the Rocky
Mountains in British Columbia by the Crowsnest and upper Elk
River coalfields. Based on inconclusive fossil evidence the
Kootenay Formation has been placed in the Jurassic, although
it may be of Lower Cretaceous age, and therefore equivalent
to the Blairmore and Luscar Formations of Alberta. The strat
are highly folded and faulted and as many as 15 to 20 coal
seams may be present with true thickness ranging from 8 feet
to 15 feet. The coal fields and coal areas of British Columb
are shown in Figure 6.

Isoclinal folding and thrust faulting have increased the
apparent thickness in many areas to as much as 45 to 50 feet.
Correlation is assisted by the presence of a basal marker –
the Kootenay Sandstone. Erosion by unconformities and
depositional thinning are common in southeastern British
Columbia.

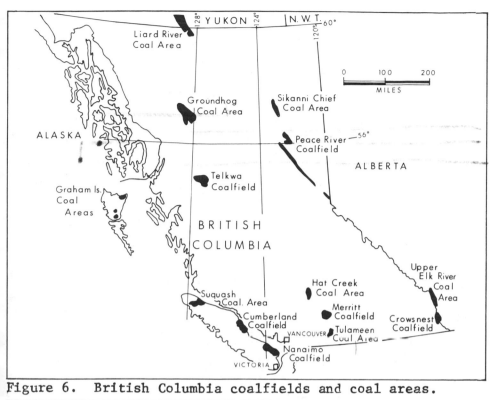

Figure 6. British Columbia coalfields and coal areas.

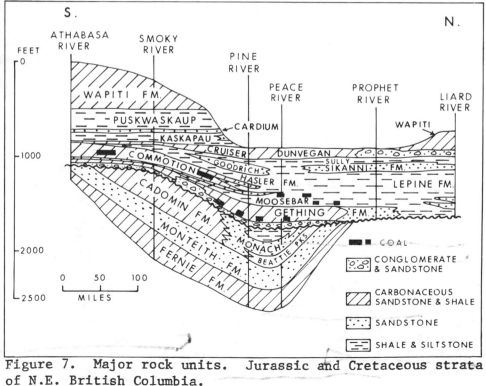

Figure 7. Major rock units. Jurassic and Cretaceous strata of N.E. British Columbia.

The coal quality is very good with both coking coal and thermal coal. It is classed as low to medium volatile bituminous in rank. The coking coal from the Crowsnest and Upper Elk River Fields is characterised by high ash content but is nevertheless acceptable to the steel works on account of its good blending qualities. Output comes from large open-pit operations of Kaiser Resources Limited at Sparwood and Fording Coal Limited on the Upper Elk River. Kaiser Resources also operate a hydraulic mine at Michel.

In the Peace River area the coal measures include the Gething Formation which is equated stratigraphically with the Luscar and Blairmore Formations in Alberta. The overlyir Moosebar Formation and Gates Members of the Commotion Formation are also productive. The fields shows considerable promise for production of coking and thermal coal with the deposits of Sukunka and Quintette Mountain having been thoroughly tested in recent years. The coal deposits in the Peace River are largely low-volatile bituminous and much of the coal possesses good coking qualities.

Thrust faults commonly follow the coal seams and become parallel to the bedding as for example at Sukunka. Although roof support difficulties are anticipated, proper application of rock mechanics will undoubtedly solve this problem in underground operations. Facies changes, non-sequence and lensing out of formations have affected the coal deposition, and as a result, it is difficult to establish reliable means of correlation in some areas of the Peace River. In the Foothills of the Rocky Mountains in British Columbia strong facies changes are evident in the Cadomin and Gething Formations (Figure 7).

As in highly productive coal fields of southwestern Alberta and the Rocky Mountain area of southeastern British Columbia, the Peace River coal seams are in many places highly disturbed by thrust faults, normal faults and intense folding. In almost every case it is almost impossible to lay out a comprehensive mining pattern to enable maximum extraction rate to be achieved.

2. Interior Region

The Telkwa and Groundhog fields occur in basins of Lower Cretaceous age in central British Columbia. They have been tested by adits and diamond drilling but do not have potenti for large tonnages. The coal in the Telkwa District is classed as low-volatile bituminous whereas the Groundhog

Coal Seam No. 9 at Exploratory Adit No. 9, Line Creek Ridge.
Crowsnest Industries Limited, Elk River, B.C.

Coal Seam No. 8, Line Creek Ridge.   Crowsnest Industries
Limited, Elk River, B.C.

field carries low-volatile coals, approaching semi-anthracite
in rank.

Tertiary coal deposits are widely scattered throughout
British Columbia where the Hat Creek, Princeton, Merritt
and Tulameen deposits occur in fairly large basins and the
coals range in rank from lignite to bituminous. Preliminary
plans have been prepared to utilize the Hat Creek coal for
large scale generation of electric power. The coal is sub-
bituminous B in rank.

3. Coastal Region

The coal deposits of Upper Cretaceous age on Vancouver
Island and Graham Island are classed as high volatile "A"
bituminous. They were the first deposits extensively worked
in British Columbia and estimated reserves are low.

Yukon Territory

The coal resources of the Yukon Territory have not been
adequately tested except for small scale mining for local
heating requirements. The deposits are extensive and widely
distributed and are largely of Lower Cretaceous and Tertiary
ages. The Lower Cretaceous coal deposits are the most important and are confined principally to the drainage basins
of the Lewes River and have been mined on a small scale at
Tantalus in the Carmacks district.

The coal is classed as high-volatile sub-bituminous in
rank. Further deposits occur to the south of this district
in the Laberge, Aishihik and Whitehorse areas, and to the
north in the Artic coast area.

The coals of Teritary age are all of lignitic rank and
occur in mineable seams in the Dawson, Ogilvie, Kluane,
Dezadeash and Watson Lake districts and also in the Bonnet
Plume district. The coal fields and potential coal areas
of the Yukon Territory are outlined in Figure 8. Preliminary
estimates of the coal reserves of the Yukon Territory indicat
a total of nearly 2,000,000,000 tons of probable and possible
mineable coal.

Acknowledgements

The writer wishes to give thanks to Dr. Hubert Gabrielse,
Geological Survey of Canada for assistance in editing the pap
and to Mr. Alan Kemp of Canex Placer Limited for drafting
the illustrations.

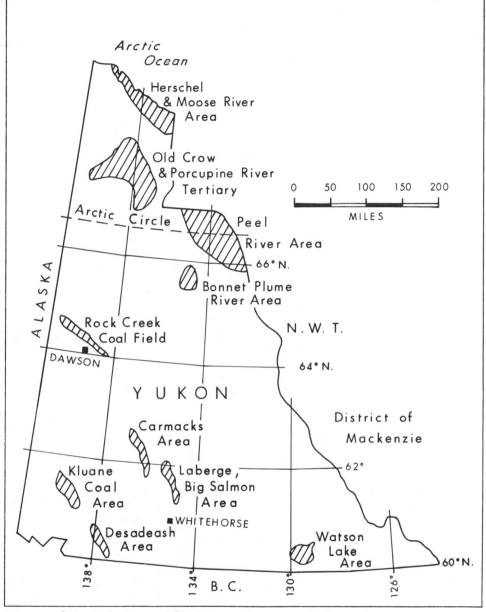

Figure 8.   Yukon coalfields and potential coal areas.

REFERENCES

1. Dowling D.B.,   "Coal Fields of British Columbia"
                         Memoir 69 Geol. Surv., Canada, Ottawa
                         1915

2. McLearn F.H. and   "Some Coal Deposits of the Peace River
    Irish E.J.W.,   Foothills B.C.," Geol. Surv. Canada
                         paper 44-15, Ottawa, 1944

3. Mackay B.R.,   "Coal Reserves of Canada" Geol. Surv.
                         Canada, Ottawa, 1947

4. Stott D.F.,   "Stratigraphy of the Lower Cretaceous
                         Fort St. John Group and Gething and
                         Cadomin Formations, Foothills of Northe
                         Alberta and British Columbia,"Geol.
                         Surv. Canada Paper 62-39, Ottawa 1963

5.   "Proceedings of the 22nd Canadian
                         Conference on Coal", Vancouver, B.C.
                         September 29, 1970 Department of Energy
                         Mines and Resources, Ottawa

6. Campbell J.D.,   "Alberta Foothills Coal," Research
                         Council of Alberta, October 1, 1970

7.   "Geological History of Western Canada,"
                         Alberta Society of Petroleum Geologists
                         Calgary, Alberta, Canada, December 1964

8. Latour B.A. &   "Preliminary Estimate of Measured Coal
    Chrismas L.P.,   Resources including assessment of indi-
                         cated and inferred resources in Western
                         Canada," Geology Surv. Canada Paper
                         70-58 Ottawa, 1970

9. McCurdy R.,   "Structural Geology of Cardinal River
                         Mine, Alberta" 75th Annual Convention
                         of Can. Instit. of Mining and Metallurg
                         Vancouver, B.C., April 17, 1973.

10.   "Proceedings of the First Geological
                         Conference on Western Canadian Coal,"
                         Edmonton, Alberta, Research Council
                         of Alberta Information Series No. 60
                         September, 1972.

REFERENCES

11. Stott D.F., "Lower Cretaceous Bullhead Group between Bullmoose Mountain and Tetsa River, Rocky Mountain Foothills, Northeastern British Columbia," Geol. Surv. Canada Bulletin 219, Ottawa, 1973

12. Review of the Alberta Coal Industry, 1973 Energy Resources Conservation Board Calgary, Alberta, Report 74E, March 1974.

Note: The Coal Reserves of Western Canada, indicated in Tables II and III are based on estimates by B.A. Latour and L.P. Chrismas and Figure 7 is based on a cross-section by D.F. Stott of the Geological Survey of Canada.

20

FINANCIAL CONSIDERATIONS IN EVALUATING

NEWLY DISCOVERED COAL DEPOSITS AND VENTURES

John K. Hammes
Vice President - Metals & Mining Department
Citibank, N. A.

INTRODUCTION

In the past several years, I have increasingly found myself involved in discussions regarding the financing of international coal projects. While most of those projects have been relatively early in the exploration phase, the discussions often assume a viable project and one that is entering the development phase. Thus, sources of loan funds and the structuring of finance for an assumed capital expenditure and cash flow profile are the principal topics. Seldom is the financing of exploration or are other financial considerations in the exploration and evaluation phase discussed with commercial lenders, and yet the financial aspects during this phase should not be dismissed as unimportant. Financing decisions may add to exploration costs, and contractual negotiations during this phase may affect the terms, cost and availability of finance should exploration be successful. Clearly, the primary objective during the exploration phase is the finding of coal reserves, but let me remind you that many known occurrence of coal and other mineral resources are not being developed as quickly as possible partly because governmental, finance and other external environmental factors introduce uncertainty or bring about delays.

Let us then examine financial consideration during the exploration phase in the following sequence. First we will look at financing the cost of exploration itself. Second we examine some of the special financial problems in evaluating international projects. Finally, we look at work contracts or concessions agreements and at coal supply contracts both of which are most often negotiated during the exploration phase and the provisions of which can have a significant impact on the development financing.

FINANCING EXPLORATION EXPENDITURES

I was recently talking with another commercial banker who had noticed the topic of my paper for this meeting and who asked how a paper could be written on the subject. Of course, he was thinking strictly about the fact that the burden of the cost of exploration is carried by the equity holder and is not a bankable risk. Therefore, the banker could not contemplate financing exploration. This is true in the sense that it is outside the risk tolerance of commercial bankers. Commercial bank lending is based either on the financial strength and past record of the borrower or on some form of security which, in the case of new ventures, is usually backed by contractual obligations of one or more financial strong parties. Financing exploration where loan repayment is dependent on cash generated as a result of the success of exploration does not meet these standards. Commercial banks and other lenders do provide finance to exploration subsidiaries of creditworthy companies but loan repayment is normally assured by the parent company.

One obvious consideration is that of raising local equity financing either through public offerings or negotiated with existing private companies. Many governments require substantial local participation in new private sector projects either by dint of direct legislation or indirectly through pressure applied by investment and administrative policies.

Unfortunately, in most developing countries the equity market is small, and this, together with the small appetite of local investors for exploration risk, reduces the likelihood of raising substantial local equity. Where private company funds are available the penalty for raising funds during the exploration stage is often an excessive dilution of ownership. Further, the exploration company will, in the end, bear the burden of arranging finance to the extent of carrying the interest of the local investors without adequate compensation. Most often the public shareholders or local investor cannot provide sufficient additional equity and credit strength to assure financing their proportionate share of development costs. In this event, the foreign company may find no alternative but to finance the entire cost of development. Usually, it will find government regulations and other pressures preventing it from redeeming equity or obtaining an adequate return for providing this disproportionate share of financing.

The incentives of resource development to governments or of assurance of critical raw material supply to consumers has induced these groups to provide finance on favorable term for project development. I am not aware of any specific programs for this type of financing of coal exploration, however governmental or consumer financing of coal exploration by non-affiliated private sector companies should not be overlooked. Even where such exploration funding programs have been formally established they often allow for substantial administrative freedom of action in determining the amount and terms of financing. In addition, such programs come and go depending on the political party in power, the current level of the economy, or the caprices of decision makers. The Brazilian agency, CPRM, which operates as an exploration company on its own account provides exploration financing up to 80% of the funding of approved programs for a variety of commodities. Finance is provided either "without risk" or "with risk". Under the first provision, the exploration company must repay all the financing with interest, and is required to put up a bond to ensure that the loan is repaid. Under the

"with risk" financing, the borrower must have a prospect
evidencing mineralization and must submit a budgeted
exploration program for approval. CPRM is repaid only
if a mine is developed and the amount repaid is determined
by a risk schedule depending on the commodity and the
location. While this sort of program benefits local
prospectors and companies, most multinational companies
engaged in international coal exploration will not find the
terms or the effort in negotiating participation in such
programs worth the financial advantage obtained.

I should comment at this point that much of this paper is
devoted to financing and financial consideration in international exploration where the exploration company is owned
in whole or part by a company which is not exploring in
its own country nor receiving substantial income in the
currency of the country where it is exploring. For this
reason, the effects of foreign exchange rate fluctuations
and government exchange regulations are a major
consideration.

If a one or two year exploration budget has been
approved for a project in a country whose currency is
expected to undergo revaluation in the near future, then
it might be advisable to buy local currency units immediately to meet the future budgeted level of expenditures.
Purchasing local currency at later dates as required to
meet exploration expenditures would result in a greater
cost if the local currency has revalued relative to the
parent company's currency. On the other hand, if the
exploration is being carried out in a country whose
currency is expected to devalue it might be prudent to
defer injection of parent funds or to borrow local currency
in anticipation of devaluation. In the latter case, local
currency would be purchased after devaluation and the
local currency loans repaid with cheaper local currency.
In reality, the complexity of forecasting exchange rate
movements, the relative borrowing costs in various
money markets, the parent company's foreign exchange
position and a host of factors make this funding a complex
and difficult process which should be left to the treasury
department and those who manage the foreign exchange

position.

Yet another consideration is the alternative of providing funds to an exploration subsidiary in the form of equity or shareholder debt. The form may be important because of the future impact of exchange controls and also because of tax consequences. Controls may preclude or limit the remittance of dividends and fees, or may be applied to the repatriation of capital. Similar restrictions may apply to the remittance of interest on shareholder loans or to the repayment of these loans. Future project financing in the event the exploration is successful may require a minimum level of true equity because interest payments are allowed by the government on a limited portion of the total investment.

Third party debt financing guaranteed by the parent company is less common but may be preferred in the event sovereign risk insurance is available for that form. Alternatively, specific risk insurance covering expropriation or currency convertibility may be available on equity invested in an exploration subsidiary. Another situation which is seldom encountered is the availability of blocked currency at a favorable exchange rate. Exploration in Indonesia has been financed with low cost rupiahs sold to exploration companies by foreign creditors at a discount from prevailing exchange rates. These rupiahs became available under a Debt Investment Conversion Scheme (DICS) which exists for the benefit of pre-1966 foreign creditors holding claims against Indonesia. These creditors, willing to make investments in Indonesia or to transfer the claims to investors in Indonesia, are paid in rupiahs in settlement of the claim.

It is difficult to formulate a check list or systematic approach to evaluating the alternatives or, for that matter, to identify all the financing alternatives for a given exploration project. However, some limited commitment of time and attention should be given to the assessment of alternatives even at this early stage in the history of a mine.

FINANCIAL CONSIDERATIONS IN EVALUATING NEW COAL VENTURES

Mine valuation has long consisted of a very scientific analysis of a set of numbers produced by artisans and sprinkled with a bit of commodity price speculation. International mine valuation is further complicated by the addition of a new set of risks and uncertainties which further impair the capital budgeting decision. Much of this additional uncertainty can be defined in terms of financial risk.

The calculation of economic returns, for example, the DCF rate of return on equity, requires that sales revenue, costs and debt repayment all be translated into a common currency (normally the currency in which the parent company reports its financial results) despite the fact that the actual flows of cash occur in several different currencies. This requires that future rates of exchange be forecast and introduces a new risk element. In addition, the reinvestment assumption which has been criticized as a shortcoming in DCF analysis becomes a more real concern in international projects because limits on repatriation of profits and requirements for additional local investment definitely limit the reinvestment opportunities.

Much has been written on time value of money theory, and as an ex-devotee of net present value, I feel compelled to suggest that less emphasis should be placed on these calculations as a project evaluation tool. The following considerations should be given more weighting by multinational natural resource companies:
1. What is the project's position on the supply curve? - where will the cost structure of a proposed project place it competitively?
2. How inflation proof is the project? - Outlook of inflation's effect on the project versus remainder of the industry.
3. Backward integration objectives - e.g. What is the hurdle rate of return for a steel company seeking a defensive investment in a captive source of

metallurgical coal?

4. Effect on shareholder reported earnings - What will the project do for the company's earnings per share or reported profits in project year 5 or year 10 ?

5. Expansion Potential - Are reserves and cash flows sufficient to support rapid expansion and thereby an attractive pattern of reported earnings growth?

6. Reduction of sovereign risk exposure concentration - accomplished through diversification into other countries.

7. Fulfilling other corporate goods - e.g. Attaining coal mining experience and/or reserve position looking ahead to coal gasification. Establishing a "missionary investment" to position the company in a country by developing a smaller project albeit one with smaller returns.

8. Financeability - Can the project be more easily financed than other alternative projects?

There are other considerations which could be added to the list including the case where no other viable alternative projects are available irrespective of rate of return. I would like to comment further on one consideration, that of the financeability of the project. Such considerations as total magnitude of capital investment, risk profile of the project, and cash flow margins impact the debt/equity ratio and total debt funding that can be achieved in the world financial markets. More than one project has passed all the hurdles prior to the development stage only to find financing unattainable on an acceptable basis to the project sponsors. I suggest the approach of estimating imaginary or model coal reserves in order to determine target criteria for exploration should be extended to include an appraisal of the difficulty in financing development of the target.

We have already touched on certain of the financial risks particularly encountered in international ventures and we might elaborate on how those might impact a project. If inflation is viewed largely as a condition brought about by monetary policy and as one having consequences closely allied to those of exchange rate movements, then inflation and exchange rate risk as well as exchange controls and other measures prescribed for correcting balance of payments problems should be included in our discussion.

Decreases in future cash flow from those projected for a project may occur because of exposure of one or more elements of the cash flow equation to foreign exchange rate fluctuation. On the revenue side of the equation, a coal sales contract might denominate its selling price in the currency of the buyer or in another currency than that in which the coal operating company maintains its account. If that currency devalued relative to the mine operator's, and adequate protection was not written in the sales contract, then the operator would suffer a decrease in selling price for coal in terms of his currency. On the other hand, cash outflow for servicing debt could increase above that forecast if finance and the subsequent interest payments and repayment of debt were denominated in a currency which revalued relative to the operator's currency. One school of thought is that these risks are all part of the general problem of assessing the effect of inflation, and that margins will be maintained as costs and prices move up together. Actually, this is not a valid assumption and the exchange exposure created by the restraints of the currencies in which sales are denominated, finance is arranged, and costs are incurred should be considered in the evaluation. Recent history points to a trend of rapid production cost increases at mines in developing countries brought about in part by inflation coupled with government and labor union policies which hinder productivity increases. This suggests one might emphasize exploration for a coal reserve suitable for surface mining development as opposed to underground development based on a reduced exposure to inflation risk because surface mining is less labor intense. On the other hand, the magnitude of the financing problem would be increased because of the higher initial investment in surface mine development.

Another government action designed to spread inflations gains and losses equitably, encourage savings and decrease the need for high interest rates is indexing. Forms of indexing include adjusting the value of financial assets and adjusting wage rates for inflation. This can lead to a decrease in liquidity and an increase in costs and taxes. Evaluating the possible effect of indexing on future streams of cash flow is a difficult if not totally conjectural exercise.

An entirely separate type of foreign exchange exposure risk is that created through balance sheet exposure. This is a risk which is reflected by accounting losses which occur when foreign currency accounts are translated into the company's base currency. These losses may be true economic losses and not just accounting losses because of a continuous deterioration in the value of local currency and a permanent impairment in the value of a project. Irrespective of whether the losses are real or not, translation losses are reflected in the parent company's financial statements, and an evaluation of a new project might include an analysis of the effect the project might have on the future exposure position. The accounting principles and method of analyzing balance sheet exposure are complex and the likelihood of this having a substantial role in project evaluation is not sufficient to justify covering this subject in greater detail.

Other government policies and related financial considerations that may adversely affect a project and should be considered in the evaluation are those which adversely affect distribution of cash flow, those which effectively lower revenues, and those which increase financial costs.

Those that adversely affect the distribution of cash flow include:
1. Restrictions on the return of capital or repayment of shareholder loans
2. Restrictions on dividend remittance
3. Restrictions on the payment of interest on shareholder loans
4. Requirements for local reinvestment
5. Rationing of foreign exchange

Return of capital restrictions might limit the percentage of capital or shareholder loans which can be returned each year. Dividends remittances may be limited to a percentage of initial capitalization, and interest on shareholder loans to a maximum percentage or to a percentage fluctuating with a selected money market rate. Local reinvestment of the depletion allowance or some other amount determined on an accounting formula basis may be required. Remittances could be allowed at less favorable exchange rates than those applying to other commercial transactions. These are but some of the ways that the realizable worth of a venture or the equity rate of return can be reduced for shareholders in a project because of limitations on the distribution of the cash flow actually generated by the project. Current government regulations and historical policy should be carefully examined so that the existence and likelihood of future occurence of any such limitations can be determined.

Lower project sales revenue and decreased cash flows may also result from government policy. For example, the requirement that physical sales first meet the local needs for raw materials when considered along with the effect of government price controls could reduce the effective sales price relative to that obtainable in other markets. This might be a consideration in evaluating a metallurgical coal project in a developing country where an expanding steel industry is leading to increased coke consumption.

Increased financial costs usually occur whenever access to money markets is restricted thereby reducing financing alternatives. For example, restricting offshore foreign currency borrowings to those of a tenor longer than that otherwise necessary will add to interest costs since rates increase with the tenor of the borrowing. Regulations requiring the maintenance of deposits with the central bank calculated as a percentage of external borrowing or on some other formula basis is not an uncommon policy. Since the company has the use of only part of the borrowed funds, this effectively increases financing costs.

Yet another restriction, that limiting the importation of certain classes or items of equipment, can have an adverse effect on financing costs and even make financing much more difficult. One of the principal sources of project finance is supplier credit which ties financing to equipment purchases. This credit is usually provided at lower interest rates and for a longer tenor than is bank financing. To the extent that restrictions on importing equipment are imposed in favor of that locally produced, a significant source of lower cost fixed rate financing is eliminated.

In view of the difficulties introduced into the evaluation process by those factors concomitant to international projects and taking into account the continuing evolution of multinational business, there is a definite need for the introduction of financial considerations early in the screening of new projects and a need to reconsider the capital budgeting process itself.

FINANCIAL CONSIDERATIONS IN CONTRACTUAL NEGOTIATIONS

There are a variety of contractual obligations that are usually negotiated during the exploration phase and prior to new mine development, these include royalty or lease agreements, joint venture or partnership agreements, operating agreements,coal sales agreements, and concession or contract of work agreements. They spell out the various relationships between the owners of mineral rights, the mine owners and operators, and the government. The terms and provisions of these agreements affect project economics, the distribution of revenue, and the manner in which project risks are shared by various involved parties. In addition these terms and provisions are often of prime importance in their effect on the arrangement of development financing. Partnership agreements for example, often spell out the responsibilities of the shareholders to contribute equity, arrange finance, and backstop other financial obligations. Operating agreements are often a key factor in a lender's evaluation because they provide assurances that operating management will be supplied the project. Shareholder and operating agreements are more easily amended than are those with coal purchasers or governments, so we will not consider these further. Coal supply contracts and work contracts are key agreements in a lender's evaluation because they contain provisions that bear greatly on the assessment of economic risk and on the ability of the lender to establish an acceptable security interest in the project. We should examine the provisions of these contracts in further detail.

Coal Supply Contracts

The energy crisis and shortages of metallurgical coal have been reflected in coal supply contracts with terms very favorable to the coal producers. Contracts with full pass through of cost increases for the account of the purchaser are common and contracts for future coal supply have been written which allow for price increases to maintain margins as a percentage of price or to maintain a specified minimum rate of return to the operator. Lenders are particularly interested in those provisions that specify price, quantity and term, quality, delivery, and assignability.

In the United States, numerous long term contracts were entered into in the 1960s. Those contracts often provided for escalating base prices for wage rate and welfare payment increases, and for material and supply increases determined from published price indices. Coal operators and lenders viewed these as significant advancements and many new mines were planned and financed which might not have occurred without these contracts. Today, we have learned that most of those contracts did not provide adequate protection against escalating costs. Cost increases resulting from higher reclaimation cost, from declining productivity resulting in part from new health and safety laws, and from inadequate coverage of fuel and other supply costs could not be passed through to the purchaser. Today's contracts recognize the inadequacies of earlier contracts and most provide that all cost increases are reflected in increased prices. Sometimes these contracts also include price escalation in the event franchise payments or initial estimates of capital cost are exceeded. Thus, if the tax depletion allowance is reduced or eliminated, the coal price is adjusted on a formula basis to pass on the increased tax to the coal consumer. Similarly, the parties to the contract may have agreed upon a capital cost estimate with a provision that increases in the cost per ton of coal capacity over the estimate be passed on to the customer in accordance with an established formula.

In international transactions the degree of complexity in negotiating such contracts is increased by the difficulty in establishing indices and standards, and by the added risk of currency fluctuations. Under the old fixed dollar standard, contracts could be denominated in dollars and were fairly simple to administer with each party bearing the risk of parity changes in his own currency. With the adoption of floating currencies, the contractual provisions of contracts dealing with exchange risk has become more difficult. The more simple schemes denominate price in either the buyer's currency, the seller's currency or a third currency. If the seller's currency is used then the buyer bears the risk, and vice versa. Each party bears risk if the contract is denominated in a third currency. Other schemes involve a mix of two or more currencies. These may be weighted so that each party bears more of his own currency risk and less of that of other currencies. Yet another approach is the denomination of contracts in SDR's. Lenders view these provisions from the standpoint of economic risk and also the risk of the availability of currency for debt service. Where the sales proceeds are denominated in a currency other than that in which debt finance is arranged there is added risk that exchange will not be available or that the exchange rate will have moved against the contract currency such that adequate proceeds will not be available after conversion to service debt.

The quantity and term specified in the contract are important to lenders because they introduce uncertainty that the output will be sold. If the contract allows considerable flexibility in establishing minimum take provisions or allows considerable latitude as to the timing or rescheduling of deliveries, then an element of uncertainty is added. Also, the provisions regarding price and quantity may be quite favorable but where a reopening of negotiations is provided for at a date prior to that when debt will be repaid, then the value of the contract as security to the lender is reduced.

The conditions of delivery are also of interest to potential lenders. Ideally, the contract contains a "take and pay" provision which requires the purchaser to take delivery of the coal upon notification of availability, and failure to take because transportation is unavailable or because of a condition of force majeure does not relieve the buyer of responsibility to advance funds. Clearly, this puts a burden of risk on the buyer. This provision should not be confused with "take or pay" provisions which also pass the production risk onto the buyer. Lenders view take or pay contracts as providing assurance that minimum payments will be advanced to the operator regardless of the availability of product and irrespective of any condition of force majeure. Such contracts are very rare but have been arranged in vertically integrated companies to support borrowings by subsidiary companies where the purchaser holds a strong economic interest in the operating company. We have seen such contracts between public utilities and subsidiary coal mining companies.

Provisions allowing assignability of sales contracts and of the contract proceeds are also important. If limited recourse project finance is to be negotiated then it is important that the lenders can take an assignment as part of the security for the financing.

It would be misleading to imply that the negotiation of coal sales contracts with all of the aspects we have examined favorable to the operator assures that financing will be easily arranged. Such provisions are obviously difficult to negotiate and may require that the operator give up the potential of increased economic reward. For example, the operator may be required to dedicate or assign the entire reserve to the purchaser thereby foregoing some of the benefits of a future mine expansion at a time when the coal could be sold at better prices. In fact, contract provisions have allowed the coal purchaser an option to purchase the reserves or the operator's economic interest in the event the operator does not perform up to expectations. Finally, whether or not the contract allows for reopening or for arbitration in the event inequity is experienced, it is certain that renegotiation will occur should inequities occur.

Work Contracts

Although most countries have general mining, taxation and foreign investment laws and regulations, many new mineral projects are developed under ad hoc agreements. These agreements between foreign investors and host country governments have been called work contracts, concession agreements, production-sharing agreements etc., and they spell out how the economic benefits of the project will be shared and the various obligations of each of the parties. A thorough description of how these have been structured is beyond the scope of this paper, however, we can examine those provisions which specifically impact the financing.

Obviously, a negotiated sharing of income which does not allow for the dedication of a large percentage of cash flow to debt service in the early years of a project's life can be a severe hindrance on the level of debt which can be raised. This aspect of the negotiations reflects the relative bargaining strengths and objectives of the parties.

Import duties and withholding taxes on interest might properly be considered part of this negotiated sharing of economic returns, however, they affect both the magnitude of initial investment and the cost of financing. Major projects have often successfully sought relief from these duties and taxes by obtaining exemption from them in work contracts.

We have previously examined the need for lenders to negotiate security in the project through the assignment of sales proceeds. Work contracts that allow long term sales agreements and the assignment of the proceeds of those contracts so that the funds flow into trust accounts in which the lenders have an interest can greatly facilitate financing. This assures lenders first call on cash flow for debt repayment and may eliminate currency convertability risk.

Negotiations for the removal of restrictions on imports and of limitations on raising foreign or local debt should also be considered. These restrictions limit flexibility in obtaining financing, and they often are not apparent early in the project or they become a problem because of laws enacted after the date of signing of the work contract. Assurances should also be obtained that necessary approvals of financing plans will be granted.

Other provisions that come under critical observation by lenders are those which relate to the government's role in project management, the provision of transportation, power and other essential inputs to production, and the governmen's financial obligations particularly if they are a joint venture partner liable for providing funds for capital cost overruns or other deficiencies.

CONCLUSIONS

Coal financing, whether of the early stages of exploration or of project development,is becoming more difficult as projects become larger and more remote, and as environmental and other uncertainties increase. This problem becomes even more complex when multinational companies are exploring and developing projects on an international basis.

This paper has discussed some of the financing considerations which should be taken into account in the exploration and evaluation phase. It's purpose is to identify problem areas and to suggest that financing considerations should be brought into focus at an early stage.

Financial Guidelines for Exploration Activities

Financial Considerations in Evaluating Newly Discovered Coal Deposits and Ventures

- Financing Exploration Expenditures
- Financial Considerations in Evaluating New Coal Ventures
- Financial Considerations in Contractual Negotiations

Financing Exploration Expenditures

- Equity Risk - Not Bankable
- Sources
  - Local Equity
  - Governments and Consumers
  - Guaranteed Bank Loans

Considerations in Exploration Financing

- Equity Dilution
- Foreign Exchange Risk
- Form of Advances
- Sovereign Risk Insurance

Financial Considerations in Evaluating New Coal Ventures

- Calculation of Economic Returns

- Foreign Exchange Exposure and Inflation
    - Transaction Exposure
    - Balance Sheet Exposure

- Government Policy
    - Exchange Rate
    - Investment
    - Price

Capital Budgeting Considerations

- Time Value of Money Theory
- Project's Competitive Position
- Defensive Investment in Captive Source
- Reported Earnings Effect
- Expansion Potential
- Diversification
- Financeability

**Financial Considerations
in Contractual Negotiations**

- Royalty or Lease Agreements
- Joint Venture or Partnership Agreements
- Operating Agreements
- Coal Sales Contracts
- Contract of Work Agreements

Coal Sales Contracts

- Price Escalation Provisions
    - Currency Denomination

- Quantity
    - Take and Pay
    - Take or Pay
    - Force Majeure

- Delivery

- Assignability

Work Contracts

- Income Distribution
- Import Duties and Withholding Taxes
- Restrictions on Finance
- Assignment of Security

Section Eleven Discussion

S. GAZANFER: In evaluation studies one would normally take
into account the price of the coal and one would think that
the price would not change during the operating life of the
project. Also one would think that the expenditures would
not change and if revenue and expenditures do change, the
difference between the two will remain the same. In a world
of high inflation in places such as United Kingdom, Turkey
and other places, inflation really plays an important part
and I would like to ask Mr. Hammes, how would he go about
evaluating a project to get the inflation rate and, also in
deciding on this current rate, how he would include the
factor of inflation.

J.K. HAMMES: Well, that is a very good question in fact.
Yes, we always certainly are looking at the revenue side of
the equation and we are looking at mineral commodity prices.
I guess we tend to look at the historic trend lines and what
we tend to call the downside band around the longterm trend
line where the price might fall and give the company a
liquidity problem, or the new project a liquidity problem.
One of the problems in coal, and certainly coal more than
any other commodity, is that financings have been tied to
marketing agreements with base prices that have been escalating. One of the problems in coal and certainly on the
international side is there isn't a pricing mechanism, so
when we look at it I am not even sure that I should say to
you, well this price is the price on which we are agreed
that we will lend, because I don't know if there is enough
history for lenders to make that decision. I am faced with
that problem right now in a financing outside the United
States, where we are talking about a contract and what sort
of pricing mechanism will be used, will it be tied for
example to the oil price, will it float with the oil price
at some discount? I cover in fact, in the paper, this question
of inflation, of varying inflation rates in the short run
and the long run, of inflation rates in construction costs,
and whether prices and costs will move in tandem. It is not
necessarily true that the prices will move in tandem with
the margins. I don't know that we have got the time or even
if I could do an adequate job of explaining to you how we
have been trying to look at various inflation rates. I was
involved in the negotiations of this large copper project in
Zaire where that became a very significant problem and, we
concluded there just wasn't any mechanism to measure. We

ended up using, in terms of some contractual obligations, an index, but it certainly wasn't an index that was relative to the country and we used it in a pricing mechanism. But, it is a problem, we look at historical rates of inflation, we look at a country in terms of its balance of payments situation, because that gives us some idea of what is likely to happen in the country. We look at perhaps stripping equipment, we look at oil and we try and really take a hard look at the components of the costs side of any project and come up with our best estimates of what they are likely to be for that particular component. It is a very tough problem.

R.H. CLAYTON: (Watts Griffiths and McQuat Ltd., Canada):I would like to ask Mr. Hammes, looking at it from a banker's viewpoint, how loans to mining projects start up. Suppose you had three projects, one a coal mining project, one copper or potash say, and another a shoe factory, and they were all say 50 million dollar projects that would return say, 15% dcf. Which one you would prefer to put the bank's money into?

J.K. HAMMES: That is not fair. You should remember first of all that my background is in mining so I have a biased viewpoint. I would try to say that the bank had to do more in the mining business. I think I have to be frank with you, what leaders in general are really looking at is not that dcf rate of return but reliability of that cash flow and how likely it is going to vary. You know that in most mining projects it is going to be tougher to predict that reliability than it is in say, a manufacturing situation. In manufacturing our merchant bank goes out with a series of telexes to the banks around the world, saying how much do you want of this and they get back responses and often they allocate, because there is more than enough who want a share in the loan. In mining we can't do that, we go out to a relatively small number of banks that we go visit and we say how about taking a piece of this - you understand the mining business - this is a good project - let us explain it to you.

A.B. LAWRANCE: (R.W. Miller (Holdings) Ltd., Australia):Mr. Hammes, in your appraisal of financial considerations, is it common for you to introduce probability factors into the critical elements?

J.K. HAMMES: Well, that is an interesting question. Just before I came to the bank I started to do some work at Kennecott, looking at some projects using the sort of approach that David Hersch developed using Monte Carlo theory and risk analysis. I suggested we ought to do a little bit more of that at the bank when I joined it, but I found that I have never gotten on with the programme, I have been too busy. What we really end up doing is picking a few key areas and I would say we do a sensitivity analysis on key areas, but it is not a very sophisticated analysis and it is usually on anything about the process that is new: technology, mining costs, revenues, but it is a fairly rough cut and not a very scientific approach to sensitivity analysis.

Modern mobile drilling rigs insure faster penetration and more flexibility in moving from site to site. Photo courtesy Portadrill

SECTION 12

21

CONSUMER COAL CRITERIA AS A GUIDE TO EXPLORATION

Richard A. Schmidt
Technical Manager
Fossil Fuel Resources
Fossil Fuel Department
Electric Power Research Institute
Palo Alto, California

ABSTRACT

Exploration for coal resources and reserves will fall short of meeting consumer requirements unless coal quality factors are considered together with those pertaining to quantity. Standard coal analyses are useful and convenient, but are not universally available; even when available, such data do not necessarily represent the behavior of coals under conditions of use. As a result, deposits of coal-bearing rocks that otherwise may be targets for intensi exploration could prove to be of limited potential. An assessment of resource/reserve data from the standpoint of criteria for optimum use in electric utility combustion and conversion is presented as a guide to future coal exploration.

INTRODUCTION

Exploration for coal reserves is a rather simple activity in concept, yet it is intensively complex and intricate of detail in actual practice. Far from being a set of cut-and-dried activities, coal exploration as an enterprise is replete with uncertainties about conditions of origin of coal-bearing strata, subsequent metamorphasis of deposits through physical and chemical processes, and the influence of geologic time upon their evolution. Despite the efforts of generations of coal miners, engineers, and (more recently) scientists, the essence of coal remains an enigma although a wealth of empirical information is available. In large measure, therefore, knowledge employed to guide coal exploration results from a history of trial-and-error attempts to solve some particular problem. This situation, it is submitted, is to be expected because of our imperfect understanding of the processes of nature that created the coal deposits which we seek to locate and exploit. As a result, successful identification of workable quantities of coal is a major accomplishment. Nevertheless, it is the quality of coal which ultimately controls its utilization. In order to meet fully consumer requirements for coal, it is essential to have in mind from the initiation of coal exploration programs those criteria most important for principal uses.

The variability of coal character and properties is such that its utilization is essentially controlled by a set of empirical factors derived from practical experience. In view of this situation, it would be presumptuous indeed for this paper to attempt more than to place in perspective several of the key parameters of coal that are of concern to consumers as a guide to those designing exploration programs. Toward this end, a review of facts pertaining to the physical, chemical, and engineering properties of coal are described, followed by an assessment of United States coal reserves and potential exploration targets in light of consumer criteria.

COAL PROPERTIES

The physical, chemical and engineering properties of coal were determined over the years as a result of practical experiences in combustion and/or coking. Much of this information is empirical, and is transferrable to other applications only with difficulty. A brief discussion of each class of properties is given below.

Physical Properties

Numerous physical properties of coal have been studied by large numbers of investigators. Among the properties determined are x-ray diffraction, ultraviolet and visible absorption, reflectance, refractive index, infrared absorption, electron spin resonance, proton spin resonance, electrical conductivity, diamagnetic susceptibility, dielectric constant, sound velocity, density, porosity, strength, reflectance, caking properties, and heating value.(1) No attempt will be made here to repeat previous comprehensive discussions of these properties. Instead, this section will be limited to a brief review of the inherent physical properties that are of principal apparent importance to consumers; density and heating value. A description of derived engineering or mechanical properties of coal is presented in a later section.

Density – The density of coal measures its weight per unit volume. Because coal is a porous substance, it is difficult to determine its volume accurately. Accordingly, most measurements of coal density are apparent densities rather than true density. The apparent density curve passes through a minimum at about 85 percent carbon; thus, bituminous coals are least dense of any of the other members of the coal series (Figure 1).

There is a general correlation between density and ash content in coals (2), although the relation differs for various ranks. For bituminous coals, the ash contents for various densities are as follows:

Ash Contents vs. Density,
Bituminous Coal (3)

Density	Ash Content, Percent
1.3 – 1.4	1 – 5
1.4 – 1.5	5 – 10
1.5 – 1.6	10 – 35
1.6 – 1.8	35 – 60
1.8 – 1.9	60 – 75
1.9 and greater	75 – 90

These data are empirically determined for each coal rank, and are employed to guide design of coal preparation and utilization methods.

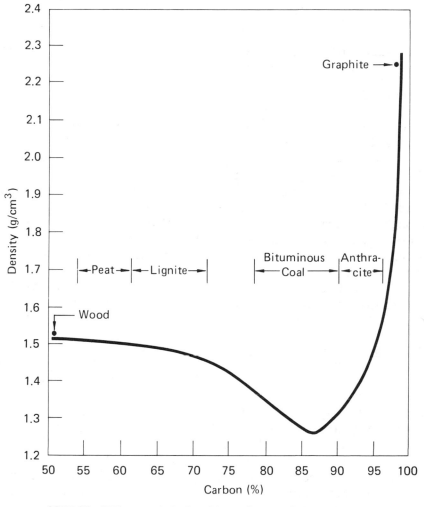

SOURCE: Williamson, I. A. Coal Mining Geology, Oxford University Press, London, 1967.

Figure 1   Relationship between apparent density and the rank of coal (as indicated by the carbon content)

Heating Value - The heating value of coals is typically expressed in British Thermal Units (Btu).(4) Heating value increases progressively with rank (expressed as carbon content) from peat through bituminous coals; a slight decrease in heating value occurs with anthracite (Figure 2). Heating value is, of course, a function of many physical coal parameters as well as its chemical properties.

The main chemical properties and their variations among different members of the coal types are summarized next.

Chemical Properties

The chemical analysis of coals has been a challenge to chemists for centuries. As a result of this previous work, analytical techniques have been developed and standardized so that results. may be correlated between different laboratories. Numerous accounts of this previous work have been prepared (5), this section cannot attempt to supplant such thorough and comprehensive compilations; rather, the presentation will be limited to a brief review of the principal chemical properties of coals that appear to be of primary importance in processing and utilization. The discussion covers classes of coal analysis, relations of principal constituents to rank, and content and occurrence of mineral matter and three elements.

Classes of Coal Analyses - Two main classes of coal analyses are routinely employed:(6) (a) Ultimate Analysis, which determines the elemental composition of the coal (particularly the carbon, hydrogen, sulfur, nitrogen and ash content; oxygen is estimated by difference. The ultimate analysis may be supplemented by measurements of trace elements present in either the organic or inorganic constituents of coal. (b) Proximate Analysis, which determines the presence of certain coal compounds (namely, moisture, volatile matter, and ash; fixed carbon content is estimated by difference). The proximate analysis is often supplemented by determination of sulfur content and estimation of heating value.

Ultimate Analysis - expresses coal composition in weight percentages of carbon, hydrogen, nitrogen, sulfur, oxygen and ash. Carbon content reflects the content of organic material, and hydrogen content includes water of mineral constitution in addition to that present in organic matter. Sulfur may occur as part of the organic matter, as sulfide minerals, or as sulfates. All nitrogen is present in the organic material.

614 COAL EXPLORATION

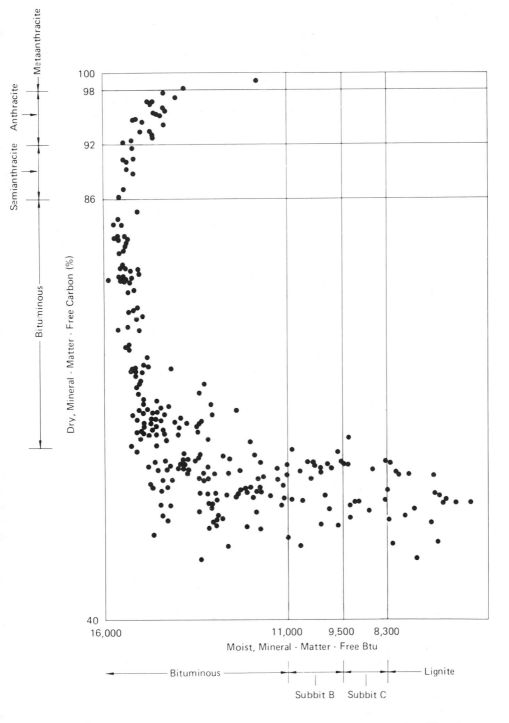

Figure 2 Typical U.S. coals graphed according to standard classification by rank

There is a continuous variation in chemical properties
with rank. For example, coals illustrate a progressive
increase in carbon content and decrease in oxygen content
with increasing rank. In contrast to the progressive and
direct increase in carbon content with increasing rank,
hydrogen content remains relatively constant at about 5
percent for lignites and much bituminous coals but decreases
to 3 to 4 percent for higher rank bituminous coals and
anthracites. The decreases in hydrogen content with increasing rank is related to the decrease in volatile matter
content.

Maximum nitrogen contents are about 1.7 percent for
bituminous coals, and less for coals of lower and higher
rank (Figure 3). Lignites, sub-bituminous coals, certain
bituminous coals, and anthracites show low sulfur contents
(less than 1 percent by weight) (Figure 4). Highest sulfur
contents are found in bituminous coals; a progressive
decrease in sulfur content with increased carbon content is
indicated for these coals. A number of minor or trace
elements are found in coals.

Analyses of trace elements present in low temperature
ash resulted in the recognition of four groups of potentially
volatile trace elements.(7)

(1)   Elements of greatest organic affinity that are
        concentrated in clean coal fractions: Ge, Be, B
(2)   Elements of least organic affinity that are
        concentrated in the mineral matter of coal: Hg,
        Zr, Zn, Cd, As, Pb, Mn, and Mo.
(3)   Elements associated with both organic and inorganic
        matter, but which tend to be more closely allied
        with the organic fractions: P, Ga, Ti, Sb, and
        V.
(4)   Elements found in both organic and inorganic
        matter, but which tend to be more closely asso-
        ciated with inorganic fractions: Co, Ni, Cr, Se,
        and Cu.

Ultimate analysis, however necessary to provide data on
the total elemental composition of coals of varying type,
is time-consuming and costly. As a result, ultimate analyses
are not performed regularly on most coals produced and used
in the United States; out of 255 coal samples for which
analyses were recently reported, ultimate analyses were
reported for only 39.(8) All samples reported indicated
proximate analyses, a more rapid and simpler procedure

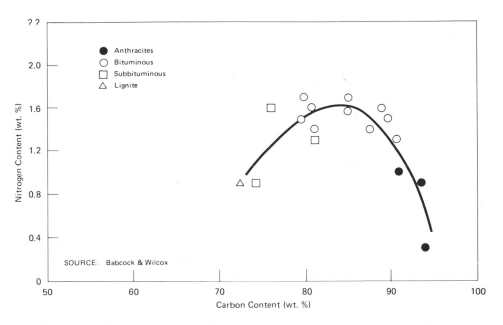

Figure 3 Relationship of nitrogen content to carbon content in coals
of varying rank

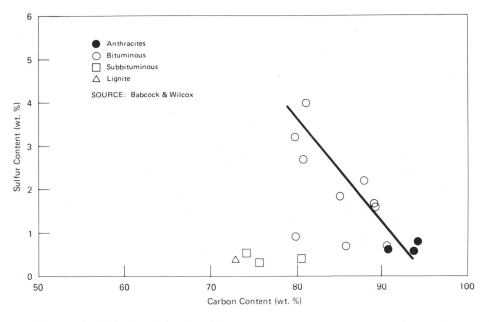

Figure 4 Relationship of sulfur content to carbon content in coals
of varying rank

which provides information about coals according to their response to heating under controlled conditions.

Proximate Analysis – characterizes coals in connection with conditions of utilization, by determining the distribution of products obtained by incremental applications of heat in a controlled atmosphere. Proximate analysis separates coal compounds into four groups (1) water or moisture; (2) volatile matter consisting of gases and vapors driven off during pyrolysis; (3) mineral impurities or ash; and (4) fixed carbon (9), the remaining non-volatile fraction of the pyrolyzed coal, obtained by difference following determination of the above groups.

The composition of the fixed carbon in all types of coal is "substantially all carbon," and "the variable constituents of coals can, therefore, be considered as concentrated in the volatile matter."(10) The heating value of the volatile matter is "perhaps the most important property as far as combustion is concerned."

Figure 5 shows the relationship of volatile matter to sulfur content for coals of varying rank. Sulfur is relatively independent of volatile matter for anthracites, lignites, sub-bituminous coals, and some bituminous coals. However, for the majority of bituminous coals included in this limited sample, the data suggest that higher volatile contents are accompanied by higher sulfur contents. Volatile matter is determined by heating the coal to 875° to 1050°C.(11) Sulfide minerals present in coal tend to decompose at elevated temperatures (12), and this could contribute to part of the volatile matter measured. Ash content appears to be relatively independent of fixed carbon content, perhaps reflecting variations in initial deposition of mineral matter in coals as well as subsequent alterations in such minerals or formation of new mineral species subsequent to deposition. In addition to various organic constituents, coals contain a variety of inorganic mineral compounds. Present in mineral matter may be (a) silicates of alkalis, calcium, magnesium, iron, and titanium; (b) oxides of iron and silica; (c) carbonates of iron, calcium, and magnesium (which may change to oxides or heating); and (d) sulfides of iron; and minor amounts of sulfates, phosphates, arsenides and others.

Engineering Properties of Coal

The physical and chemical properties of coal described

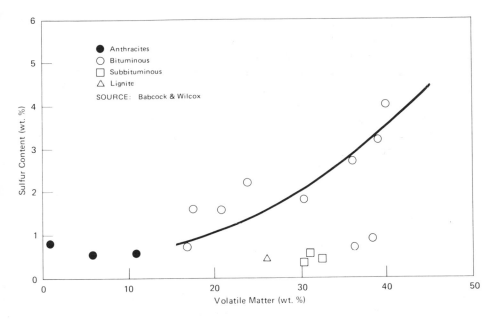

Figure 5 Relationship of sulfur content to volatile matter content in coals
of varying rank (proximate analysis)

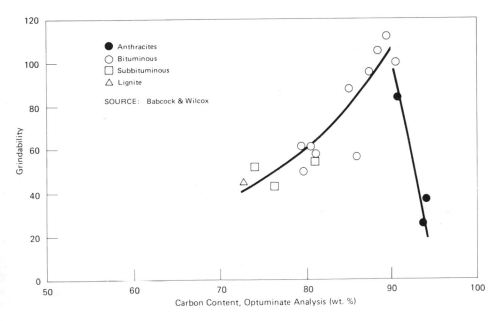

Figure 6 Relationship of grindability of coal to carbon contents for coals
of varying rank

earlier serve to distinguish among various coals and aids in establishing preliminary criteria in connection with design and performance parameters for coal utilization equipment and facilities. It is noteworthy that standard coal properties are typically determined on rather small samples of coal in contrast to the large volumes consumed by a commercial facility. Additionally, test conditions rarely duplicate those of actual operations, and the applicability of coal data obtained through such measurements to full-scale operation may be incomplete. Although the basic coal properties are useful, it is also essential to know more detailed data on coals under conditions of use. It is the engineering data for actual operations (largely empirical information for practical coals in well-defined utilization conditions on equipment that is derived from a history of trial and error iterations) that are of primary importance in assessing potentials for coal utilization. This section will present an overview of selected information on coal production factors; an introduction to sampling methods; engineering properties; effects of preparation on properties, and effect of coal properties on utilization conditions.

The trend in coal mining technology is toward more highly powered equipment that is capable of cutting any materials encountered. This more highly mechanized equipment produces a product that is both smaller in size and diluted by admixture of non-combustible material.(14) It was found that conventionally mined coal contained less total fines and less dust of aerofloat size than that produced by continuous mining equipment. Furthermore, generally speaking, higher rates of methane emission are found with finer coal particles.(15) From the standpoint of utilization, therefore, finer coal produced by continuous mining methods has lost more of its methane content than coal from the same seam mined by conventional methods. As noted in an earlier section, the heating value of coals is related in part to the content of combustible volatile matter; it would be worthwhile to investigate whether some part of the decline in heating value of utility coal in recent years is related to this phenomenon (and the growth of continuous mining).

Generally speaking, the end result of modern coal production is a product that is small in size, an intimate mixture of combustible and noncombustible material, and from which much of the inherent volatile matter has been lost. Coal utilization processes, accordingly, are presented with a material whose natural variability in essential properties has been accentuated and compounded in a relatively random manner as a consequence of its extraction from the earth.

Sampling

A number of sampling procedures are employed to determine coal characteristics. The procedures seek especially to identify deleterious constituents which are either inherent in particular coals or which become introduced through extraction. This information is indispensible for design and operation of facilities to use naturally-occurring coals, as well as for the establishment of coal treatment practices that seek to remove all or most of the undesirable constituents (or, failing that, to suppress them to a tolerable level).

The targets in coal sampling and analysis are to determine its primary characteristics under utilization conditions. Experience to date with various combustion systems indicates that coal grindability, moisture, and ash properties are controlling parameters. The following discussion deals with these factors in greater detail as they affect coal grindability, separation of impurities, and ash characteristics.

Size Reduction - The main approach to treatment of raw coal (16) is comminution, with the objective of liberating non-combustible constituents for greater ease in separation and segregation. The principles of coal breaking and crushing are well-described, together with a practical description of typical equipment.(17) The performance of this equipment is affected by the size and nature of the feed coal. As noted above, this will vary for different coals, different mining methods, and different end uses for which the coal is intended.

The hardness or softness of the coals is an important parameter which determines the relative ease of reduction of particle size. These characteristics are measured through a standard procedure that determines the relative ease of pulverizing coals in comparison to a standard coal having a "Hardgrove Grindability Index" of 100. Typical values of this index are shown in Figure 6. The grindability index is greatest for most bituminous coals and least for lignites and anthracites (bituminous coals are softer).

The grindability index varies greatly from seam to seam (and among samples from the same seam) because of the variations in impurities that are commonly present. Also, because impurities tend to segregate in certain size fractions, the grindability index of any given coal can vary from size to size. Furthermore, any segregation of coal from mineral matter during handling or other processing will result in changes to the grindability index.

Further, the grindability index for lower rank coals is influenced by moisture content. The index passes through a minimum at intermediate-range moisture contents (i.e., these are harder to grind). Thus, partial drying of high moisture coals may result in a product which is more difficult to grind. While it may be intuitively desirable to dry such coals, such efforts could result in even further utilization problems and costs by increasing the difficulty of grinding.

Clearly, the Hardgrove grindability index is merely an empirical tool that represents only those coals (or fractions) that are sampled. It is subject to significant variability, and if misinterpreted could result in serious operating impacts. While research is in progress to perfect hypotheses of coal breakage and distribution of particles, there seems to be no practical alternative to continued use of the index for at least the immediate future. One reason for this statement is the fact that equipment manufacturers publish Hardgrove grindability indices for their equipment.(18) The performance of each piece of equipment is thus reflected in empirical terms as well, and it would appear to be most difficult to translate into quantitative terms (assuming that a comprehensive expression could be developed). There seems no escape, therefore, from the continued use of empirical data on coal grindability. In fact, the present impetus toward rapid acceleration in coal utilization will place a premium on such knowledge and encourage others to develop expertise in such practices, for the simple reason that they are known to be effective.

The effect of grinding on coal quality cannot be generalized because of the differences in coal characteristics. For example, the ash content/ash fusion temperatures of one coal may be lowered by grinding and separations, while those of another may be increased. The principal benefit of grinding coal is to separate ash or mineral matter to realize reduced handling and shipping costs as well as to achieve higher boiler efficiencies.

Separation

Removal of impurities liberated from coal through grinding and size reduction is typically accomplished by mechanical methods employing gravity concentration. Most common impurities are heavier than coal, and will sink in an appropriate medium while the lighter (principally coal) fractions will float.

In addition to removal of ash-separation processes also are concerned with removal of sulfur-bearing minerals. The potentials for sulfur reduction through mechanical methods have been investigated for several coals by the U.S. Bureau of Mines.(19) It was found that less than 30 percent of the samples tested could be reduced to 1 percent or less total sulfur, although reductions of 50 percent were found in more than half the samples.

Finally, it is important to recognize that coal preparation to separate unwanted materials evolved mainly as a method to reduce transportation costs, only secondarily being concerned with quality factors.(20) Thus, the applicability of available preparation technologies to separation of sulfur and ash needs to be re-assessed to accomplish a new objective that is different from the original purpose. While reduction in coal transportation costs remains of obvious importance, the use of separation methods to gain better control over coal behavior under utilization conditions is likely to become of even greater impact. This point is illustrated by the following discussion of the characteristics of coal ash.

Ash Characteristics

Considerable attention has been given to the problems created by ash in coal-fired combustion systems. It was noted that "Most of the routine coal analyses, such as proximate, calorific value, etc., are of little value in estimating the severity of ash deposition."(21)

Because the nature and amount of ash in coal is of major concern to the design and operation of utility facilities, a number of engineering procedures have been developed to determine various empirical parameters relative to ash behavior.(22) Many of these parameters are oriented toward specific types of coal-fired units. A given coal may behave differently in different units because of inherent design conditions. As a result, the parameters must be

used in connection with a particular coal under stated
utilization conditions or the results will not be reliable
or meaningful.

Physical - Perhaps the most common method to determine
ash properties is to measure ash fusion temperatures. The
purpose of this test is to provide an indication of how
coal ash will behave in the furnace. This test is strictly
empirical, employing loosely-defined softening and fluid
points which are observed during deformation of a standard
cone-shaped coal sample under heating. The following four
temperatures are reported:

- Initial Deformation Temperature, at which the
  first rounding of the cone apex appears.
- Softening Temperature, at which the cone has been
  reduced in height so that height equals width of
  the base.
- Hemispherical Temperature, at which the height is
  reduced to half the width of the base.
- Fluid Temperature, at which the fused mass has
  spread out in a nearly flat layer.

Generally speaking, the ash fusion temperatures are less
for coals of lower rank (both under reducing and oxidizing
conditions). However, the variability in coal and ash
content and character is such as to require analyses on
individual coals.

Chemical - Most of the parameters compiled by ASME are
expressed in terms of ash composition (Table 1). The table
shows that each of the 17 empirical engineering parameters
listed can be determined from a knowledge of ash analysis,
another reason why more request sampling and determination
of coal/ash composition should be undertaken.

Coals may be treated as two broad groups based on ash
analysis; (a) coals with "Bituminous-type ash": ($Fe_2O_3 >$
$CaO + MgO$); and (b) coals with "Lignite-type ash": ($Fe_2O_3 <$
$CaO + MgO$). This criterion applies to all U.S. ranks of
coal regardless of source; generally speaking, however,
Eastern coals have bituminous-type ash while Western coals
have lignite-type ash. Because Eastern coals have been
used more extensively than those of the West, many (if not
most) of the empirical parameters have been developed for
utilization of these resources. It is not clear that these
parameters will be applicable to Western coals; indeed, the

Table No. I
Derivation of Fouling and Slagging Parameters for Coal-Fired Boilers

Parameter	Equation
1. Total Coal Alkali	$\% \ Na_2O + 0.6589 \ (\% \ K_2O) \ \times \ \dfrac{90 \ ash}{100}$
2. Total Ash Alkali	$\% \ Na_2O + 0.6589 \ (\% \ K_2O)$
3. Total Acid	$SiO_2 + TiO_2 + Al_2O_3$
4. Total Base	$Fe_2O_3 + CaO + MgO + K_2O + Na_2O$
5. Base/Acid Ratio	$\dfrac{Fe_2O_3 + CaO + MgO + K_2O + Na_2O}{SiO_2 + TiO_2 + Al_2O_3}$
6. Ferric/Lime Ratio	$\dfrac{Fe_2O_3}{CaO}$
7. Dolomite Percent	$\dfrac{CaO + MgO}{Fe_2O_3 + CaO + MgO + Na_2O + K_2O} \times 100$
8. Ferric/Dolomite Ratio	$\dfrac{Fe_2O_3}{CaO + MgO}$
9. Silica/Alumina Ratio	$\dfrac{SiO_2}{Al_2O_3}$
10. Silica Ratio	$\dfrac{SiO_2}{SiO_2 + Fe_2O_3 + CaO + MgO}$
11. Slagging Factor	Base/Acid Ratio x S
12. Fouling Factor	Base/Acid Ratio x Na_2O

"slagging factor" and "fouling factor" are not applicable to Western coals. Probably, it will be necessary to develop a set of new empirical parameters to deal with Western coals in combustion facilities. It seems likely, moreover, that a new set of parameters will probably be developed for coal conversion processes using coals from each region.

Typical data for selected parameters are shown by type of fouling and slagging behavior in Table 2. While it may be possible to obtain a rough idea of the behavior of individual coals by reference to such data, careful determinations are required to properly ascertain the likely performance of a particular coal.

These engineering parameters are of concern to both designers and operators of utility plants, as noted above. However, the present state of knowledge is such that it is impossible to estimate values for the parameters in terms of capital or operating and maintenance costs. Instead, these parameters generally are expressed in coal contracts or equipment specifications as limits for acceptable delivery or performance. Each manufacturer has its own set of approaches to these parameters, based on its practices for collection, organization and utilization of basic and applied data on coals and their utilization, contributing to the proliferation of parameters as new equipment is designed. "Parameters used to judge the fouling and slagging potential of coal ash are confusing because they are numerous and because their theoretical significance has not yet been thoroughly established."(23)

A further problem is that the inter-relationships between the several parameters are investigated infrequently, and potentially valuable information about coal behavior remains obscure. In an attempt to arrive at a better understanding of coal ash behavior, it was suggested by Weingartner and Ubbens that a two-component phase diagram could be constructed for acid and basic constituents in coal as a function of T_{250} temperature (temperature at which the ash viscosity curve reaches 250 poise). Figure 7 shows an example phase diagram constructed in this manner. While it is recognized that the actual system involved is much more complex than the simple two-component system depicted, this approach nevertheless has the promise of integrating information that is seemingly unrelated and moves in the direction of a more quantitative evaluation of the processes acting upon coal ash. It is clear that much work remains to be done to test this approach through further observation and through

Table No. II

Summary of Parameters Regarding Fouling and Slagging

| Parameter | Low | Fouling Type | | Severe |
		Medium	High	
$Rf = \dfrac{Base}{Acid}$ x % Na_2O	<0.2	0.2 - 0.5	0.5 - 1.0	>1.0
Na_2O content (%)	<0.5	0.5 - 1.0	1.0 - 2.5	>2.5
Total alkali on coal (%)	<0.3	0.3 - 0.45	0.45 - 0.6	>0.6
Chlorine on coal (%)	<0.2	0.2 - 0.3	0.3 - 0.5	>0.5
Ash sintering strength				
at 925°C = M Pa	6.89	6.89 - 34.47	34.47 - 110.32	>110.32
at 1700°F = psi	1000	1000 - 5000	5000 - 16,000	>16,000

Summary of Parameters Regarding Slagging - Slagging Type

		Slagging Type		
T_{250}				
°C	>1275	1400 - 1150	1245 - 1120	<1200
°F	>2325	2550 - 2100	2275 - 2050	<2200
$Rs = \dfrac{Base}{Acid}$ x % S	<0.6	0.6 - 2.0	2.0 - 2.6	>2.6

Source: ASME Research Committee on Corrosion and Deposits from Combustion Gases.

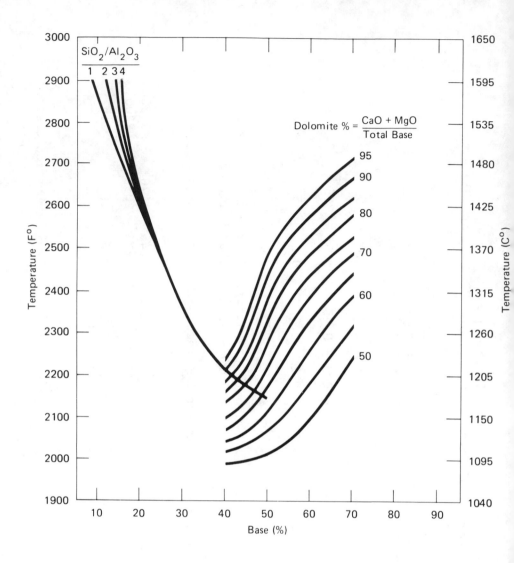

Figure 7 T_{250} Temperature versus ash composition

specifically designed experiments. It may be expected that revisions may be necessary as further data are obtained (through the same process of trial and error that led to our present set of parameters). Still, this is an encouraging start that appears quite promising in beginning to place engineering data on coal utilization on a sound theoretical basis. With such information, it may be possible to establish more effective separation and preparation methods so as to achieve coals of desired properties.

Consumer Criteria to Guide Coal Exploration

The preceding data illustrate the variability of coal properties, any of which could prove to be critical to utilization. There is no substitute for detailed analyses of coal to determine essential properties with precision; unfortunately, as noted, such analyses are performed infrequently.

A further limitation is that the variability in coal properties and the variety in coal utilization systems tends to frustrate attempts to employ simplifying concepts -- there are just too many exceptions. In consequence, there has been a tendency to deal either in special cases (which are not generally applicable) or in broad generalizations (which are not representative of real conditions). This has been particularly true during the development of coal-fired power generating plants in the United States. The evolution of power plant technologies has led to plants that employ pulverized coal; have capacities in the range from 400 to 800 MW; operate at pressures between 2400 and 3500 psig, superheat between 925°F and 1024°F, and reheat between 955°F and 1024°F. In addition, these plants are equipped with complex systems to minimize production or release of nitrous oxides and sulfur oxides, as well as high efficiency electrostatic precipitators or filters for particulate control. While there is still considerable latitude for design to meet the requirements of individual power systems, the relative commonality of modern power plant features enables the following gross identification of coal properties which are important to operations (Table 3).

Table 3
Consumer Criteria to Guide Coal Exploration

Coal Property	Consumer Preferred Value
1. Heating value	Uniform; consistent with specifications
2. Sulfur content	Less than 0.6 lbs. per MM Btu
3. Ash content	Less than 20 percent
4. Moisture content	Less than 15 percent
5. Alkali content	Less than 0.4 percent
6. Grindability	Hardgrove Index greater than 50
7. Ash fusion temperature	Less than 2200°F
8. Fouling index	Less than 0.5
9. T_{250}	Greater than 2300°F
10. Slagging index	Less than 2.0

The above assumes that sufficient reserves of coal having most of the preferred qualities will be available to furnish reliable supplies for the entire useful life of the power plant, avoiding the necessity for costly redesign and refurbishment to accept coals with a different combination of properties. A typical power plant of 1000 MW capacity will require nearly 4 million tons over a 40 year lifetime. Coal consumed at a power plant in the future will probably be cleaned, with losses of up to 20 percent experienced in the cleaning process; therefore the total coal production to meet the power plant demand will be about 200 million tons. With an estimated coal reserve recovery factory of 50 percent, the total coal reserve required to support a power plant of 1000 MW capacity throughout its lifetime would be approximately 400 million tons. Clearly, it will be no simple task to locate such large reserves of coal, especially those having a large number of the preferred properties.

Assessment of Reserve Data

The above data suggest that large coal reserves of coals having rather well-defined properties are required to meet electric power demands. An intensive assessment of U.S. recoverable coal reserves is in progress at EPRI. The following discussion is a summary of part of that work. For purposes of this paper, the estimated U.S. recoverable reserves of coals with differing rank was compared, by state, against average sulfur content per million Btu in an attempt to illustrate the application of one of the consumer criteria in recognizing exploration targets. In actual practice, of course, it will be necessary to take into account several of the criteria simultaneously.

Considering only recoverable reserves greater than 200 million tons (approximately the amount of production to supply one 1000 MW power plant for its lifetime) and sulfur content less than 0.6 pounds per million Btu, the following exploration targets are identified:

(a) Anthracite: Pennsylvania
(b) Bituminous: Colorado, New Mexico, Utah, Virginia, and Wyoming
(c) Sub-bituminous: Alaska, Colorado, Wyoming
(d) Lignite: Montana

If sulfur limitations are extended to 1-2 pounds per million Btu, the following additional exploration targets are identified:

(a) Bituminous: Alabama, Maryland, Montana, Tennessee
(b) Sub-bituminous: Arizona, New Mexico, Montana, Washington
(c) Lignite: North Dakota

If sulfur limitations are ignored entirely (implying that some other coal property is of principal importance to a particular consumer) the following additional exploration targets are identified:

(a) Bituminous: Illinois, Indiana, Kansas, Kentucky, Ohio, Oklahoma, Pennsylvania, West Virginia
(b) Lignite: Texas

Clearly, this summary assessment is only exemplar of an approach to the detailed analysis of the properties and quantities of individual seams required in order to design practical exploration programs and to evaluate their results in order to meet future coal requirements of principal consumers.

SUMMARY

A review was presented of data on coal physical, chemical properties, and engineering properties related to coal utilization. Much of the information that exists about coal is empirical, derived from trial-and-error experiences related to some particular application. The variability of

coals and related materials is so great as to frustrate
most attempts to derive unifying concepts of any but the
broadest kind. This places a premium on the collection and
transfer of empirical knowledge about coal character and
behavior in specific conditions of use. Present impetus
toward development of relatively standardized approaches to
coal-fired power generation facilities enables recognition
of a set of coal properties important to meet consumer
operating objectives. These data can be used, together
with available estimates of coal resources and reserves, in
the design of practical coal exploration programs and in
the evaluation of their results. Implementation of systematic approaches to coal exploration and development appears
to be essential if future coal demands are to be satisfied
with conservation of finite coal resources.

REFERENCES

1. See extensive description by Tschamler, H. and E. deRuiter, "Physical Properties of Coals" in Chemistry of Coal Utilization, Supplementary Volume, H. H. Lowry, Editor, John Wiley and Sons, New York, 1963, page 35-118.

2. Tschamler and deRuiter, op. cit.

3. Babcock & Wilcox, Steam, 1972, page 85.

4. A British Thermal Unit is the quantity of heat required to raise the temperature of one pound of water 1 degree Fahrenheit.

5. See, for example, the comprehensive syntheses of data prepared by Ode, W. H., "Coal Analysis and Mineral Matter," and Dryden, I. G. C., "Chemical Constitution and Reactions in Coal," both in Chemistry of Coal Utilization: Supplementary Volume; H. H. Lowry, Editor, John Wiley and Sons, New York, 1963, page 202-231 and 232-295, respectively.

6. U.S. Bureau of Mines, Dictionary of Mining, Mineral, and Related Terms, 1968, page 872.

7. Ruch, R. R., H. J. Gluskoter, and N. F. Shimp, "Occurrence and Distribution of Potentially Volatile Trace Elements in Coal," Illinois State Geological Survey, Environmental Geology notes, No. 72, August, 1974.

8. See "Analyses of Tipple and Delivered Samples of Coal Collected During Fiscal Year 1971," U.S. Bureau of Mines Report of Investigations No. 7588, 1972, page 20.

9. Actually, "the fact that coal on distillation gives a residue of impure carbon (coke) and volatile products (tar and gas) suggested the idea that coal contained free carbon, associated in some way with bituminous matter. The terms, "bituminous coal" and "fixed carbon" (for ash-free coke) are survivals of this idea, which is, however, entirely incorrect. The process of distillation results in an entire destruction of the original coal and the rearrangement of its elements -- neither the coke nor gas are contained in the coal as such." Boulton, W. S., Editor, Practical Coal Mining, Gresham Publishing Company, London, 1913, Vol. 1, page 7, 63.

10. <u>Steam: Its Generation and Use</u>, Babcock & Wilcox Company, New York, 1972, page 5-14.

11. Ode, W. H., op. cit., page 208.

12. Hard Book of Chemistry and Physics, page B-99.

13. Moore, op. cit.

14. Schmidt, R. L., W. H. Engelman, and R. R. Fumanti, "A Comparison of Borer, Ripper, and Conventional Mining Products in Illinois No. 6 Coal," U.S. Bureau of Mines, Report of Investigations No. 7687, 1972. See also, Stutzer, O., <u>Geology of Coal</u>, University of Chicago Press, 1940, page 261.

15. Kissell, F. N., J. L. Banfield, R. W. Dalzell, and M. G. Zabetakis, "Peak Methane Concentrations During Coal Mining," U.S. Bureau of Mines Information Circular No. 7885, 1974.

16. <u>Raw coal</u> refers to coal that has experienced initial, rough segregation of mining-induced wastes; coal that has not been so handled is termed run-of-mine, and may contain a variety of unwanted materials (e.g., wires, bits, tubing, cardboard, wood clips, etc.)

17. See, Leonard, J. W. and D. R. Mitchell, <u>Coal Preparation</u>, AIME, New York, 1968, Chapter 7.

18. <u>Keystone Coal Industry Manual</u>, 1974, McGraw-Hill, New York, page 199.

19. Deurbrouck, A., "Sulfur Reduction Potential of the Coals of the United States," U.S. Bureau of Mines Report of Investigation No. 7633, 1972.

20. Symonds, D. F., G. Norton, and G. B. Bogdanow, "Some Aspects of the Economics of Coal Preparation for Coal Conversion Processes" presented at the Engineering Foundation Conference on Coal Preparation, Rindge, New Hampshire, August 15, 1975.

21. Attig, R. C. and A. F. Duzy, "Coal Ash Deposition Studies and Application to Boiler Design," Proceedings, American Power Conference, Chicago, April 22, 1969.

22. Winegartner, E. C. (Editor), "Source Book of Procedures and Definitions of Fouling and Slagging Parameters for Coal-Fired Boilers," ASME Research Committee on Corrosion and Deposits from Combustion Gases, 1974.

23. Winegartner, E. C. and A. A. Ubbens, "Understanding Coal Ash Quality Parameters," presented at 1975 AIME Annual Meeting, New York, February 16-20, 1975. Preprint No. 75-F-32.

22

Memorandum of the Delegation of the Public

Power Corporation (Greece) on Greek Lignite

A. Papadopoulos Manager, Division of Mines
E. Doganis, Assistant to Manager

During Carboniferous and Permian times, the first large period of coal formation, what is known as Greece today, was lying under the sea, therefore no deposits of hard coal exist in the Country. There are of course a few exceptions of no economic significance.

However, during the Tertiary and Quaternary, the second large period of coal formation, Greece was a land, conditions were favourable and as a result every district of the Country was endowed with major or minor coal deposits, particularly lignite (see Table and Map at end of Chapter).

A few years after the second war power generation and transmission in Greece was totally revised and put on a new basis. For this purpose a state enterprise was founded, the Public Power Corporation (PPC), and among its targets was to utilise hydropower and fossil fuels of Greece, the latter being the base load. The first lignite fired steam electric plant was constructed at Aliveri near an old underground mine which was redeveloped.

Today PPC owns and controls, through the Division of Mines, the three major Greek Lignite mines, i.e., Aliveri in the island of Euboea (underground), Megalopolis in Central Peloponnese (open cut), and Ptolemais in West Macedonia (open cut). These three mines produce almost 100% of Greek coal, a total of 17,750,000 metric tonnes in 1975. Out of this figure 1,200,000 metric tonnes were used for non electric purposes, (fertilizers, briquettes, lime).

Under the current development programme of PPC the annual production of lignite will rise to 42 million tonnes by 1990, out of which 28 million tonnes will be produced at Ptolemais, a production which is nearing the limiting capacity of the Ptolemais mining fields.

As one can see in the Table, the measured reserves amount today to 3,600 million metric tonnes, an impressive figure for Greece. However, besides the fact that the figures refer to geologic reserves, due to limitations characteristic to large scale operations, there can only be approximately utilised 50% of this potential.

These limitations are:

- The exclusion of lignite deposits that cannot be mined by low cost open cut methods. The relevant criterion is the ratio of "barren material (m^3) to lignite (t)", a ratio which under today's conditions varies between 5:1 and 7:1 (extreme cases 12:1).

- The exclusion (at least for contential Greece) of lignite deposits with a total energy content (Gross Calorific Value) less than 0.5×10^{14} Kcal. Smaller deposits cannot support a long term operation of large capacity steam electric stations.

The so called energy crisis has triggered research for lignite, which besides hydropower is the only conventional energy source. Lignite research in Greece is carried out by PPC, (Division of Mines) and Institute of Geological and Mining Research (a State organisation) which operates in close cooperation. However, it should be noted that the whole programme is financed by PPC.

There were examined last year:

- sixteen individual lignite deposits.

- There were drilled over 33,000 meters of bore-holes.

- There was located an unknown lignite deposit in the Amyndeon area, West Macedonia, with measured reserves of 250.000.000 m.t. out of which 100.000.000 m.t. are considered exploitable under present conditions.

- There were increased exploitable reserves of Ptolemais basins by 70.00.000 m.t.

The current research programme includes:

- Supplementary drilling in the basins of Aliveri, Megalopolis and Ptolemais.

- Geologic and deposit research at Amyndeon, East Florina, Serrae, Kyme and Crete.

- Reconnaissance in 10 prospective areas.

Parallel to geologic research, due attention is paid by the State to the non-electric uses of lignite. Along this line there is an intention to found a "Lignite Institute", a State research and development organisation, which will study the utilisation of lignite in its whole spectrum of uses.

TABLE

GREEK LIGNITE RESERVES

DISTRICTS	RESERVES (Metric tons $\times 10^6$)	
	Measured	Probable and Inferred
1 PELOPONNESE		
Korinth and Achaia basins		0.1
Pyrgos-Olympia basin		40
Megalopolis Fields (5)	470	30
Sparta-Afissos basin		A few thousand tons
Pellana (Laconia) basin	5	
Messinia basin		0.1
2 MAINLAND GREECE—THESSALY—EPIRUS		
Oropos-Malakassa-Rafina area		10
Kalogreza-Neon Eraklion-Peristeri area		2
Lokris area		0.1
Pelasghia-Myloi (Fthiotis) area		0.1
Elasson-Larisa basin	1.6	90
Aetolia and Akarnania basins		7
Prevesa basins	0.5	
3 WEST MACEDONIA		
Ag. Anarghyri-Amyndeon area	250[6]	
Vevi-Amyndeon area	62	40
Vegora-Amyndeon area	13	30
Akhlada area	40	100
Florina greater area		130
Ptolemais Fields (5)	2000[7]	500
Prosilion-Trighonikon basin (Kozani)		10-20
Kozani-Servia basin	508[1]	
Pieria basins	0.5	15
Agras river basin		25
4 EAST MACEDONIA		
Serres basin	6	100
Drama basin		0.15
Paranestion basin		0.2

DISTRICTS	RESERVES (Metric tons X 10^6)	
	Measured	Probable and Inferred
5 THRACE		
Aemonion-Kotyli basin[2]	0.24	0.16
Konotini basin		Less than 0.1
Alexandroupolis area[3]		1
Orestias basin[3]		1
Soufli area [3]		A few thousand tons
6 ISLANDS		
Crete		
Almyri Panaghia	0.5	
Kandhanos	3	2
Euboea		
Aliveri greater area [5]	3[4]	2
Kymi area	4	8
Central Euboea basin	2.5	4.5
North Euboea	0.09	
Rodhos		
Apolakia basin	5	
TOTAL	**3593.9**	**1150**

(1) Probably exploitable reserves amount to 293 x 10^6t.

(2) Uranium bearing coal

(3) Small lenticular scattered deposits.

(4) Remaining exploitable reserves

(5) PPC owned mine

(6) Exploitable reserves amount to 100 x 10^6 t.

(7) Exploitable reserves amount to 1150 x 10^6 t.

MAP OF GREEK COAL DEPOSITS

 Lignite

Peat

MEMORANDUM OF THE DELEGATION OF THE PEOPLE'S
REPUBLIC OF MOZAMBIQUE
ON THE COAL DEPOSITS OF THAT COUNTRY

J. NUNES NETO
MINING ENGINEER (1st)
MINISTRY OF INDUSTRY & COMMERCE
PEOPLE'S REPUBLIC OF MOZAMBIQUE

As you all know Mozambique is one of the countries in
Southern Africa which possesses a considerable number of
minerals (some of them considered strategic) and mines in
operation. Naturally "the world-wide impetus to find and
develop new coal resources" caused by the oil crisis,
presents an "exploration challenge" to us. Mozambique is
one of the small producers of coal in the area. The fact
that we are a small producer of coal in Southern Africa
makes the topical subject of this Symposium important to us.

Our contribution therefore will be focussed on general
information on coal qualities and quantities as they are
found in the limited area under operation in our country.
Of course, other aspects which are deemed important to
prove the point or to illustrate it will be mentioned, for
our coal has some special features.

At the present moment we have the Moatize coal deposit in
production. The Moatize mining zone occupies an area of
about 200 kms^2, 30 km long and 6 km wide on average.

<u>Situation</u> It is situated in the Province of Tete. This
area is sparsely populated and cultivated and has a very
dry climate. The rainy season is from December to April,
the highest intensity falling in the months of December and

January. The rainfall during these five months rarely surpasses 750 mm total. The air temperatures in the period from October to April are very high, exceeding during long periods 40°C in the shade. During the months of May to September the climate is bearable. There is a permanent water course crossing the zone in the N.W. third - Ruvugue and another temporary course crossing following the line N.E. - S.W. - the Moatize - which is a subsidiary of the first and which during the rainy season is torrential.

Geology The coal bearing structure is situated in the Karoo series which in this area is represented by mudstones, sandstones, and coal with more or less shale. The area is relatively flat, the difference in level in the coal bearing field being rarely more than 10 metres. The seams dip on average 15 degrees and form assymetrical synclines, and sometimes anticlines. The dip of seams has a tendency to decrease with depth, the greatest dip being on the outcrop. The area is crossed by a complex system of faults with important overlaps which sometimes limits mining both underground and opencut. Some of these faults are filled by dolerite dykes, which by contact metamorphism burnt the coal near to these dykes. We also find dolerite dykes whose extrusion was not through the faults. The Karoo formation in the area has a total thickness of 300 to 400 metres and six different coal seams are found.

These coal seams have a different thickness going from 1.5 metres for the top seams (the sixth) to 40 metres for the fourth. We find coal and shale in all seams. The coal seams are separated by layers of sandstones and/or mudstones about 40 to 50 metres thick. The chronological succession from the top to the bottom of the known coal seams is the following:-

(6) Andre
(5) Grand Falaise
(4) Intermedia
(3) Bananeiras
(2) Chipanga
(1) Sousa Pinto

Numbers 2 and 6, in which underground mining started over 30 years ago, are well known. For marketing reasons the mining of the Andre seam (6) was temporarily stopped and the seam Chipanga (2) has been the only one mined during the last 25 years.

<u>Chipanga Seam</u>   This seam is a complex of mudstones and
coal with 30 to 32 metres of thickness. (A typical section
of this complex is attached where the layers of coal,
mudstones, and shale/coal alternating in the thin beds are
indicated (figure 1). This seam is relatively regular and
occurs in nearly all the mining zones. The drillings over
the mining zone show that the bottom layer is the most
**regular in the seam, and even of all the seams in the
Moatize area.** This fact, together with its highest quality,
lead us to concentrate mining of the Chipanga seam only in
its bottom layers for the past 30 years.

The bituminous coal obtained from this seam has the
following characteristics:-

–	Inherent moisture	1%
–	Volatile matter	18%
–	Ash content	20%
–	Sulphur	1%
–	Calorific value gross 6,800 K.cal/Kg.	

From this run of mine and by dry screening we separate two
coals of grades 0-11 mm. and over 11 mm. with the
following characteristics:-

	0/11 mm.	+11 mm.
Moisture	1%	1%
Volatile matter	19/20%	17/18%
Ash content	14/15%	21/23%
Sulphur	0.95%	1%
Calorific Value	7,200 K. cal/Kg.	6,600 K.cal/Kg.
Swelling index	$7\frac{1}{2}$%	1 to $1\frac{1}{2}$

In the quantitative percentage of respectively 52 and 48%.

As previously stated this seam shows many faults and some
dolerite dykes. Also in some areas water is found in
underground mining. This seam releases a large percentage
of methane gas. The coal is friable because it has a high
percentage of vitrain which contributes to make it a good
coal. It is also subject to spontaneous combustion
certainly helped by the presence of nodules of pyrites
and sulphur in the upper layers. The upper layers of coal

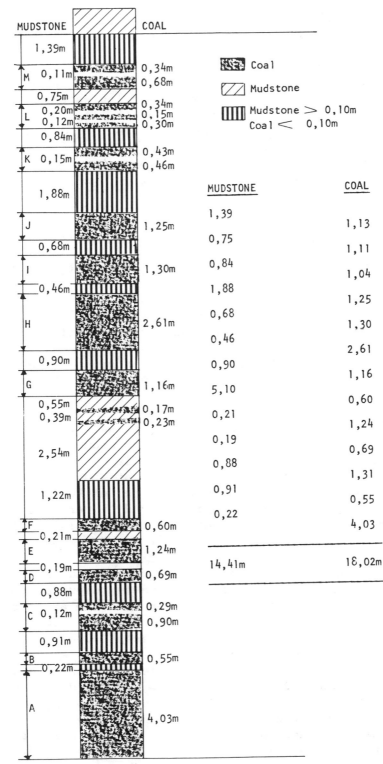

Figure 1. Typical section of the Chipanga Seam.

are slightly shaly but the quantity of the vitrain is
maintained and can equally produce a good coking coal in the
smaller grades.

Mining of the Whole Chipanga Seam Studies on the grading
and selective washing of all coal beds indicated in the
Chipanga seam show the economical possibility of obtaining
coals with the following characteristics:-

Volatile matter	19 to 20%
Ash content	10 to 13%
Swelling	9 to 8%
Sulphur	Less than 1%

For all the above studies the idea of open cut mining
occurred. The idea would be:-

(a) Open cut mining to a determined depth of all the seam,
     improving by washing the indicated susceptible beds.

(b) Below this depth mining should continue underground
     only on the 4.5 m. bed which constitutes the bottom
     of Chipanga seam.

With this programme in mind, various areas within the
Moatize mining zone were selected which seemed to be
susceptible to open cut mining, not only by their dimension,
but also by their regularity. The first studies considered
that for an average dip of 15°, and the thickness and
characteristics of the seam to be mined, a vertical depth
of 80 metres would be an economical target. This depth
would be reached with a ratio of 1:7 (coal/overburden).

Therefore, areas were selected that on open cut mining and
down to 80 metres depth, could guarantee reserves of about
200,000,000 tons of run of mine. In the same areas the
underground mining of the 4.5m. bottom layer from 80 metres
down to a depth of 300 metres, provides 40,000,000 tons
more by the method of room and pillar. It is evident that
if open cut mining is carried on below 80 metres the total
reserves of 240,000,000 estimated up to this depth by the
previous reckoning, will be increased by 2,500,000 for
each metre depth that the open cut mining may reach
economically. Therefore, it was decided, on a first
approximation, that the establishment of open cut mining
capable of producing 3,600,000 tons of run of mine divided

into 3 distinct fronts, each with an approximate annual production of 1,200,000 tons run of mine, which after washing would produce about 2,000,000 tons saleable coal annually, would be the desirable size for the project.

Analyses of Moatize Coal Besides the above encouraging data, it is important, in our view, to point out an element of controversy with regards the basis of the evaluation of the coal. In analysing the Moatize coal we were surprised to learn that some of the components of the ash are minerals of high value, of which the most important was Germanium.

COAL EXPLORATION IN INDIA

P. K. GHOSH

Coal Controller, Ministry of Energy, India.

Planning for coal exploration and its success depend basically
on proper and scientific appraisal of the geology of the
area concerned. In this context, the geological map is of
paramount value.

In India, the importance of coal was first realised during
the late twenties of the last century when coal from Newcastle
mainly bought as bunker coal, was found to be costly for
shipping. The Geological Survey of India was thus officially
established in 1851 with the charter ' to survey the Indian
Coalfields'. After 25 years of its organised existence i.e.
by 1876, the survey of the major Indian coalfields was
completed. This work by the Survey not only laid the foundation for exploration and development of the coal mining
industry in India but also led to many a discovery leading
to the concept of "Gondwanaland" based on various fundamental
findings in the field of Gondwana geology. Recent exploratory
operations have led to the discovery of a number of thick
and workable seams in many virgin coalfields - the most
spectacular find being the discovery of the 134 - metre (440
ft.) thick bituminous coal seam in the Singrauli coalfield
in Madhya Pradesh (see Figure 1). This is possibly the
thickest bituminous coal seam in the world. With this new
find, this coalfield is going to be one of the leading
producing fields of the country.

The Geological Survey of India is the fourth oldest organisation of its kind in the world. Thus, India especially on
the basis of her contribution in the field of Gondwana
Geology, can rightfully claim to be one of the pioneering
countries in the field of coal exploration.

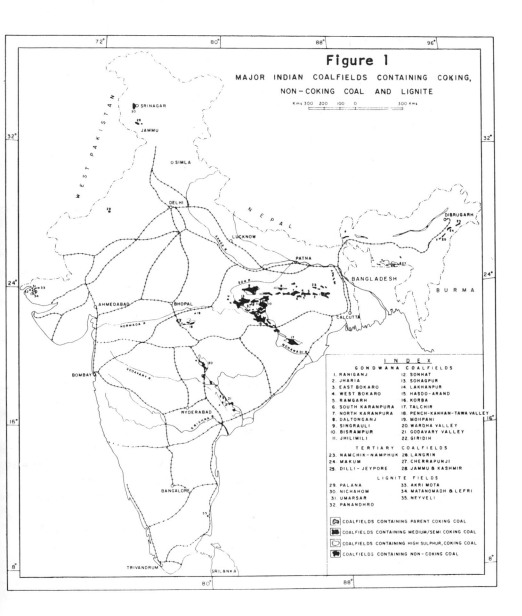

Figure 1

MAJOR INDIAN COALFIELDS CONTAINING COKING, NON-COKING COAL AND LIGNITE

Kms 300 200 100 0   300 Kms

INDEX

GONDWANA COALFIELDS

1. RANIGANJ	12. SONHAT
2. JHARIA	13. SOHAGPUR
3. EAST BOKARO	14. LAKHANPUR
4. WEST BOKARO	15. HASDO-ARAND
5. RAMGARH	16. KORBA
6. SOUTH KARANPURA	17. TALCHIR
7. NORTH KARANPURA	18. PENCH-KANHAN-TAWA VALLEY
8. DALTONGANJ	19. MOHPANI
9. SINGRAULI	20. WARDHA VALLEY
10. BISRAMPUR	21. GODAVARY VALLEY
11. JHILIMILI	22. GIRIDIH

TERTIARY COALFIELDS

23. NAMCHIK-NAMPHUK	26. LANGRIN
24. MAKUM	27. CHERRAPUNJI
25. DILLI-JEYPORE	28. JAMMU & KASHMIR

LIGNITE FIELDS

29. PALANA	33. AKRI MOTA
30. NICHAHOM	34. MATANOMADH & LEFRI
31. UMARSAR	35. NEYVELI
32. PANANDHRO	

COALFIELDS CONTAINING PARENT COKING COAL

COALFIELDS CONTAINING MEDIUM/SEMI COKING COAL

COALFIELDS CONTAINING HIGH SULPHUR, COKING COAL

COALFIELDS CONTAINING NON-COKING COAL

COAL EXPLORATION IN INDIA 649

Once the geology of the coalfields was known, exploratory operations, followed by exploitation, was resorted to. Demand of coal being limited and fluctuating activities on exploration by the Geological Survey of India waxed and waned from time to time - the Survey being the only exploratory organisation in existence then. After Independence in 1947, the position however, considerably changed. With emphasis being laid on Iron and Steel and other basic industries including power generation during various Plan periods, the tempo of exploration was considerably intensified, both regional assessment and detailed exploration. Regional assessment by drilling (one borehole per sq. mile or 2.6 sq.km.) was carried out mainly to have preliminary data on the quantity and quality of coal locked up in various coalfields. Detailed exploration in selected blocks led to development of a number of mines in the Public Sector prior to 1971 when both Public as well as Private Sector undertakings were operating the coal mining industry. In October, 1971, all coking coal mines were nationalised. This was followed by nationalisation of the non-coking mines in January, 1973. Since then the entire coal mining industry has come under the Government. Now the development of this industry lies fully in the hands of the Government.

EXPLORATION

Geology: In India Coal resources are confined to two geological horizons - Gondwanas (coal measures are of Permian age) and Tertiary. About 98 percent of the reserves are, however, within the Gondwanas represented by more than 74 separate basins covering an area of about 45,125 sq.km. (17,356 sq. miles). The fields, varying in size from 1 sq. km. (0.39 sq. miles) to as much as 1,500 sq.km. (579 sq. miles), are primarily confined to Peninsular India. More than 98 percent of production comes from the Gondwana fields containing coking as well as non-coking bituminous to sub-bituminous coals. Tertiary coals are, on the other hand, ligno-bituminous in nature and are mainly confined to the North-Eastern and North-Western part of the country. There are also lignite deposits, the most important of which is at Neyvile in the South which is the only deposit now under exploitation.

Gondwana coals are found in three stratigraphic horizons, viz., Karharbaris (Upper Sakmarian), Barakars (Lower Permian) and the Raniganj (Upper Permian). Out of the total area of the Gondwana coalfields, Karharbari coal measures occupy 350 sq. km. (135 sq. miles) whereas the Barakars and the Raniganj represent 14,220 sq.km. (5,489 sq. miles) and 1,236 sq.km. (477 sq. miles) respectively.

Tertiary coals are confined to two stratigraphic horizons, viz., Eocene and Oligocene.

Lignites of India are of various age groups and are found to be associated with sediments extending from Eocene to Lower Pleistocene in age.

Coal Seams: In the Gondwana fields, specially those in the eastern region in the States of West Bengal, Bihar and Orissa, a number of thick coal seams within a short span of vertical strata are found as will be seen from the following:

> Raniganj Coalfield: 13 major seams varying in thickness from 0.9m. to 38.1m. (2.5 to 125 ft.) within 640m. (2,102 ft.) of Barakar sediments with coal to non-coal ratio as 1:4.8. In the overlying Raniganj coal measures, 12 major coal seams varying in thickness from 0.9m. to 13.7m. (2.5 to 45 ft.) are confined within 1,036m. (3,398 ft.) of sediments with coal: non-coal ratio as 1:15.4.

> Jharia Coalfield: In the adjacent Jharia coalfield, 18 coal seams varying in thickness from 0.9 to 22.5m. (2.5 to 74 ft.) have been recorded within 609m. (2,000 ft.) of Barakar sediments with coal: non-coal ratio as 1:2.5. The coking coal resources of the country are primarily confined to the above coal measures of the field. In the overlying Raniganj coal measures, three major coal seams varying in thickness from 0.9 to 3.7m. (2.5 to 12 ft.) are found within 561m. (1,840 ft.) of sediments with coal to non-coal ratio as 1:39.0.

Tertiary coals are usually the lensing type. Maximum thickness of a seam recorded is about 33.0m. (100 ft.) in Makum coalfield in Assam. Tertiary coal measures are generally highly disturbed tectonically. These fields being located in hilly and forested areas are not easily approachable. All these have posed serious problems in exploration in the region containing such coals.

Exploratory Operations: Once the blocks/areas are selected for exploitation, detailed exploratory operations are carried out in a planned manner, of which drilling occupies an important phase of investigation. In areas which are covered,

geophysical methods are adopted for delineating the extent
of the coal measures and/or determining the basement profiles
etc. as the first step prior to drilling.

During the pre-Independence era (prior to 1947), besides
geological mapping, following by pitting and trenching,
little was done by way of drilling except by some of the
leading coal mining concerns. This was but natural as the
properties being small, individual entrepreneurs were not
inclined to invest much on exploratory operations specially
in view of manual mining involving minimum investment. For
regional assessment, the Geological Survey of India, however,
carried out a number of drilling operations during this
period again to be adopted in a planned manner covering a
wider field of operation after 1947. Simultaneously with
regional drilling, the Geological Survey of India had also
been engaged in detailed exploration in a number of blocks
which had subsequently been developed into working mines in
the Public Sector.

Drilling: The main aim in an exploration programme is to
obtain information, as accurate as possible, on the subsurface behaviour of the coal seams regarding their quality
and layout along with tectonic features. Although modern
geophysical techniques are making an inroad in such exploratory operations, drilling continues to be the main field of
activity for this purpose. Thus the meterage of drilling
completed, in a way, reflects the magnitude of such operations
In India, drilling along with other geological investigations
attained a new tempo with the launching of different plans.
So far from 1956, the first year of the Second plan from
whence emphasis had been laid on sub-surface exploration by
drilling until August, 1976, a total amount of drilling done
in many widely located blocks of various coalfields is of
the order of 996,054m.(3,269.049 ft.) the details of which
are indicated in Table I. In such operations, drilling had
generally proved the full sequence of the coal measures.
The deepest hole so far completed is 1,216m. (4,000 ft.) in
depth.

TABLE 1
DRILLING PROGRESS (1956–75)

	Second (1956–61)	Third (1961–66)	1966–Aug 1975	Total
(i) Meterage drilled by Geological Survey of India	76,317	11,250	247,043	334,610
(ii) Meterage drilled by Indian Bureau of Mines	284,647	186,797	Nil	471,444
(iii) Meterage drilled by other organisations				160,010

Total: 966,064

i.e. 3,269,049 ft.

The total target of drilling of all the organisations for the year 1976-77 is of the order of 193,000m. (633,426 ft.).

ASSESSMENT

For assessment of resources there has been a standing committee known as the Committee on Assessment of Resources which had worked out certain procedures known as I.S.P. (Indian Standard Procedures) for the categorisation of reserves quality-wise and quantity-wise as indicated below:

Reserve Calculation   In India, Coal reserve is calculated under three categories based on the following criteria,

(i) Proved: Not exceeding 200m. (660 ft.) from outcrop/ borehole/trenches/mine workings.

(ii) Indicated: Usually 1000m. (3,300 ft.) from point of observation, but 2000m. (6,600 ft.) for beds of known geological continuity.

(iii) Inferred: 1000 or 2000 m. (3,300 to 6,900 ft.) based on geological evidences.

Thickness: Coal seams 1.2m. (4 ft.) and above are considered
Partings greater than 5 cm. (2 inches) are excluded. The
classification of Gondwana coal by grade is given in Table 2.

Tertiary coals are invariably high in sulphur content (average
2.5 to 3 percent) but low in ash (2 - 12%). The following
criteria is adopted for reserve calculation.

Ash	Sp. Gravity
0 to 5%	1.30
5 to 10%	1.34
10 to 15%	1.38

Reserves: Following the above criteria, gross reserves (in
million tonnes) as estimated by Geological Survey of India
based on their work and that of other exploratory operations
carried out by different agencies, such as National Coal
Development Corporation, Coal India Ltd., Indian Bureau of
Mines, Mineral Exploration Corporation, various State Governments etc., are indicated in following Tables 3, 4 and 5.
Total gross reserves of coal in India for seams 1.2 (4ft.)
in thickness and down to 609 m. (2,000 ft.) depth is of the
order of 83,670 million tonnes.

The recoverable reserves have been arrived at after deducting
the locked up coal in barriers etc. and allowing for losses
due to:

a)   geological disturbances, intrusives etc.,

b)   mining and,

c)   washing.

Grade: A different classification is followed for marketing
of coal. There is not much of a difference between these
two classifications except that the latter is somewhat more
specific. Commercial classification is thus more in common
usage not only for marketing but also for the estimation of
gradewise production. Broad details of this grading are
given in following Table 6.

Recently useful heat values in kilo-calories per kilogram
have been included under each grade of non-coking coal.
For low moisture coals these are Grade IIIB - exceeding
3,800 but not exceeding 4,710 through 5,940 to 6,350 for
Grade I to Selected A as 6,620. For high moisture coals

TABLE 2

GRADE CLASSIFICATION (Gondwana coals with sulphur content less than 1%)

	Low to medium volatile coals		High volatile or high moisture coal	
	Sp.Gr.	Air dried moisture up to 2%; V.M. not more than 35%	Sp. Gr.	Air dried moisture more than 2%; V.M. more than 35%
Class I	1.42	Ash Content under 17%	1.40	Ash & Moisture content under 19%
Class II	1.47	Ash Content 17 to 24%	1.45	Ash & Moisture content 19 to 28%
Class III	1.57	Ash Content 24 to 35%	1.55	Ash & Moisture content 28 to 40%
Class IV	1.70	Ash content 35 to 50%	1.70	Ash & Moisture content 40 to 45%

TABLE 3

RESERVES OF COAL AND LIGNITE IN INDIA (Millions of tonnes)

CATEGORY	P	R	O	V	E	D	IND	INF	TOTAL
CLASS	I	II	III	IV	UC	Total			
GONDWANA COALS	1,513.78	7,008.60	7,105.78	1,856.89	3,091.98	20,577.03	31,882.51	30,312.12	82,771.66
TERTIARY COALS	-	-	-	-	-	161.21	191.71	549.06	901.98
LIGNITE	-	-	-	-	-	1,868.56	202.00	28.70	2,099.26

TABLE 4
RESERVES OF METALLURGICAL COAL (included above in Gondwana Coals)
(Millions of Tonnes)

COAL TYPE	PROVED	INDICATED	INFERRED	TOTAL
PRIME COKING COAL	3,251.89	1,586.26	460.73	5,298.88
MEDIUM COKING COAL	3,793.33	4,275.20	1,308.03	9,376.56
BLENDABLE COAL	1,206.16	2,600.98	914.79	4,721.93
TOTAL:	8,251.38	8,462.44	2,683.55	19,397.37

TABLE 5
RECOVERABLE RESERVES OF METALLURGICAL COAL (Millions of Tonnes)

COAL TYPE	COAL FIELD	1964 ESTIMATE	REVISED ESTIMATE
PRIME COKING COAL	JHARIA	1,360	1,604
MEDIUM COKING COAL	JHARIA	440	427
	RANIGANJ	280	241
	E. BOHARO	1,195	560
BLENDABLE	RANIGANJ	738	477
TOTAL:		4,013	3,309

TABLE 6
COMMERCIAL CLASSIFICATION

A. Non-Coking Coals

High Moisture Coals

Ash & Moisture	Under 17.5%	– Selected Grade A	
	17.5% – 19%	– Selected Grade B	
	19% – 24%	–	Grade I
	24% – 28%	–	Grade II
	28% – 35%	–	Grade III

Low Moisture Coals

Ash	Under 15%	– Selected Grade A	
	15% – 17%	– Selected Grade B	
	17% – 20%	–	Grade I
	20% – 24%	–	Grade II
	24% – 28%	–	Grade IIIA
	28% – 35%	–	Grade IIIB

B. Coking Coal

Ash	Under 13%	–	Grade A
	13% – 14%	–	Grade B
	14% – 15%	–	Grade C
	15% – 16%	–	Grade D
	16% – 17%	–	Grade E
	17% – 18%	–	Grade F
	18% – 19%	–	Grade G
	19% – 20%	–	Grade H
	20% – 24%	–	Grade J
	28% – 35%	–	Grade K

TABLE 7
GRADEWISE PRODUCTION OF COAL AND LIGNITE IN INDIA
(Figures in million tonnes)

Grade	1972-73	1973-74	1974-75	1975-76
Non-Coking coal				
Sel.A	3.38	2.33	2.02	2.20
Sel.B	4.30	4.11	4.14	4.75
Gr.I	28.38	29.00	30.29	33.10
Gr.II	7.93	7.77	9.73	10.02
Gr.IIIA	1.97	2.00	1.69	2.20
Gr.IIIB	4.50	4.32	7.72	8.00
NG/UG	10.14	12.87	11.86	12.00
Sub Total: (non-coking)	60.60	62.40	67.45	72.27
Coking coal				
A	0.25	0.29	0.24	0.15
B	0.72	0.61	0.87	0.90
D	1.08	0.89	1.39	2.02
E	1.59	1.75	2.14	2.64
F	4.38	4.02	5.25	5.30
G	2.35	2.21	3.27	3.43
H	2.04	1.89	2.29	3.03
HH	4.11	4.06	5.43	7.50
Sub Total: (coking) (A to HH)	16.62	15.77	20.93	25.00
J	–	–	0.03	1.12
K	–	–	–	0.96
Sub Total:			0.03	2.08
Total: (Coking)	16.62	15.77	20.96	27.08
Lignite	2.89	3.32	2.94	3.03
TOTAL: (Coal & Lignite)	80.11	81.49	91.35	102.38

Production Target:
(In million tonnes)

Years	Coking	Blendable	Non-Coking	Total
1976-77	26.0	1.5	80.5	108
1977-78	28.5	1.7	92.8	123
1978-79	32.0	2.0	101.0	135

heat values are: Grade III - exceeding 3,770 but not exceeding 4,570 through 5,170 to 5,930 for Grade I to Selected A - exceeding 6,135.

Production: After the complete nationalisation of the coal mines by January 1973, Government had been laying stress on enhanced production of coal to meet the growing demand in the country. In the context of the energy crisis that engulfed the world in 1973, India has to re-orient her energy policy with greater emphasis on the solid fuel resources to replace petroleum and petroleum-products to the maximum in the field of energy. The challenge thus thrown to the nationalised Coal Mining Industry had been fully met so much so that at present there is a surplus of coal which could possibly be channelised for additional export. From figures as indicated below (Table 7) it would be observed that there has been an annual increase of about 10 million tonnes during the period 1973-74 (81.49 millions including lignite) to 1975-76 (102.38 million tonnes including lignite). With the comfortable position regarding the current availability of coal, production targets for the future years are now under revision.

INDEX

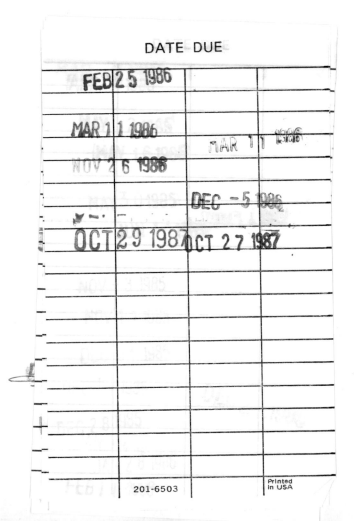

DATE DUE

FEB 2 5 1986		
MAR 1 1 1986		MAR 1 1 1986
NOV 2 6 1986		
	DEC − 5 1986	
OCT 2 9 1987	OCT 2 7 1987	

201-6503   Printed in USA